Reality, Universal Ontology, and Knowledge Systems:

Toward the Intelligent World

Azamat Abdoullaev
EIS Encyclopedic Intelligent Systems Ltd, Cyprus, Russia

IGI PUBLISHING
Hershey • New York

Acquisition Editor:	Kristin Klinger
Development Editor:	Kristin Roth
Senior Managing Editor:	Jennifer Neidig
Managing Editor:	Jamie Snavely
Assistant Managing Editor:	Carole Coulson
Copy Editor:	Maria Neithercott
Typesetter:	Sean Woznicki
Cover Design:	Lisa Tosheff
Printed at:	Yurchak Printing Inc.

Published in the United States of America by
IGI Publishing (an imprint of IGI Global)
701 E. Chocolate Avenue
Hershey PA 17033
Tel: 717-533-8845
Fax: 717-533-8661
E-mail: cust@igi-global.com
Web site: http://www.igi-global.com

and in the United Kingdom by
IGI Publishing (an imprint of IGI Global)
3 Henrietta Street
Covent Garden
London WC2E 8LU
Tel: 44 20 7240 0856
Fax: 44 20 7379 0609
Web site: http:/www.eurospanbookstore.com

Library of Congress Cataloging-in-Publication Data

Abdoullaev, Azamat.
Reality, universal ontology, and knowledge systems : toward the intelligent world / Azamat Abdoullaev, author.
p. cm.
Summary: "This book provides cutting-edge research on reality, its nature and fundamental structure, represented both by human minds and intelligent machines.--striving to describe a world model and ontology; organized human knowledge; powerful reasoning systems; and secure communication interoperability between human beings and computing reasoning systems promising the profound revolution in human values and ways of life"--Provided by publisher.

Includes bibliographical references and index.
ISBN-13: 978-1-59904-966-3 (hardcover : alk. paper)
ISBN-13: 978-1-59904-967-0 (e-book : alk. paper)
1. Reality. 2. Ontology. 3. Expert systems (Computer science) I. Title.
BD331.A29 2008
110--dc22
2007044145

British Cataloguing in Publication Data
A Cataloguing in Publication record for this book is available from the British Library.

For my sons, Alexander and Iskander, as the future citizens of the Intelligent World

Reality, Universal Ontology, and Knowledge Systems:

Toward the Intelligent World

Table of Contents

Preface

All men by nature desire to know.
Aristotle

He who does not know what the world is does not know where he is.
Marcus Aurelius

If I have seen further, it is by standing on the shoulders of giants.
Isaac Newton

Of all sorts of intellectual pursuits, nothing appears of the greatest challenge than giving a formal account of the world understandable both by humans and computing machines. Art, religion, science, and technology in their specific ways seek to model, explain, or represent reality. But only one human activity is thoroughly and systematically committed to the grand cause of inquiry into the world, the universe, existence, or being, known as **a universal, global, master, ultimate, foundation, general, common, top**, or **standard ontology**, the hardest, most exacting and most rewarding ambitious search for the essential knowledge and truths of the whole existence. Being alone in its penetration of reality, such a unifying science addresses the whole universe of entities, its elemental categories and primary rules with their true formal and cultural representations. Particularly, how the general realities (entity classes and relationships) are transformed to the specific classes of things and embodied in distinct levels of reality as specific ontological layers:

- The material entities and interactions in the physical universe (physical reality)
- The constructs and associations in the mental world (mental reality, phenomenological reality, personal realities, or common sense world)
- The words, expressions and sentences in natural languages (linguistic reality)
- The cultural objects, processes, and relationships in a social realm (social reality, a product of social interactions and actions, as hyperreality, proxy, simulacrum reality as a media-created world; where representation is identified with reality itself, such as pornography, casinos, celebrities, human clones, androids, etc.)
- The coded representations and sign structures, static and dynamic, in the information universe of reasoning computing machines (cybernetic semantic reality)

Being a theoretical groundwork and knowledge base of basal sciences and technologies, real ontology is the heart and soul of all human learning, wisdom and reasoning, unveiling the common nature, mechanisms and meanings of things and their relationships diversified as the key levels and kinds of reality: natural, mental, social, cultural, technological, or informational. As the source science of all knowledge bases and domain theories, it develops the most general theories and representational languages about reality giving the most profound and consistent account of things, entities, or beings, their properties and relationships.

All kinds of science, basic or applied, hard or soft, theoretical or empirical, fundamental or descriptive, natural or humane, to some degree partake in the general theory of entities and relationships as special sciences, or domain specific ontologies. All types of knowledge, theoretical, formal, experimental or practical, presuppose essential, ontological knowledge of things. Implicitly or explicitly, ontological principles can be found among mostly general theories, mostly universal axioms and laws, and in mostly interesting scientific problems. As underlying ideas, ontological categories, classes, concepts, notions, and terms lead the list of the great ideas making the very substance of the grand elemental conceptions. For they are the abstractions by which thought knows the world and minds think things, the terms in which we formulate major principles and facts of reality, the notions in which we make definitions, put fundamental questions, and solve decisive problems. Ontological ideas constitute the very framework of mental contents and cognitive processes as the heart of mental life. They reside in languages, natural or artificial, as the mind in the body, as pungency in pepper; since the syntactic and grammatical categories and semantic classes are inherently tied to world things. Our human language is pervaded with ontological categories, in terms of which we describe the fundamental constituents and properties of reality and explain the complex dynamics of the nonlinear world of things. All great human actions and intellectual achievements, all our rational practice of choice and moral codes are intrinsically guided by ontological rules and principles as primary and unvarying truths of reality.

The large order and broad goal of ontology is to produce the explanatory schemas of all being and reality, giving the guiding principles and rules for a wide variety of special truths and particular facts. The ontological verities come up as the basic laws of reality occupying the highest level in the hierarchy of truths and meanings: mental, logical, mathematical, semantic, verbal, scientific, empirical as well as moral, ethical, esthetic, and religious. So the quest of underlying truths, universal and necessary, is the ultimate goal of the fundamental ontology and ontological theories, aimed to uncover the general knowledge and universal laws applicable to all existence in its basic levels, parts, and domains. These are not all the substantial implications of the general knowledge of reality. Never-foreseen before technological artifacts and engineering systems are lining up beyond the current horizons of knowledge and technology.

With the beginning of the third millennium, unprecedented and profound impacts are increasingly forced not only on human life and culture but also on the dominancy of human beings as the only creatures of sapience and the end-all of evolution. Ontology by means of its radically new ontological technologies and applications underpins the most sophisticated information technology such as large-scale knowledge systems and intellectual technologies as the World Wide Intelligent Web. Providing the unifying reality modeling schemes and languages for intelligent technology, ontology is bridging the real world and the information universe, giving the dynamic world modeling fundamentals, principles, constructs, representations, and algorithms for building ontological technology, a radically new class of intelligent technologies and knowledge systems.

Thus the global ontology as a comprehensive formal abstraction and conceptual model of reality yields a dramatically new type of generic engineering, **ontological AI technology**, world-centered intellectual systems and knowledge machines. It then becomes the central cause of an unexampled global social transformation, an Intellectual Knowledge Society. That profoundly advanced stage of human societies may be defined as *an intelligent civilization (culture) sustained by knowledge infrastructures and machines, where a personal, economic, societal and environmental development is mostly driven by ontological AI technology*. Unlike the traditional technologies, it employs the fundamental world knowledge, elemental categories, axioms, rules, laws and principles, for the design and construction of the most advanced information and communication systems and intelligent machines.

To lay down the theoretical foundations of the knowledge society technologies, it is set forth a unifying account of entities and relationships as the reality representation and reasoning (RRR) framework in the form of **unified framework ontology** or **universal formal ontology** (UFO), the overlap between **universal ontology** and **fundamental mathematics** The encyclopedic world model is developed as Reality Modeling Framework or **World Description Framework** (WDF) equally applicable for both types of intelligent entities, natural and artificial. Being a unifying framework, a universal theory, a general system, a comprehensive model, and a theoretical account of reality, the UFO maps the corresponding elements and structures of the real world onto the mental representations and operations coded as the computing software classes and processes; thus functioning as the common semantic language both for computers and persons. The unique combination of the ontological and mathematical fundamentals results in the consolidative account of the general classes of things: substances and objects; states and properties; changes, actions, processes, and events; relationships, connections, and associations.

Some fundamental solutions and significant answers coming from the convergence of two master sciences (formal mathematics and substantial ontology) are as follows. The all-entity description framework allows for a reference system of knowledge standards underlying the key concepts of scientific knowledge, the basic constructs of minds, the major categories of languages, the metadata models and reasoning schemas of information systems. Crucially, the UFO also provides a unified (causal) description of the nonlinear world of heterogeneous systems (physical, chemical and biological, cognitive and conceptual, economic and social, engineering and informational) revealing the fundamental principles and rules of their behavior. This presents the entity reference as a representational and inferential framework for the rising generation of intelligent applications equal to reason about different pieces of reality in causal and predictive ways.

As an integrative theoretical system, the UFO makes the largest semantic context of use and meaning within which the major classes and relationship patterns of physical, mental, or cultural worlds as well as the natural language semantics can be described and explained. Again, the natural language is proved to be the most effective representation and reasoning language not only for persons but also for computing machines as able to process and communicate semantic information about the world and its domains in NL forms. This capacity opens up the possibility of linguistic intellectual systems such as encyclopedic intelligences and global knowledge web resources (as the UFO driven Real Web) and so bringing order, meaning, and intelligence to the heterogeneous content of the WWW.

As an original contribution, the book is proposing the unifying account of the dynamic world merging into a single frame the significant elements of various conceptual models, schemas, and data languages such as upper ontologies, the ER extended data models, data integration systems, and Web ontology languages. Representing the content of reality, its meanings and natural language constructions to formal reason of humans and to programming machines will make possible the powerful knowledge technologies and intellectual systems. All is along the strategic lines of automated knowledge society aimed ultimately at an intelligent world:

Reality or the World → World Knowledge {(Universal Ontology of Everything + Mathematics + Semantics + Science + Real Logic → Natural Language } → Applied Ontology → Informatics and Computer Science → Ontology AI Technology {Semantic Web → NL Engineering and Intellectual Technology → NL Knowledge Machines (systems, applications, agents, robots, software programs, tools) → Encyclopedic Intelligences → Real World Web → Global Intelligent Cyberspace } → Global Knowledge Society → Intelligent World (Universe, Reality)

Accordingly, the core subject matter is basically the world, its nature, model, formalization, content, structure, dynamics, meaning, representation, and how it may be effectively represented by human minds and intelligent machines. The world is analyzed as the largest unbounded environment by the agency of ontological representations and mathematical descriptions of its key constituents and relationships. This approach affords to develop the emerging Knowledge Society technology as the real world knowledge systems and applications capable to efficiently represent and reason about everything that exists, changes, and related. The entire research is performed with the basic goal to design a standard ontology (a master schema of things in the world) capable to uniformly organize human knowledge, to create powerful reasoning systems and to secure communication interoperability between diverse species of intelligences, natural and artificial.

To demonstrate the above, the content of the book is arranged into 13 parts. A brief description of each of the parts is as follows.

In Chapter I, **Introduction**, we introduce the central meaning and theme of the whole work. The general science of reality, universal formal ontology, together with its theoretical principles and technological applications, is alpha and omega of advanced Knowledge Technology, the fundament of the Intelligent Civilization. The Great Chain of the Intelligent World is traced as: Reality → World Knowledge → Knowledge Technology → Knowledge Society → Intelligent Reality.

The architecture of reality representation and reasoning systems as the emerging Knowledge Society technology is outlined as:

UFO (fundamental rules, classes, objects, processes, and relationships of reality) + Upper Ontologies (Generic Conceptual Models) + Domain Knowledge Base (Domain Ontologies) + Domain Data Sets + Formal Schema Language + Programming Implementation

A strategic plan for building the Intelligent WWW is shown to follow the developmental stages for the reality-centered ontology technology:

INTELLIGENT WWW: THE WORLD > THE WORLD WIDE WEB > THE SYNTACTIC WEB > THE SEMANTIC WEB > THE PRAGMATIC WEB > THE ONTOLOGICAL WEB

In Chapter II, **Ways to View the World**, it is developed a generic theory of things in the world as a reality representation and reasoning system acting as the world description or reality modeling framework. This part of the book points that the type of ontology relevant to the standard case is the one modeling the universe as the totality of entities organized under few primitive classes which are embodied as a multitude of contingent particulars, objects, properties, occurrences, and connections. In such a global model whatever is and happens can be named an entity or a thing or a being, where all beings, things, or entities are arranged into a comprehensive hierarchy of entity types: the prime classes, their subordinate kinds and innumerable individuals. Here the few fundamental kinds of entity are uncovered, namely, substances, states, changes, and relations, make the principal ontological classes of things in the total reality as well as in its particular domains, realms, regions, and levels. Considered apart from the concrete embodiments and circumstances, the fundamental kinds of things determine the top classes of entity, while all sorts and manner of particular objects, qualities, events, and links make up their instances, cases, and exemplifications. Such an integrating theoretical account provides the highest classes of ontological predicates as the conceptual templates under which all other constructs, notions, terms, and predicates can be ordered and distributed whereby unifying a multitude of domain specific ontologies. This postulates that any entity data framework should first appeal to a formal account of reality, to a consistent and exhaustive classification reflecting the real structure of the world with the inherent ordering of its entities.

In Chapter III, **The World Code**, employing the system of symbols and techniques of mathematical foundational theories, vocabulary of higher algebra, lattice theory, set theory, and category theory, we propose an account of the world in onto-mathematical terms of entity variables. As a result, the key assumptions and principles of the Mathematics of Reality (as Real, Ontological Mathematics) are formulated upon which the dynamic formal theory of complex structures and causal systems can be evolved. The chapter indicates that the functional schemas of realities, the categories of the world, the lattice of the classes of entities, the tree of universal properties, and the material equation of things, take priority over the formal constructs as the lattice of theories, the categories of concepts, or the tree of axioms. Also, we sketch the outlines of a new sort of universal mathematics involving ontological principles, definitions, and axioms, namely we talk about the reality-centered, ontological mathematics as the standard model of the world, its distinct levels, realms and domains (natural, mental, cultural, informational).

In Chapter IV, **What Makes Reality**, we analyze the fundamental classes of the world. Leaving relationships for a special analysis, we give a formal description of the primary entity types, objects, states, and changes, as the template classes of the universe of discourse about anything and everything. It is proved that to have a well-developed ontology of computable classes, to construct a high quality Web ontology language, we first need to formulate the universal entity framework supporting all the major classes in the world hierarchy of entities. To design such a machine-centric type system requires a formal inquiry into the world's objects, states, and changes, as the classes of ontological predicates specifying the structure of any realms of thing or any knowledge domains and bases. Having in view the close perspective of practical application of the entity type system as the unified modeling language, this part also examines and formulates the distinguishing features of objects, states, and changes, their definitions, axioms, properties, classification, and rules, so that to identify the common rules of reasoning about them. This part indicates that the entire gamut of knowledge about the world can be expressed in a single ontology language common to scientific knowledge, extant upper ontologies, and web ontology languages.

Chapter V, **What Orders Reality**, indicates that a relation is a real associative entity, like an interaction or space-time relationship, and that this entity-universal constitutes the vital part of the world description frame and hence the core of any general ontology language. We constructed an axiomatic relational model of reality involving algebraic relation theory. A consistent system of ontological axioms is given as displaying the inherent properties of relational entities, which are as follows:

- Holding between or among things (as components)
- Being endued with order or direction
- Being able to reverse the order
- Being capable to be realized (instantiated, exemplified, or embodied)
- Being comparable as identical, like or unlike

It's shown how a universe of relations can be consistently classified, so that the universal taxonomy of relations covers the rich listing of the WordNet relation types running from the part-whole relations to causality to spatial relations to temporal relations to social, human and business relationships to magnitude relations to semantic, linguistic and formal relations.

Chapter VI, **What Organizes the World**, presents a universal schema for representing N-ary relations connecting an arbitrary number of kinds of things modeled as entity variables ranging over classes and individuals. It demonstrates how the n-relational ontology patterns should be defined in the web ontology languages, which reduce relations to mathematical functions. In the simplest case of entity-to-entity relationships, all the possible relative orderings are systematically arranged in the Entity-Entity Relationship Matrix (EERM) structure:

O : O	O : S	O : C	O : R
S : O	S : S	S : C	S : R
C : O	C : S	C : C	C : R
R : O	R : S	R : C	R : R

Given the nature of correlatives, the relational matrix formula allows for (and generate) a full extent of specific relation types, as in:

- Object-to-object relations for spatial (material) substances and non-spatial (abstract) objects
- State-to-state relations for attributes or properties, qualities and quantities
- Change-to-change relations for natural, mental and social events
- Relation-to-relation meta-relations as analogies and proportions of various sorts

In Chapter VII, **What Determines the World**, it is emphasized that of all sorts of relationships the relation of causes and effects comes to be the most substantial connection in reality; the formal analysis of which, of its nature, structure, and meaning can be given within the general framework of relations in reality. It is established that whatever significant is predicated of the relationship entity also is affirmed of causality as its distinguished species. Hence, all the ontological axioms revealing the inherent properties of relational entity apply to causal relations, namely:

- Holding between two and more change types
- Endowing with ordering or direction of actions
- Reversing the causal order of actions or effects
- Instantiating in concrete facts and particular events

While resting on the common theory of relations, there appears the possibility to give a unified account of causality and causation covering the various causal models and perspectives: functional, instrumental, counterfactual, probabilistic, factual, etc.

Chapter VIII, **How to Reason About the World**, indicates the following crucial point: as far as the relations are among the prime constituents of any universe of discourse, the knowledge of relations, particularly the ontological relation of causality, constitutes the basic core of world knowing and reasoning. It is shown that any consistent reasoning upon the world, its particular classes, parts and features, is to be founded on: (1) the ontological relationships of substances, states, changes, and relations; (2) the meta-relations of 'parthood' (part-to-whole/whole-to-part relations specified as classificatory relations like subordination, membership and instantiation) and comparison (contrast, identity, resemblance, difference, and analogy). It turned out that the species of reasoning are ultimately determined by the kinds of relationships. As a result, a whole new opportunity opens up, that of ontological reasoning, realistic inference rules, or entity-driven inference system (an organized system of ontological axioms and inference rules) enabling a new class of intelligent artifacts: general reasoning applications and ontology-based search engines.

Chapter IX, **How the World is Signified**, is largely about building the human natural language as a common knowledge representation and reasoning language for advanced information systems, AI software, and knowledge technologies and applications such as the upcoming intelligent web and versatile reasoning services. This part of the book discusses how the entity and relationship classes may serve as the ultimate semantic values to the word classes and expressions and sentence constructions. It is thus devoted to a formal ontological analysis of natural language, to the role the world knowledge plays in adjusting natural language for meaningful processing by knowledge machines through establishing the semantic relationships between the entity types and the language units. As far as the world knowledge resides in the natural language as intelligence (mental powers) in the human brain, any rich natural language can be well regarded as a linguistic extension of ontology specifying its verbal classification of things.

In Chapter X, **How to Represent the World**, the natural human language (English) is constructed as the most effective RRR language. NL is approached from many sides: physical, physiological, biological, psychological, linguistic, logical, mathematical, and semantic. However, we show that the only one plan of attack may bring out the underlying structure of linguistic knowledge, the fundamental nature and meaning of its parts and lexical units (combining forms, morphemes, words, terms, phrases, expressions, sentences, or utterances). It is that which is based on ontological knowledge and real world semantics. We prove that the number, meanings and relations of word classes (both content words and function words) are tied to the classes of things, to the relationships of their meanings. Or, the syntactic, semantic, and logical structure of language, along with the meaning of its major elements, is ultimately determined by the typology and relationships of the top world things, the referential objects of linguistic signs.

In this chapter, it is also demonstrated that the language as a collection of expressions can be classified with respect to the principal classes of things: substances, objects, or agents; states, properties, qualities and quantities; changes, actions, activities, events and occurrences; or relations, dependencies, and connections. An onto-semantic analysis of the parts of speech and basic sorts of sentences is performed on the assumption that there are virtually several sorts of words and expressions: entity referring terms; substance naming words; state expressing words; change indicating words; relation signifying words.

As far as there are four distinct name types of things like the names of substances, states, actions, and relations, there are four kinds of adjectives, substantive, stative (expressing states), active (expressing actions), and relational (expressing relationships), and four types of verbs indicating being or existence (to 'be' or to 'exist'), states, changes, or relations. As a consequence, it is confirmed that any linguistic statement asserting that something exists (or does not exist) firstly signifies either substance (stuff and identity) or state (attribute, property, quality, and quantity) or action (change, process, and behavior) or relationship (association, connection, and link); while, secondarily, there come the derivative features like temporality, spatiality, conditionality, or modality.

In Chapter XI, **Natural Language Intelligences**, we formulate the principles of designing the NL driven knowledge and reasoning applications, like as NL question/answering systems, domain reasoning engines, representation tools, web search mechanisms, and many other knowledge artifacts, empowered with human language understanding. It is pointed out that most important practical consequence of the WDF comes to be the NL encyclopedic intelligence, titled the Virtual Aristotle Machine (VAM), whose progenitor more than 2200 years ago had envisioned an 'instrument' that 'could accomplish its own work, obeying or anticipating the will of others, like the statues of Daedalus, or the tripods of Hephaestus' (Aristotle, Politics, section 1.4). The principal component of such a universal ontology language machine is a question-answering system possessing the world knowledge and large lexicon enabling deep causal reasoning in any open domain in terms of human language. It is demonstrated how the basic kinds of things can determine a full set of question classes (and answer types) to meet human inquiry in any knowledge domain. Unlike the demised Halo project of Digital Aristotle (http://www.projecthalo.com/) using the traditional KR&R formal logical technologies for modeling, inference, query encoding, answer justification, the VAM is to be designed by applying onto-semantic techniques involving real world modeling, entity (or content)-based reasoning, NL query, and causal explanation and prediction.

Chapter XII, **The Knowledge Society Applications**, examines the critical issues for creating the VAM as a RRR machine: the semantic properties of world representation and reasoning formalism; how the world knowledge base content should be structured; interpretation of the VAM in terms of automata theory as a universal Q&A intelligent transducer, or a general-purpose knowledge automaton subsuming the extant logical models of intelligent agents. The VAM is expected to critically contribute into transforming the Syntactic HTML Web into the Semantic Ontology Web as the next generation of the Internet's World Wide Web. To tell apart the Virtual Aristotle from the Digital Aristotle as well as from traditional KR&R systems, we focus on the attributes of the VAM as an **all-knowing natural language machine**. The main thrust of the Part is that to develop the Virtual or Digital Aristotle capable to intelligently extract and process the meanings of symbolic representations, we need to move away from the historical knowledge engineering tools and technologies blindly banking on formal logical languages. Due to the algorithmic (formal) description of natural language within the general ontological meaning framework (as underlying semantic context), we could establish the principles for designing the encyclopedic knowledge and reasoning machines as a new class of intelligent information (computer) systems understanding human natural language. Empowered to learn, acquire, process, accumulate and communicate the world knowledge in NL forms, the Virtual Aristotle Machine might crucially contribute to the advent of **Global Intelligent Cyberspace**.

In Chapter XIII, **Reality Classification System**, the knowledge production line of the UFO is described. The world categorization system is proposed as Universal Standard Entity Classification System, or USECS®, published as a downloadable Adobe Acrobat 7.0 Document, with the memory size of 14.7 MB and the number of words about a million. To flesh out the categorical skeleton, the rich content of the largest online lexical reference, WordNet 2.1, was accommodated according to the UFO's classification system. Besides, the number of upper taxonomies and web ontologies are included, which can also be encompassed provided that their categorical schemes are properly reclassified.

By its nature, USECS® is nothing else but the world reference system serving as an all-entity classification, the global directory of entities and relationships, or a standard code of basic meanings and definitions supported by common ontological foundation. The UFO product line may carry the following convertible connotations such as World Stock List; World Knowledge Directory (Tree); Universal Scheme of Things; World (Universal) Directory of Meanings; Meaning Searcher, Sense Finder; Human Knowledge Digest, or the Global Nomenclature. The USECS® may be viewed as a general resource of meanings supplying all the essential information about natural, mental, social, cultural, and virtual domains. In its substance, such a general nomenclature, terminology, or language of things is nothing else but the digest (epitome, compendium, outline, abstract, or synopsis) of human learning covering all the key domains of reality, subject areas and fields of study.

In an endeavor to ultimately create **the entire reality classification code**, the USECS project is particularly committed to digesting, systematizing, and encapsulating human knowledge into a summary scheme retaining the most essential things, so that the frame of human learning can be integrated in the human minds or the global knowledge base of reasoning computing applications.

Why This Book is Must Reading

There are few things that distinguish the book as a valuable and significant reading preferable to others of the same topics regardless of some complexness of materials and sometimes unavoidable involution of scientific style. The reasons why readers need to read this book in its entireness are at least three: to learn how to formally describe the world; how to create a single global ontology of anything; and how to establish the ontological and semantic fundamentals of real world knowledge systems.

The loyal reader will be rewarded with the principal answers to the most thrilling and challenging intellectual issues of humanity: whether and how a consistent account of reality can be obtained, which field of knowledge is capable to deal with the whole reality and its formal representation, and how powerful intelligent machines can be brought into existence.

First of all, the reader will discover what is the nature and meaning of the world, existence, or reality, and that the universe is all that exists, physically or mentally, materially or conceptually, the totality of all substances, states, changes and relationships. That there is single underlying ontological dynamic reality coming in diverse forms and levels: physical space-time reality; personal reality; phenomenological reality; commonsense world; semantic reality; social reality, etc.

Second, the reader will let be known that the only science concerned with a comprehensive formal representation of reality, its domains, levels, and constituent entities, which can be effectively used to formulate real world computable models, causal algorithms, and reasoning strategies about the world, is nothing but a global ontology. In the course of reading, the reader will resolve for whether the whole activity is about:

i. The inquiry of reality, its forms, kinds and properties
ii. The general concepts with their formal logical relations
iii. The taxonomy of terms (words) and their semantic relationships

Thus, the reader can finally decide on one of three perspectives, reality-centric, concept-driven, or name-based: **the world ontology** (realistic and veridical); **the concept ontology** (conceptual and notional); **the word ontology** (linguistic and nominal). And the stakes are rather high. The way we are going to choose might affect the validity of the whole cause of Knowledge Technology and Knowledge Societies.

From the pragmatic view, the reader will find out that emerging knowledge systems and real intelligent technologies, as semantic web applications, thinking machines, etc., are all designed to be using the world representations as ''sensible'' signs (physical signals, codes, or words) in order to process and communicate information about things, processes, facts, rules, laws, feelings, ideas, thoughts, or concepts. Unlike the human brain, in the intelligent machines the symbolic codes signify things directly without the agency of concepts, constructs, notions, categories or abstractions. This means that the nature of mechanical meaning is dependent on the types of symbols and the kinds of things these symbols stand for, and that knowledge machines are devoid of mental experience or meaningful mental constructs. The symbols processed by the mechanical intelligence are the signs of entities and hence they get their significance without the mediation of the conceptions of human intellect. Hence the significance of symbols is rather to come directly from the real entities denoted and their relationships connoted, thus promising to leave off most conceptual troubles so long discommoding human beings.

As the last caution against a possible misreading; to get the essential meaning of the book, the reader should try and avoid mixing the scopes and subject matters of Ontology and Logic, under the headings of which the computing ontology and semantic web activities are currently performed. The real semantics or meanings of any symbolism or notations or signs is defined by ontology; for this is the only knowledge domain studying the Being of Everything which is, happens and relates. If Ontology is a real science, Logic is a formal science, because the latter is concerned only with the formal parts of Discourse about Anything or Everything. Logic considers the elements (the terms, propositions, inferences or syllogisms) of the whole discourse aside from their reference to the world (or their real meanings and significance). As a formal science, Logic deals with the formal patterns of discourse common to all sciences. So, Ontology deals with the matter and content, with the real components of discourse about anything, while Logic is an art, an instrument of intellectual reasoning, at least a formal science, and it has directly nothing to do with reality, real significance or real meanings, a coherent representation of which is the ultimate target of Intelligent Web.

So, the content to read is all about three cardinal things:

i. What is the nature and meaning of reality, the real world, existence, the universe, being, or entity, as all that exists and changes
ii. How it can be described, represented, or formalized by the unifying standard ontology, fit both for humans and machines
iii. How to design real artificial intelligence (general AI, advanced AI, global intelligence, or superintelligence) with a potential to supersede the best human minds

Acknowledgment

The ultimate goal of our research is to develop the world science as reality representation and reasoning framework and the universal schema of things in the world, revealing the nature of ontology, its subject matter and scope, categories and principles, knowledge inputs and technological outputs. While striving for this hard aim, the author involved many people to publicly discuss the most challenging questions and issues referring to the subject of ontology and intelligent technology, invoking constructive comments, criticisms, and advices. Much gratitude is due to all the active members of the on-line Internet fora: "Standard Upper Ontology Working Group"; "The Ontology and Taxonomy Coordinating WG", ONTAC; the Semantic Web Interest Group of the World Wide Web Consortium, and Ontolog Forum, a Virtual Community of Practice on Ontology, Ontological Engineering and Semantic Technology, for the inspiring on-line debates of the crucial topics of ontology, meaning, knowledge representation, semantic web, and intellectual technologies.

Most gratitude should be expressed to my wife, Marina, for her persistent encouragement and not being jealous of the labor of spiritual love.

Chapter I

Introduction: Toward the Intelligent Civilization of Ontological Technology

Today, computers anchored to electronic measuring devices, transducers, and telemetric systems are increasingly transforming into worldwide integrated information processing networks. And ontology science and engineering is getting a long-awaited and fully deserved recognition of a critical factor in the 21st century information and communications technology (ICT); especially, in building knowledge-driven intellectual technologies, meaningful machines and reasoning systems, the engine and driving force of the Global Knowledge Society (Novik, Abdoullaev, 1991).

Providing the unifying modeling schemes and languages for sophisticated information technology, the cornerstone of knowledge societies, ontology is bridging the real world and the information universe, giving the dynamic world modeling fundamentals, principles, constructs, representations, and schemes for building a radically new class of intelligent technologies and knowledge systems, **Ontological Technology**.

What are Ontology and Computing Ontology?

Being a fundamental form of knowledge representation about the world, or any part and domain of it, ontology defines the basic constituents and elements of reality: entity kinds, categories, or classes with their constituent relationships, inherent properties and possible instances. Or, in the fundamental sense, it is an all-comprehensive account of reality and realities and their invariant interrelations; for it concerns with the entity-relation types in the world in particular.

Mapping the real world features into the mental constructs encoded as objects types (data structures) and procedures in the information systems, ontology comes as the most comprehensive world model and conceptual schema of things. Accordingly, computing ontology is a world knowledge base (realm of abstraction or conceptualization or model) formally defining a domain (universe of discourse, area of knowledge or practice), its constitutive elements and structure, and yielding the rules of reasoning about its organic relationships, presented by machine-processed formal specifications, notations, languages, and codes.

Being essentially, about theoretical grand schemes and master plans, ontology affords widely shareable commitments of how to view the world, giving a single conceptual environment for fundamental intelligent reasoning about the world. Also, it yields a universal language in which intelligent agents, human and machine can represent the world, expressing and communicating their semantic representations and knowledge. Since ontological classes and rules are, metaphorically speaking, the meat and potatoes of human knowledge and reasoning, they constitute an ideal single representation and reasoning platform for various knowledge representation technologies.

If computing is concerned with computable structures and processes, ontology relates to general structures and patterns of relationships permeating the key branches of computer science: knowledge engineering in AI, conceptual modeling in information systems and databases, and type systems and domain modeling in programming languages design. It can be claimed that the most important and breakthrough technological works currently are in ontology research, design, engineering, and large-scale intelligent applications constructing thereof (Abdoullaev, 1989).

Crucially, the computing ontology is promising the developer not only the tools for organizing information (data) but also the generic mechanisms of reasoning over data (strategic rules). In the inherent nature, computing ontology is basically a formal representation of reality and its domains, levels, and complex entities and is used to formulate computable models, causal algorithms, and reasoning strategies about the world, its parts and aspects. Being the heart and soul of diverse coded classification systems, conceptual schemas and reasoning models, it organizes all the things into a hierarchy of entity groupings defined as classes, categories, universals, resources, kinds, or types. The basic unit of the classification system may be taken as a category (as in philosophy), a kind (as in empirical sciences), a set (as in mathematics), a class (as in logic), or a type (as in computer programming).

Today particular domain theories and models, designated as ontologies, are actively used in most basic fields of computer science and technology, namely, artificial intelligence, computing networks, informatics, software engineering, programming languages, and computational linguistics. It is increasingly realized that knowledge systems ought to be founded on the real world models involving ontology-controlled syntax (Web content structure, axioms, and rules), semantics (Web content meaning), and pragmatics (agents, attitudes, intentions, values, actions, communication, measurement) replacing the formal representational languages implying the logic-restrained syntax and formal semantics.

So it is essential to differentiate two classes of representational languages and technologies: general and specific. On the one hand, there is Reality (or World) Representation and Reasoning [(RRR) or (WRR)] system describing the universe of things that involves foundation ontology as a content and problem-solving theory providing strategic knowledge and reasoning mechanisms about the world. The RRR system is equivalent to Reality Modeling

Framework or World Description Semantic Frame or the Universe of Discourse of Everything, all in the same sense of the unified framework ontology or universal formal ontology (UFO) (Abdoullaev, 1989, 1997, 1998, 1999). On the other hand, there are knowledge representation and reasoning (KRR) systems, which mostly involve formal logical languages, models, and schemes. While the former tends to a unitary representation of reality, the latter is diversified as various formal languages and technologies, such as formal logic operating with individual entities and their relationships; rule-based systems using attribute-object-value triples and plausible inferences; frames operating in terms of prototypical objects and taxonomies; or object-oriented languages using classes (object types) and objects (instances).

If KRR languages belong to the foundational mechanisms of AI technologies, the RRR language is everything for constructing a unifying knowledge structure compounding the particular models and theories about diverse portions and parts of reality, all within a single theoretical account. In other words, there is a global world schema covering the whole reality, material and mental and cultural, with its things, beings, and relationships, and there is a plurality of domain ontologies dealing with the specific regions, parts, domains, or realms of reality. That means that there is Ontology in the primary sense, foundational (intensional), and there are ontologies in the secondary senses, extensional and applied.

The basic meaning of "intension" consists in a fundamental account of reality, its realities and their associative orderings; hence, in the first place, the global model of UFO is about the key features of reality, its underlying entity-relationship classes, principles, and rules. In the second place, it studies how the realities [world things and relationships] may map (or project) to the concepts and associations in the mind, to the coded representations and structures in machines, and to the words and sentences in natural languages. So, computing ontology also is about the representation of the world, its entity states, changes, and relationships, but in machine-processed codes and forms. The foundational ontology, looking at things from an objective and universal viewpoint, provides a consistent set of axioms and rules of inference about reality implemented as general purpose reasoning engines for intelligent application programs, Web services and processes. Thus intensional (fundamental) ontology provides a basic set of ontological commitments about the largest unbounded environment, the world, its parts and relationships, specifying the very content of formal logical systems and representational languages (Davis, Shrobe, and Szolovits, 1993). Despite the wise admonition, most current research projects creating ontology models and languages for knowledge technologies are still guided by the logicist and formalist programs employing first-order or higher-order logics and intensional logics long struggling with the meta-logical issues of completeness, consistency, and decidability.

The evidence that the KRR languages of the classical AI paradigm are inefficient for real computing applications is very compelling. The Vulcan's project Halo, recently folded because of failing to meet the expectations and promises of the initiators, may well serve as an object lesson (Project Halo, 2004). By initial design, the project was targeted to build a Digital Aristotle, a large knowledge representation and reasoning system consisting of extensive world knowledge, organized as computer-readable rules, and an inference engine for managing those rules (Friedland, Allen, Witbrock, et al. 2004). The venture to create a question-answering knowledge application driven by upper-level descriptive taxonomies and logic-based knowledge representation and reasoning technologies (Friedland & Allen, 2004), (Friedland et al., 2004) was doomed to failure for the same reason. First of all, it is because of the critical deficiency of the unified representation of the world consistently and uniformly organizing the classes of things with their lawful associations, both in human-understandable and machine-readable forms. Again, only the RRR language can enable an educational dialogue system with broad scientific knowledge and deep causal reasoning, as the Digital Aristotle is being designed to be.

Other illustrative examples may be found among the ongoing research projects initiated within the AI and Web communities and aimed to construct domain-independent ontologies and ontology languages to build advanced knowledge and reasoning applications.

SUO, an ambitious project started under the auspices of the Institute of Electrical and Electronic Engineers (IEEE), was intended to develop a standard upper ontology (http://suo.ieee.org) of top-level concepts, definitions, and relationships processed by computers (IEEE SUO, 2000; 2004). Such a standard was targeted to formulate a general model of things for enabling intelligent information search and retrieval, semantic data interoperability, automated inferences, and natural language processing. To achieve this broad goal, several starting candidates were sequentially brought together with a view to merge into a single reference system: IFF Foundation Ontology; Lattice of Theories; OpenCYC; SUMO; 4D Ontology, MSO. The project was originated by the IFF (Information Flow Framework) foundation meta-ontology based on a mathematical category theory and the SUMO (SUMO, 2003) targeting to sort out more than 20k common notions of objects and processes.

Later, the CYC ontology was added, making the core of the largest general knowledge base and commonsense reasoning engine, and containing more than 100k terms, 10k predicates, and 1M assertions with the first-order predicate logic as its representational language (OpenCYC, 2003).

The number of candidates was increased with the Shell's data model of things (West, 2003) as a sample of 4D ontology, named Lifecycle Integration Schema, and the multisource ontology (MSO, 2003) aimed to unite some of these taxonomies. The project's objective was to construct an ontology library of distinct modules by mapping, merging, and by the integration of broad vocabularies and nomenclatures and domain terminological sources (Lattice of Theories, 2003).

Regardless of much intellectual efforts, high expectations, and hot online disputations, the implementation of the standard upper ontology as covering all types of domain ontologies is far away from a practical fulfillment. Such non-achievement is chiefly due to ignoring a decisive fact that ontology is neither a catalogue of the world, nor a taxonomy, nor a terminology, nor a list of objects, but rather the general framework, or organizational knowledge structure for all sorts of catalogues, taxonomies, terminologies, and vocabularies (Poli, 1996). Still the SUO project carries a high utility and great potential for knowledge systems and intelligent applications. But to accomplish the project, its general taxonomies should be integrated into the entity reference frame in which all things are orderly placed, much like the space-time continuum for physical events, a kind of universal standard entity classification system (USECS, 2005).

The same observation applies to another outstanding undertaking initiated under the World Wide Web Consortium (W3C) Semantic Web Activity, where the leading role is again assigned to ontology, now in the cause of transforming the current Web into the Semantic Web (W3C, 2004).

The traditional Syntactic Web is *a worldwide information space of resources and services and Web agents interrelated by hypertext links, which architecture is based on three principles:*

identification of resources by global identifiers; representation of resources states, or data formats; and interaction protocols (Architecture of the World Wide Web, 2004).

In its turn, the upcoming Semantic Web is *a worldwide knowledge space of intelligent contents and resources (programs, databases, Web pages, models, and sensors) and services communicating by reasoning agents via the standard ontology language encompassing the Internet markup languages, schemas, and logics*. So the emergent meaningful Web is planned to develop radically novel features with respect to the traditional Web, chiefly relying on Web ontologies and ontology engineering (Semantic Web Topic Hierarchy, 2007). The Syntactic Web/Semantic Web distinction is basically seen in using the common ontology languages as furnishing the necessary facilities for meaningful processing of the Web content by machines. Since, by its very nature, the Semantic Web is mainly about building a computational form of global ontology, "a single consistent Ontology of Everything," encoded as a universal semantic platform and language regardless of its current technical difficulties with scaling up (Shadbolt, Hall, Berners-Lee, 2006).

The key technologies of the Semantic Web usually are structured as a seven-tier cake of specifications: the URI naming scheme; the Extensible Markup Language (XML) syntax (XML, 2004); the RDF (Resource Description Framework, 2004) vocabulary language for coding the Web data; the Web Ontology Language (OWL) vocabulary; description logic; proof; trust (W3C Web Ontology Language, 2004). This architecture guides more specific goals, topics, and projects: Web ontology formalisms and languages; Web-based intelligent software agents with semantic markup languages; ontology-based semantic portals; reasoning and knowledge representation on the Web; intelligent Web services and products; semantic integration of information resources on the Web. And the essential component of the whole undertaking is an ontology language (built atop of URIs, XML namespaces, and the RDF vocabulary) viewed as an engine of interoperable technologies, web search and retrieval, knowledge management, software intelligent agents, and reasoning applications through providing machine-understandable processing of the Web content. Unlike the SUO projects more focused on the nature of things and the content of information, the OWL formal language had a better fate of being recommended as a standard semantic markup language for Web ontologies (OWL, WC3 Recommendation, 2004).

It is however symptomatic of the whole enterprise that any advanced computing projects purposed to build either a foundation ontology standard, or construct an ontology-based Web, or create the Digital Aristotle, tend to postulate some sort of extensive world models expressed in formalized ontology languages. So, implicitly or explicitly, the general classification schemes and conceptual models are involved in the tasks performing the most ambitious and challenging computing projects. And the search for a computing representation capable to cover all the possible entities and their relationships, either as a common upper ontology, a common ontology language, or in the form of global knowledge application, currently is established as an activity of great engineering import with many candidates aspiring after the Holy Grail and Soul of intelligent knowledge technologies.

Current research projects in this area might look brighter and much promising, but one annoying and persisting obstacle. For the time being, the Semantic Web and AI research communities are marked by a deep division with respect to the fundamental issue of the whole field: whether it is possible and how it is possible to build a standard upper ontology of comprehensive coverage and a standard language with a machine processed syntax and unambiguous semantics. Above all, this discordance occurs on account of a widely spread ambiguity of the very nature of ontology, the meanings of ontology knowledge, ontology design and engineering.

There is today an explicit general consensus that the "models of the world"' turn out to be an unavoidable and decisive part of any large-scale knowledge applications. But, at the same time, there also is a widely implicit consent not to explore the nature of fundamental (philosophical) ontology, as if computing descendants were illegitimate and not directly related offspring. As a result of such disregard of its noblest line, the term *ontology* is apt to many interpretations, readings, versions, and renditions.

In information sciences and engineering, an ontology (instead of Ontology) is claimed to be "an explicit specification of conceptualization," "a theory of content," "a theory (a system) of concepts/vocabulary used as building blocks of information processing systems," "a set of agreements about a set of concepts," or "the representation of the semantics of terms and their relationships." Alternatively, it is interpreted as "the class hierarchy in object-oriented paradigm," "a complete schema of the domain concepts," "an entity-relationship schema with *subsumption* relations between concepts." Sometimes, one can meet such definitions as "conceptual patterns," "concept heterarchies or hierarchies," "a body of conceptualizations," "schemata," or "metadata scheme," "a common set of terms," "a controlled vocabulary of terms," "a representation vocabulary," or "a body of knowledge." At best, in the context of computer science, information and communication technologies, the science of entities is reckoned to be:

- A set of generic or philosophical concepts, axioms, and relationships for domain ontologies (IEEE SUO, 2000, 2003)
- A taxonomy of world terms/categories comprising definitions, hierarchical relations, and formal axioms (Mizoguchi, 1998)
- A set of definitions of classes and their relations, as well as individuals and their properties (OWL 2004; 2006)
- A catalog of the types of things (representing the predicates, word senses, concept and relation types of some formal language) organized by the class-subclass taxonomical relation (Sowa, 1997, 2000)
- Metadata schemas with machine processable semantics (Horrocks, 2003)
- Content theories about the kinds of objects, their properties and relationships possible in a certain knowledge field (Chandrasekaran, Josephson, & Benjamins, 1999)
- The total of a taxonomy and a set of inference rules or a document (or file) formally defining the relations among terms (Berners-Lee, Hendler, & Lassila, 2001).

It appears that in artificial intelligence, the Semantic Web, and software engineering, the science of things and their patterns of relationships is commonly regarded as an extension or an external layer of logical calculi and formal languages. As a result, such a formal logical ontology is commonly specified as consisting of the following logical elements: concepts (classes, objects, or categories) with their characteristics (attributes, slots, functions, roles, or properties) and relations (generalization and specialization, functions) restrained by logical axioms (assertions) and exemplified by the instances of classes and properties (Gomez-Perez & Corcho, 2002).

Such confused state of affairs may be partly attributed to the longstanding disagreements, disputes, conflicts, polemics, and arguments over the scope and nature of the subject area even among its greatest scholars. For, when taken as pure and abstract knowledge, the general study of entity in all levels and kinds of reality is formulated as differently as:

- The science (account) of entity (or being) in general
- The knowledge of the most general structures of reality
- The theory of the kinds and structures of things in every domain of reality
- The study of entity types and relations
- The most general theory concerning reality, being, or existence
- A collection of absolute assumptions
- The study of change
- The science of all possible worlds and everything conceivable
- The study of semantic values of natural and formal languages and ontological commitments about the world.

Such total ambiguity or rather equivocation causes the researcher to decide: whether the whole activity is about the inquiry of entity, its forms and properties, or just about some general concepts with their formal logical relations? Or, are we supposed to deal with "the nominal" ontology of terms and their semantic relationships instead of "the world" ontology of entity types and their external relationships? Then one has no choice but to choose one of three perspectives: either *the world ontology* (realistic and veridical) or *the concept ontology* (conceptual and notional) or *the word ontology* (linguistic and nominal). As human history witnesses once and again, big troubles in human life stem from the erroneous views and false world models. It appears the upcoming intelligent systems are not going be an exception; for just as wrong ontologies are fatal to the way of life of the human race, so they are destructive of the knowledge artifacts that are built to represent and operate information about the world.

To handle the art of ontology engineering and computing applications, first of all, we need to consider the fundamental issues: What kind of science (knowledge domain) is ontology? What sort of stuff makes the subject matter of it? What are the most general kinds of entity in reality? How are these things related? How could entities and relationships be truly modeled, represented, and expressed? And how should a standard world specification language for global knowledge applications such as the (ontological) Semantic Web be constructed?

Apart from the technical and instrumental issues as growing pains of computing ontology, it emerges that here lies a virtually insurmountable hindrance, unless the matter is given a more profound and fundamental approach. As much as traditional engineering, ontology engineering is expected to go through the same life cycle stages: research, development, design, construction, production, operation, and management. The first, the research phase, is the most challenging, science-intensive and crucial stage; for it involves disclosing a consistent set of well-defined fundamentals for an applied or engineering ontology.

Such fundamental principles and rules come from **theoretical ontology** consisting in a systematic inquiry of reality and its properties with the help of conceptual tools of science, mathematics, and logic. Evidently, this research should commence from the clear identification of the scope and range of the subject matter (as the universe of discourse), its major principles and methods of inquiry, enhanced with the analysis of their validity, and only then the issues of goals, roles, practical uses, and engineering applications may come to the surface.

The purpose of this book is to propose a general framework as a unifying theoretical system and universal language by compounding the classical models and theories about the nature and pattern of reality within a single standard account. For, as we pointed out, there is Ontology in the primary, intensional sense, giving a general knowledge about the things in the world via a definitive set of basic categories, meanings and definitions; and there are ontologies in the secondary, extensional senses, referring to particular problem domains. The basic meaning and definition consist of a fundamental representation of reality; thus it concerns with the entity and relation types in the world at the first place. Then, ontology considers how the world content, its objects, states, qualities, quantities, events, processes, or relationships, may be represented or mapped by different static and dynamic sign (symbol) systems; namely:

- Mental signs (thoughts and associations in the mind)
- Computing codes (data and structures in machines and software agents)
- Codified knowledge representation (theories and models in sciences, mathematical and formal symbolisms)
- Linguistic signs (letters, words and sentences in natural languages)
- Cultural symbols (socially established symbolic institutions, actions, rules, norms, codes, universals, and social relationships).

In its primary and fundamental sense, this most intellectually thrilling study has been investigated by great minds for many ages.

Without proper respect of the legacy works on ontology, the most productive insights and findings, our new models are doomed to be complementary, confusing, misleading, or conceptually trivial. In other words, any ontology reference frame has to embrace the fundamental categories and principles underpinning the great ideas and scientific knowledge to date. Since while pursuing the final cause of universal ontology, we always risk of falling into a fundamental flaw by leaving out the standard resources by established authors. First of all, we mean the resources systematically collected in the great books of the western world and digested in the two-volume subject-matter index (Adler, 1990). Headed by Being or Thing or Entity, the great ideas are there alphabetically listed as follows: Angel, Animal, Aristocracy, Art, Astronomy and Cosmology, Beauty, Cause, Chance, Change, Citizen, Constitution, Courage, Custom and Convention, Definition, Democracy, Desire, Dialectic, Duty, Education, Element, Emotion, Eternity, Evolution, Experience, Family, Fate, Form, God, Good and Evil, Government, Habit, Happiness, History, Honor, Hypothesis, Idea, Immortality, Induction, Infinity, Judgment, Justice, Knowledge, Labor, Language, Law, Liberty, Life and Death, Logic, Love, Man, Mathematics, Matter, Mechanics, Medicine, Memory and Imagination, Metaphysics, Mind, Monarchy, Nature, Necessity and Contingency, Oligarchy, One and Many, Opinion, Opposition, Philosophy, Physics, Pleasure and Pain, Poetry, Principle,

Progress, Prophecy, Prudence, Punishment, Quality, Quantity, Reasoning, Relation, Religion, Revolution, Rhetoric, Same and Other, Science, Sense, Sign and Symbol, Sin, Slavery, Soul, Space, State, Temperance, Theology, Time, Truth, Tyranny and Despotism, Universal and Particular, Virtue and Vice, War and Peace, Wealth, Will, Wisdom, World.

Making our meaning plain, to achieve the great target of all-entity representation and reasoning system capable to yield unifying knowledge standards, we surely cannot ignore the conceptions systematically arranged as the Great Ideas. Besides, it is crucial for the common reference system to be validated with the essentials of human learning as represented in the *Outline of Knowledge* (Propaedia, 1994) and *Knowledge in Depth* (Macropaedia, 1994), to be additionally supported with the thesaurus of *Webster's Comprehensive Dictionary* (Webster's, the New International) and large lexical online resources (WordNet 2.0, 2004). For, to be an integrative reference schema true to science and engineering, the system must be able to bring into a single unifying description such complex entities as space, time, matter, energy, life, human, body, mind, society, culture, art, technology, religion.

In pursuit of such a tall task demanding profound scholarly learning, intellectual dedication and consecration to fundamental study, it is good to stick to several well-established and self-evident working principles.

Traditionalism, no one large classification scheme will have application prospects without being grounded on the classical ontological writings and works. Aristotle's standard reference works in the first place, to which all the great minds seeking for a broad scheme of things tried to conform for many ages (Aristotle, Logic, Physical Treatises, Metaphysics).

Fundamentality, until the fundamental ontological categories will not be cleared up and specified as the general type system, all attempts of erecting a standard will be impracticable with inherent flaws insuring their failure in applications programs.

Mathematicism and scientism, the knowledge standard should be constructed as a universal system of classes, definitions, axioms, and rules consistent with the sciences (natural, psychological and social) and mathematics, so that to render the completeness of analysis, consistency of meaning, and correctness of inference.

Hierarchy and systematism, there are three types of ontology: universal, upper-level, and regional or domain-specific ontologies; the union of which results in a complete, unifying reference system.

Universality, the reference standard is to be developed as a universal theory and a universal language at the same time. Since it can be employed as a language by the specific theory, like theoretical sciences (relativity) are using mathematical language (topology), and all of them–a general ontological language.

Professionalism, expertise, and sophistication, any computing professional, an AI expert or Web researcher or a programming developer, however brilliant, can hardly create effective entity models avoiding a professional ontology scientist.

Adhering to the methodological rules, we research the nature and meaning of reality, its classification or constitution, its basic contents, kinds and levels, by giving a systematic analysis and formal representation of the world's composition and structure, properties, states, dynamics and behavior, and all possible relationships. All this is done with a view to work out a standard world schema for programming knowledge and reasoning systems and human beings, so that to secure semantic interoperability between and among the members of these seemingly unlike species of intelligences, existing and emerging. So the main subject of the book is no more and no less but the largest existing environment, that is, the world or reality or the universe or existence. The keynote, burden, and motif of the work is the practical possibility of a unifying RRR system describing the basic kinds of things in the world (with its special domains) and authorizing the major types of reasoning processes (strategies and procedures) about reality. Or, such an integrated RRR mechanism is promising:

- To organize all substantial information about the world
- To assign semantics or meanings to representational languages, natural or artificial
- To acquire and learn new facts (data and information) and rules and principles about changing surroundings
- To prescribe architectural schemes of reasoning systems and knowledge agents
- To make and perform knowing decisions and intelligent actions.

Crucially, the work formulates the integrated account (of classes of entities or things or beings or resources) proven to be fundamental to the construction of knowledge and reasoning applications not only as a descriptive account but also as an explanatory and predictive scheme. Explaining and describing the central generic entities in terms of mathematical and natural languages, the book uncovers the way to deliver an intellectual artifact as intricate as the **Virtual Aristotle**, which can be equivalently defined as the know-all language machine or the encyclopedic AI, or the all-purpose semantic instrumentality, or the universal ontology reasoning engine as the quintessence of the UFO-based knowledge technologies. This required developing the UFO as a general RRR model capable to incorporate the basic content of scientific theories, lexical taxonomies, (upper) ontologies, and data languages for Web ontologies like as OWL and its extensions (Horrocks, I., Patel-Schneider, van Harmelen, 2003; OWL, 2004, 2006; OWL-S, 2006).

The Standard Ontology for Machines and People

To understand the core points of UFO, it may be of use to outline how the global schema may smoothly embrace the unified data models such as the entity-relationship (ER) model (Chen, 1976; Chen et al., 1998; Thalheim, 2000). The information model is known as widely utilized in information sciences and engineering, in the recommended Web data standards like as RDF and OWL, widely used by Web developers for semantic and reasoning applications. Some basic assumptions of the ontology reference also can be comprehended in terms of the ER schema, an original source of the whole new class of ontological languages and technologies in computing applications. The popularity of the ER model may be explained by its tying in the real world described as the totality of entities and relationships and their properties, where distinct entities and relations were generalized as an entity set and a relationship set, respectively. This permitted to employ the conceptual power of mathematical set theory and relation theory regardless of how these fundamental theories appear abstract and formal.

In the ER information modeling, all individual things were ordered by entity classes E_i or relationship classes R_i, while a property (attribute) was represented as a function f mapping from an entity (relationship) set into a value set (or a cross product of value sets). As a result, in the ER conceptual schema the formula for the world (or a domain of interest) W can be rewritten as the mathematical structure:

$$W = < E_i, R_i, f >$$

This ontological approach has been implicitly taken as a paradigm by many subsequent data modeling languages, including the Semantic Web ontology, where the task of the (concept) ontology also is defined as the descriptions of general things (classes), their relationships, and the properties (or attributes) that they may possess.

In our study, one principal novelty is introduced. The ontology standard schema language can be constructed by lifting up the ER model to the level of the whole world (or reality) viewed as the single universe of discourse composed of entity classes and relationship classes and instantiated by the entity (relationship) species and individuals. It clearly is demonstrated that a comprehensive formal theory of entity with relationships is something indispensable, something that is required before building any general ontology languages or information and data models of reality.

But unlike the ER model, the RRR model is based on the central thesis: the entire reality, the universe, the world as the whole entity, which parts and domains are aimed to be represented by scientific theories and data models, might be naturally split up into the following prime entity kinds (quantified as entity variables):

- **Substance, stuff, or substrate** (objects, material or nonmaterial, spatial or nonspatial) **O**
- **State** (properties, quantities, qualities, and attributes) **S**
- **Change** (actions, activities, events, or behaviors) **C**
- **Relation** (links, associations, connections, and bonds) **R**.

In the world representation schema, all the kinds, types and instances are comprised between the null class of Nothing or Nonentity, and the universal class of Thing or Entity, **T**. We can now return again to the ER model as essentially enriched with the entity theory founded on the world-formula formulated in terms of real classes expressed by real variables, which specific values correspond to individual objects, instances, examples, cases, happenings, etc.:

$$\{\mathbf{W}, W_I\} = <\mathbf{T}, \{\mathbf{O}, O_j\}, \{\mathbf{S}, S_k\}, \{\mathbf{C}, C_l,\} \{\mathbf{R}, R_m\}, \{\mathbf{f}, f\}, \square >$$

In other words, the basic classes (or universals and types) are instantiated by concrete instances (particulars, tokens, individuals, objects, events) in the realm of causality, time and space, all symbolically expressed either by linguistic items or by mathematical values of variable quantities or by objects (instances) and programming classes (types) and procedures (functions).

Essentially, universality of the RRR language is supported by its capacity to cover also the SW formal languages, systems, and vocabularies, where the key constructs, individuals, classes, and properties, are notional and abstract terms without real content and substance, that is, without reference to reality. Such partial world description is inherited from AI knowledge representation and reasoning (KRR) technologies:

- Prototypes and taxonomies without specifying what sorts of things are the prototypes (frames)
- Plausible inferences without specifying the content of the inferences (rules)
- Networks of related concepts without specifying their real content (semantic networks)
- Individuals and relations (predicates) without specifying their nature and meaning (formal logic).

As a matter of fact, to be a real representational language with the inbuilt ontological classes and rules, semiotics, reasoning logic, and efficient computation, a general formal language must be focused on the world with its domains and levels. First of all, it is centering on its basic construct, Thing or Entity or Being, viewed as equal to all things existing in reality, and generalized as the class of all entity classes. And the most fundamental categories and natural divisions or ontological classes of Thing are the class of Substance (Object), the class of State (Quantity and Quality), the class of Process (Change or Action) and the class of Relationship. Each of the classes is organized as a hierarchy of subordinate classes (kinds and types) and individual things (instances, particulars, and concrete entities), as individual objects, specific states, unique events and particular connections. That unambiguously implies that the formal notions of *definition*, *class*, *property*, *proposition*, and *reasoning* should be filled up with real contents and semantics and closely tied up with the ontological predicates of entity, substance, object, quality, quantity, relationship, change, cause, effect, time, space.

So the status, validity, and expressivity of any general representational languages and technologies are chiefly determined by the ways of treating the world, its cardinal features and operative representational constructs. The most effective conceptions scientifically proved to be the following constructs:

- Thing, with its kinds, types, varieties, and instances
- Class, with its subordinate classes, sets, and members
- Property, with its basic, essential, intrinsic or accidental, nonessential, and extrinsic properties.

Crucially, the Class/Thing distinction makes all the difference and needs some preliminary commentary; for its a central issue in all current activities and efforts of building upper ontologies (SUO, UML, GFO, etc.) as well as SW ontology languages (RDF, RDFS, OWL, OWL1.1, etc).

In terms of the XML namespace prefix syntax, the notions of Class and of Thing in different top-level ontologies are branded as `{rdfs:Resource; owl:Thing; UML:thing; GFO: entity; ISO:object; DOLCE:entity, etc.}` or as `{rdfs:Class; owl:Class; UML: class; GFO:universal; DOLCE:universal, etc.}`.

In fact, there is no any fundamental difference between Thing, generic, or Entity, generic, or Class, universal; they are marked by equality in principle. In the particular sense, *Thing* defined as an individual entity having separate, distinct existence, and *Entity* as an individual thing having separate, distinct existence. To the contrary, in its fundamental sense, the notion of Thing or Entity or Being denotes all distinct things (entities, beings), things (entities, beings) of every kind and type, or refers to the class of everything, organized as a universal hierarchy of entity classes with higher super-classes (kinds) and lower subclasses (types).

Currently there usually are three main choices widely practiced: defining *Thing* either as an individual, or a class of individuals, or the universal class (i.e., the class of all classes). What in terms of quantities may be interpreted as a fixed value (constant), an individual variable, or a class variable, respectively. The narrow view of thing as an individual entity with a specific identity has its long history of "a primary substance," "a bare individual," and so forth, borrowed by the first-order logics. Because of this, the extension of the construct-like owl: Thing has only individual things, being void of other essential meaningful dimensions of kinds, types, and species. In the biological classificatory system, this corresponds to the level of species whose members share a set of essential features and bound by a membership relationship between an individual and its class. Any species as a collection of individuals, say, the totality of human beings, can be subjected to further divisions and subdivisions, such as man and woman, white or black or yellow or red, the aged or the young, the poor or the rich, the working class or the professional class; underworld, lower class, middle class or higher class, and so forth. Yet they are not (genetically) essential classifications, and we still are in the domain of individuals, for even infinitely increasing the number of individuals never ends up with a new class or species or kind. Therefore, we used to mark two types of difference, in kind or in degree.

The right and more fundamental position is then to consider Thing (or Entity) as the class of classes (the set of subsets) at least; at best as the class of all classes (the universal set of all sets), hierarchically ordered by inclusion (containment) relationships (or whole-part relationships). Since, as the class variable, Thing will have as its values lower classes and subclasses as well, or the type of variables whose values also are variables. This all univocally implies the fundamental identity of general formal ontology languages:

UFO:Thing ≡ UFO:Entity ≡ UFO:Resource ≡ UFO:Being ≡ rdfs:Resource ≡ rdfs:Class ≡ owl:Class.

As for the properties of things, another significant construct of SW standard languages, the general construct of Property also is harmfully misunderstood: owl:Property is divided into two basic types, owl:ObjectProperty (mapping individuals to individuals) and owl: DatatypeProperty (mapping individuals to datatype values).

In fact, there are monadic and dyadic properties; essential and accidental; atomic, transient, complex, or emergent; particular and general, and so forth. But mostly important to tell the formal properties (attributes) from the ontological properties, which are classified as:

1. The property of being a substance (object), substantial properties
2. The property of being a state (quantity or quality), quantitative and qualitative properties
3. The property of being a process (change, action, operation), dynamic, functional, operational properties
4. The property of being a relationship; relational properties *per se*.

Thus, owl:Property is narrowed to the property of being a formal (functional) relationship, direct and inverse; without explicitly identifying the nature of relations between the components, like as spatial, temporal, causal, whole/part, syntactic, semantic, pragmatic, etc., all covered by UFO:Property, both essential, monadic and relational, dyadic properties. Moreover dealing with only two main disjoint types of property: owl:ObjectProperty and owl:DataProperty discard the possibility of a class-to-class mapping, or many-to-many relationships, including the pragmatic issues of measuring commensurability between magnitudes (entity variables) and multitudes (numbers, a set of numerical values, or quantities).

Now the reader can see that instead of data, information, knowledge, concepts, constructs, or terminologies, the real subject matter of computing ontology must be nothing but entities in reality with their patterns of relationships; for its main role to represent reality, its kinds of things (universal, classes, types) with their instances (objects and events) in the actual space-time continuum as machine processing symbolic structures and encodings (Smith, 2004). Therefore, modeling the total of entities in reality, the unifying formal ontology starts with the largest thing in existence, the world or reality **W**, by mapping onto the global knowledge and reasoning representation system its contents, state space $\mathbf{W} = \{W_I\}$, dynamics ($W \times W$), n-ary relationships ($W \times W \ldots \times W$), with all the key kinds, domains and levels. The backbone modeling system is marked by a self-consistent representation of the world's dynamics as the total continuum of self-transforming processes ($\mathbf{F}$: $\mathbf{W} \times \mathbf{W}$), when the world state (or space) is mapped into itself by a transformation function or operation or correspondence or map **F**. Generalizing the concept of multidimensional transformational phase space in theoretical physics, **the world state space** is marked by the {many, one}-{many, one) transformation modes (or mappings), increasing or decreasing its dynamic diversity, variety, or uncertainty (information) as **the largest domain of knowledge** and **the universe of discourse of anything**. For the whole natural world, there is applied the many-to-many mappings that correspond to complex and dynamic networks of processes, physical, chemical, biological, mental, social, or informational (like the Intelligent Web processes).

So the veridical mapping or realistic projection or true representation of the substantial world describes all sorts of real circularities and nonlinearities, the key features of complex nonlinear systems as physical systems, organisms, intelligent systems, organizations, economies, and social systems, studied in cybernetics, mathematical dynamic systems,

systems science, nonlinear physics, life sciences, and social sciences. A crucial importance is commissioned to causal circularities as the network of mutually related real processes encompassing various feedback cycles and loops, positive and negative, and thus driving all complex phenomena in the natural domain, conceptual universe, cultural region of social reality, or virtual world of computing machines. So, *the dynamic nonlinear substantial world is to be realistically modeled as the entire networks of worldwide causal loops*, the real circles of causes and effects moving the whole universe and consisting of a multitude of interacting changes (processes, causes, agents, factors, elements, or variables) reciprocally and convertibly causing (and mutually sustaining) each other (Abdoullaev, 2000). For all the forces of nature (mechanical motions and forces, gravity, heat, magnetism, electricity, radiation, chemical force, biological actions) are interrelated and mutually dependent, reversing the ways of actions by reciprocally causing each other.

Crucially, the universal dynamic loop topology structure is applied not only to the whole world entity but it also goes for any specific field, realm, region, part, or isomorphic representation of the world, for any domain of interest ranging from set theory to physical devices domain, viewed either as a set of components or a set of specific physical processes.

As a global master model, the **standard ontology** of the world (or everything) is then to be constructed as a comprehensive causal account of reality where its entities exist as interacting systems of various kinds and levels: natural, psychological, social, cultural, or computational. Evolving this basic assumption, we display the type of ontology acting as a coherent and consistent supporting structure for the diverse levels of reality:

- The material realm of natural realities
- The psychological realm of mental entities
- The social reality of cultural forms and processes
- The computational level of informational entities.

In such an extensive consideration, whatever exists and happens (or conceived as existing and occurring) can be named an entity, a thing, or a being, so that all beings, things, or entities are to be arranged into a great hierarchy of kinds, types, subtypes, and individuals, formed into distinct but interdependent levels of the universe.

To sum up the significant issue, a few fundamental kinds of entity compose the principal kinds of things in the universe (of discourse), which are substances and objects (spatial and non-spatial or abstract), states (attributes, properties, qualities, or quantities), changes (actions and processes), and relations (entity-to-entity). Then, having the underlying ontological categories and principles, we can work out a universal formal account of reality having all the necessary resources to incorporate conceptual models, lexical taxonomies, general ontologies, and formal ontology languages into a single, consistent and comprehensive framework of things.

Knowledge Society and Ontological Technology

A high practical goal of the standard ontology as a unified account of reality and the universal representation language is to produce the world knowledge standards, which meet both the requirements of machine intelligences and human minds. The underlying thesis is to provide the entity type system with descriptive, explanatory, and predictive features to be used as a single consistent frame for the emergent class of effective knowledge systems and reasoning applications. As said, such a task may be reached by incorporating into a common ontology library of things all the relevant constructs from existent upper ontologies, classification schemes, conceptual data modeling languages, and the web ontology languages. Having the ontological entities as the determinative components in meaning (semantic relationships) allows for natural language as the most general knowledge representation and reasoning language. This means that the natural language enables us to lay down the principles and rules for designing and developing encyclopedic knowledge and reasoning machines as a new class of intelligent information systems understanding human natural languages. Equipped to learn, acquire, process, accumulate and communicate the human knowledge in linguistic forms, the encyclopedic RRR applications can constitute the core of coming **Global Intelligent Cyberspace** (the intelligent Web): a worldwide computer networks of intelligent NL machines and human beings, the technological base of **Intellectual Knowledge Society**.

Advanced information and communication infrastructure is increasingly viewed an essential foundation for knowledge-based information societies, particularly in creating digitally new forms of social life: e-government, e-business, e-learning, e-health, e-employment, e-environment, e-agriculture, and e-science. Building a Knowledge Society is currently a social conception and pragmatic topic occupying the top agenda in the meetings of heads of governments debating an optimal strategic policy and seeking for a silver bullet solution. According to the Plan of Action proposed on the World summits on the Information Society, "ICT applications can support sustainable development, in the fields of public administration, business, education and training, health, employment, environment, agriculture and science within the framework of national e-strategies" (WSIS, 2003, 2005, 2005).

There are many implications and connotations carried by the emerging knowledge societies: lifelong learning, mass education, progressive schools, extensive research funding, modern teaching methods, advanced political system, innovative industries, online universities, universal access to information systems, electronic voting, enlightened governance, knowledgeable administration, e-world government, onto-technocracies, or robot-states, etc. But there is one principal feature distinguishing the most advanced phase of human society: the explosive growth (generation and production) of scientific knowledge and electronic forms of its codification, processing, transmission by information and communication technologies. And the heart and soul of information infrastructure and its innovative information processing/communication systems is intellectual technologies, ontology-driven knowledge machines, which are designed to make the cornerstone of information infrastructure of the upcoming learned societies with widely enlightened citizens.

Then *a knowledge society is equal to a state of technological civilization where a personal, economic, political, scientific, social, cultural, and environmental development is advanced by knowledge infrastructures and intellectual technologies.* The urgent necessity of a knowledge-based society is basically caused by the fact that the natural unity of world knowledge is badly destroyed by the modern system of scientific, formal research and education. For example, the harmful consequences of the greenhouse effect (or global warming) were explained by Svente Arrenhuis in 1896, who established a causal relationship between the increase (doubling) of the carbon dioxide level (via burning fossil fuels) and the atmospheric temperature (raise by 5 degrees). Only because the fragmented learning could not be available to the right people in the right place at the right time, the phenomenon should be reopened again after a hundred years. Division of labor and increasingly narrowing specialization are making knowledge irreparably fragmented and scattered, with its legally restricted dissemination and public access, and thus producing a confusing multitude of disunited scientific and technological communities and associations, often duplicating the same R&D programs even within the same community.

Due to modern information technologies and particularly the Net, there is a sea wealth of information recorded in gigantic databases. This may be mastered only by more advanced information technology, knowledge machines with the built-in common ontology framework, a single code of fundamental standards, principles, rules, and laws, suggesting a broad, integrated model of all the world things. What may alter the whole knowledge production system in an essential way, ending up in a radically new type of human society, the intellectual knowledge society, driven by entity-centric semantic technology.

To answer the historical challenge of building a knowledge-based economy, the 2000 Lisbon European Council set up a new strategic goal for the next decade, "to become the most competitive and dynamic knowledge-based economy in the world" (Lisbon European Council, 2000). Unlike the industrial economies driven by traditional physical machines and mechanisms, the knowledge economy is to be propelled with the advanced ICT, which core is constituted by the knowledge processing machines and semantic communication systems driven by ontological technologies.

The grand vision of the Europe of Knowledge, Knowledge-Intensive Society, Knowledge Base of Europe and Technological Know How, acts as a spiritual dynamic on the participants of the European Commission's Framework Programmes for Research, Technological Development and Demonstration. As an essential part of the overall strategy, the EU initiated large R&D projects in Information Society Technologies in the realm of Semantic Web. In order to lay down the knowledge infrastructures of the upcoming Information Society the EU's Research Council and the European Parliament allocated 3.8 billion Euro for Knowledge Technologies within the 6th European Union Framework Programme (FP6) for Research and Technological Development, with a budget of 17.5 billion euro (CORDIS/FP6, 2006). Within the FP6 Programme, all the Web-based knowledge technology projects are largely concerned with ontology research, design, learning, and management.

For instance, the Knowledge Web (network of excellence) is engaged to transfer ontology technology from universities to industry (Knowledge Web, 2006); the Data Information and Process Integration group is contracted to contribute to the infrastructure of Semantic Web services, promising an integrative Web service modeling framework via ontological technology (DIP, 2006). In its turn, the Semantic Knowledge Technologies network, part of ESSI cluster projects, is signed to produce ontological software and tools for Semantic Web services (ESSI, 2006). Again, the integrated research activities are performed under the costly collective delusion that ontology-based Semantic Web and services technologies can be constructed without having the single RRR Web platform. Realizing that and trying to avoid duplicating each other, recently the formation of the European Semantic Systems Initiative cluster was announced, targeted to deliver knowledge technologies by means of Semantic Web modeling ontology and languages (ESSI, 2006). It should be mentioned here that the high cost of an academic head game of missing a scientific concept of Semantic Web as an ontologically united global information system, may be equal to the total budget of the integrated project, thus making the common error of confusing public aid spending with advanced information technology delivery and eroding the public trust in science.

It would do only good for the public and the governments if the strategic planners of Knowledge Society information infrastructures show some prudence in funding the semantic projects that are too abstract and notional, without a firm ontological grounding in the real world settings. And the policy makers and IT agencies of the governments shooting for knowledge societies should be more focused on funding the socially promising research programs based on an ontological (dynamic) modeling of a complex world. Like those projecting the strategic knowledge management systems and global information framework technology (Global Information Framework, 2005) or proposing the unifying framework architecture for intelligent systems (Sowa, 2002). For one can only wonder how, for instance, the DIP program, coded as FP6–50 74 83, can deliver "a comprehensive Web Service Modeling Framework" (Fensel and Bussler, 2006) without having first developed the World Modeling Semantic Framework and hence the general model of services and products and business processes (USECS, 2005). After all, this is not just a speculative academic research, but an IST (Information Society Technologies) project with a total budget of € 16,3 M for three years only.

To avoid bad misuses, frauds, self-regulation, poor deliverables, or even no publication at all allowed by 6th FP, the whole scientific research system in the 7th FP is planned to be overhauled: professional expertise and advice, quality publications, monitoring and control, trust, independent reviewing, ethical behavior, public transparency, and financial audit. To further advance the goal of a knowledge-based Europe, the European Commission 7th Framework Programme for Research, Technological Development and Demonstration is starting January 1st, 2007 (CORDIS/FP7, 2006). The whole FP 7th Budget amounts to 54b €, with the indicative breakdown for Cooperation—32 365 € million; Ideas—7 460 € million, People—4728 € million, Capacities (research infrastructure)—4 217 € million, Euroatom—2751 € million, and the Joint Research Council—1751 € million. Under the Cooperation the following thematic priorities are highlighted: Health, Food, Agriculture and Biotechnology, Nanosciences and new Production Technologies, Energy, Environment, Transport, Socioeconomic Sciences and Humanities, Space and Security. A lion share here goes to the information and communications technologies (ICT) programs.

It is becoming evident that a strategic goal of a digital, knowledge-based society requires an intensive R&D of ontology-based knowledge systems and intellectual technologies, directly performed within the ICT programs and indirectly within other projects such as New production Technologies; Security and Space, Health (Building the Europe of Knowledge, 2006). And it is an encouraging sign that the 7th Framework Programme is planned to build on much better principles then the 6th FP. First of all, it presupposes an abundant financing of individual projects under the rubric *Ideas* as an investigator-driven frontier research by individual teams, judged by the European Scientific Council, where the author has submitted the Standard Ontology Project. Though, still the question may be raised, who will be in the Council? From our initial experience, the fate of a project application might be blocked by any officer appointed by the EC to police the eligibility criteria. However brilliant proposal could be, if the officer finds some formal reason to block it, say, a less than the minimum number of participants indicated in the proposal call, the proposal is halted for better times. The EC FP red tape might create aversive conditions for outstanding researchers, thus breaking off a real scientific and technological advancement in the EU. To head off this state of affairs, the application procedure should rather focus on the content of the proposal instead of its formal status.

To further avoid the fundamentally wrong research projects involving ontological and semantic technology and knowledge systems, a fundamental (or multibillion) problem must be resolved: to answer what is really passing as the intellectual technologies, the semantic-based knowledge systems, and the Semantic Web, to understand their nature and real meaning. Particularly, we must make out which Web (or architectural pillars) most fits the matter, the formal Semantic Web (i.e., the Syntactic Web) or the real Semantic Web (the Intelligent Web).

The Real Web is characterized by the Web hierarchical structure anchored in the real world environment via the ontological groundwork:

```
<Global Intelligent Cyberspace, or Real Semantic Web> :: = <Ontological
Framework, WDF, or RMF> <Semiotics> <the World Wide Web, or the Syntac-
tic Web>

<Ontological Framework> :: = <Unified Framework Ontology, WDF, or RMF>
<Upper Level Ontologies> <Domain Ontologies>

<Semiotics> :: = <Pragmatics> <Semantics> <Syntax>

<Pragmatics> :: = <Users> <Web Agents> <Intentions> <Actions> <Communi-
cation> <Proof> <Trust> <Truth>

<Semantics> :: = <Signs, Data Type, Natural Language Expressions> < Con
structs> <Meanings>

<Syntax> :: = <Logical Framework> <Rules> <OWL Ontology> <RDF Schema>
<RDF M&S> < RDF> <XML/SGML> <Namespaces>

<the Web>  ::= <Resources, Representation, Identification, URI, Unicode>
<Interaction, Software Agents, Hypertext links, Protocols, HTTP> <For-
mats, HTML, XHTML>
```

The central idea of the Real Web is rather simple: without a common standard ontology as a common code of meanings and rules, there is no base and foundation for the whole enterprise of ontological semantic technology, powerful programming tools and applications employing Semantic Web technologies (Semantic Web Tools, 2006). To be consistent, its creators should fully appreciate a hard truth that the Semantic Web with its technologies and tools postulate "a complete top-down ontology of everything" (Updegrove, 2005). And the world modeling framework or reality description model as the unifying ontology of Web content constitutes its common semantic framework. It is tantamount to comprehension that the Intelligent Web is to be designed as a worldwide RRR system of knowledge resources having the ontological groundwork with the built-in semiotic theory of meaning, rather than the Web involving a formal semantics and formal logic languages.

Above all, "the Internet is a giant semiotic system" of signs and symbols used by the agents to represent entities in the world and their intentions concerning them (Sowa, 2000). Accordingly, the knowledge-based Web is a **worldwide dynamic information space** with the inherent ontological foundation and semiotic models of resources. So, instead of relying on the RDF/OWL logical syntax and formal semantics, a consistent way to build the Semantic Web will be by means of real semiotics covering the relationships of signs, agents and things and so involving syntax, semantics, and pragmatics. Digesting the argumentation, any extensive knowledge system, like intelligent Web, is to be grounded on the World Modeling Framework with ontology-controlled syntax (Web content structure, axioms, and rules, representational languages), real world semantics (Web content meaning) and pragmatics (agents, attitudes, intentions, values, actions, communication, and measurement).

While sticking to the current Semantic Web cake structure (URI, Unicode; XML, Namespaces; RDF, RDFS; OWL Ontology; Rules; Logical Framework; Proof; Trust) (Berners-Lee & Fischetti, 1999), the Web applications will be without a firm semantic and ontological groundwork, but wholly banking on the descriptive capability of the SW languages. Though, this descriptive capacity comes from a legacy set of predicable relations of "definition," "class," "property," "sameness," "difference," and "inheritance" (Aristotle, Topics), and it is little to do with this specific sort of formal languages. The necessity of revising the SW technology push-down stack is even recognized by its active contributors pointing to the inadequacy of RDF syntax and semantics and urgency of having "a common semantic foundation" (Patel-Schneider, 2005). All this considering, the UFO as the world modeling language is designed to override the SW formal logical languages telling the application developer little if nothing about the actual nature of its polar constructs: individual, class, property, and relationship, their basic meanings, kinds, and instances (USECS, 2005). As a consequence, the SW languages cannot be effective for building reasoning systems, if only as a syntactic part and ingredient of the World Modeling Semantic Frame, a knowledge base and foundation of real-life Web applications. And the golden rule of any general formal language is:

In order to be tied in the real world domains, a language's constructs should be reified within the foundation context of entity and relationship classes (universals, kinds, or types) and instances (individuals, tokens, particulars).

Otherwise, the general modeling language is unreal, without substance and reality and truth and real world effect, and hence credibility. Therefore the OWL semantic language as a general ontological language is essentially defective, as devoid of ontological bottom, while offering the alluring but illusory promises, as much as the secret intelligence operations pretending to be more effective than the complex political negotiations. Scientifically and commercially, it is a serious confusion to use such unreal (formal logical) language without any essential reference to reality as "ontology infrastructure for the Semantic Web" (Bechhofer & Horrocks, 2003).

In practice, a game plan for building the Intelligent WWW should necessarily include the following developmental stages for the emerging Knowledge Society technology to be accomplished:

> INTELLIGENT WWW: THE WEB > THE SYNTACTIC WEB > THE SEMANTIC WEB > THE PRAGMATIC WEB > THE ONTOLOGICAL WEB (Worldwide RRR System)

Then the extent and depth of ontological language used for representing the Web data, services, and processes must match the master plan of building the worldwide RRR system combined of Real Entities (things in the world), Constructs (representations and rules), and Signs (sign and symbol systems, static Web data and dynamic Web agents). Or, the basic formula of the Web ontology language might be specified as follows:

> WWW UFO Language = UFO (world entities, relationships, axioms and rules) V SUO Taxonomies V Domain Ontologies V Semiotics (Web content as the largest sign system, Syntax, Semantics, Pragmatics) V Logics (Logical Frame, SW languages)

The base of the Web RRR system is the UFO representational language acting as a common ground between the SW formal systems and top-level ontologies. As such, the Web ontology language system supplies a core set of ontological template classes together with their principal relationships, all to be used as the knowledge standards in developing a new class of ontological machines: all-purpose 3R systems with the inbuilt ontological models of the world, of its domains, structures, and processes. In other words, **the world description framework** can provide computing machine intelligence framework (CMIF), a unified knowledge and reasoning structure uniformly modeling the whole universe of real things, their classes, kinds, sorts, and levels, from a single unifying perspective. When reconciled with a flexible modular framework for intelligent systems (Sowa, 2002), the CMIF may serve as an integrated intelligent environment for various domain-specific AI applications, existing and to be invented:

> Natural language processors (parsers, taggers, text-matching, syntax checkers, disambiguator, machine-translation systems, question-answering systems, dialog managers), theorem provers, inference engines, learning programs, classification tools, statistical tools, neural networks, pattern matchers, problem solvers, planning systems, game-playing programs, knowledge acquisition tools, modeling tools, robot guidance systems, intelligent search engines, smart Web browser, ontology editors, etc.

Supported by the unitary ontology, the CMIF can cover the extant class-based representation formalisms used in software engineer, databases, Semantic Web and artificial intelligence: entity-relationship models and diagrams, semantic networks and rule paradigms, UML class diagrams, and OWL Web ontology language. Being comprehensive in its scope of representation and mapping of real world structures, processes, and relations, the computing intelligence frame may constitute the uniform information architecture for encyclopedic artificial intelligences like as the Virtual Aristotle.

In all, the major objective of the book is to formulate the principles of UFO (or the world description framework or the reality modeling context or the universe of discourse of everything) as the ontology base for the real Semantic Web relying on real natural languages and thus replacing the formal Semantic Web as heavily dependent on unreal logical languages. It is to be shown that a formal Semantic Web is a poor abstraction of the real Web asking for a firm conceptual foundation as the world modeling framework with its essential components, the real ontology and semiotics, coinciding with reality. Then, having the UFO-based unifying semiotic frame as the world knowledge and reasoning platform opens up a possibility for a large variety of Knowledge Society artifacts (products, processes, services, systems, technologies, and applications):

- The real Semantic Web ontological languages
- Human knowledge integration systems and semantic networks
- Generic reasoning mechanisms and units
- World data processing physical cognitive systems, from smart transducers and intelligent sensors to all-purpose space robonauts
- Natural language understanding software packages
- Knowledge-intensive embedded reasoning systems
- Intelligent content systems for the Internet, as online smart encyclopedias, catalogues, taxonomies, vocabularies, and terminologies
- Generic reasoning platforms and intelligent search technology for the Semantic World Wide Web, interactive home TV, and so forth
- Encyclopedic intelligent applications, as e-business smart technologies, global Web trading systems, forecasting business systems, business management e-consultancy, and so forth
- Integrated intelligent solutions for e-Government, e-Learning, e-Science, e-Health, and e-Business
- Business intelligent technology providing generic industry solutions and autonomously implementing commercial processes, activities, transactions, and trading operations (e.g., logistic management systems, inventory management modules, enterprise resource planning agents, export-import brokering agents, etc.).

Before passing to the main content, the author would like to note that the development of the projected ideas took many years. This book is all about the infinite theory of the world, total and all-embracing. It is the summit of almost 20 years of studies revolving around the most complex entity in the world, **Reality**, its basic aspects, categories, fragments, fundamental kinds, major properties and underlying mechanisms, so that to construct the standard ontology fitting both human beings and computing machines (Abdoullaev, 1989, 1992, 1997,

1999, 2000; Novik & Abdoullaev, 1991; Encyclopedic Web Intelligence Internet site, 2005; Ontology Internet site, 2006; E-Forums: SUO, ONTAC, Semantic Web, 2006).

References

Abdoullaev, A. (1989). *Knowledge base of encyclopedic artificial intelligence.* Moscow: VINITI, the USSR Academy of Sciences (in Russian).

Abdoullaev, A. (1992). Theoretical (mathematical or symbolic) ontology as a foundation of encyclopedic intelligent systems. In R. Trappl (Ed.), *Cybernetics and Systems; Proceedings of the 11th European Meeting.* Singapore: World Scientific Publishing Co.

Abdoullaev, A. (1997). Creating machines with universal intelligence. In *Proceedings of the IASTED International Conference on Intelligent Information Systems* (pp. 115-122). CA, Los Alamitos: the IEEE Computer Society Press.

Abdoullaev, A. (1998). The machine intelligence based on MetaLingua. In *Proceedings of IASTED International Conference on Artificial Intelligence and Soft Computing* (pp. 375-379). Anaheim, Calgary, Zurich: IASTED/ACTA.

Abdoullaev, A. (1999). *Artificial Superintelligence*. PDC, Bowling Green State University, Ohio, USA.

Abdoullaev, A. (2000). The ultimate of reality: Reversible causality. In *Proceedings of the 20th World Congress of Philosophy.* Boston: Philosophy Documentation Centre. Retrieved April 20, 2006, from Paideia Project On-Line, http://www.bu.edu/wcp/MainMeta.htm

Abdoullaev, A. (2006). Ontology, semantic technology, and knowledge society. Retrieved December, 3, 2006, from http://www.eis.com.cy

Adler, M. (1990). The Syntopicon: An index to the Great Ideas. (Vols. 1-2). *Great books of the western world* (Vol. 1). Chicago: Encyclopaedia Britannica, Inc.

Architecture of the World Wide Web. First Edition. (2004). Retrieved April 15, 2006, from http://www.w3.org/TR/2004/PR-webarch-20041105/

Aristotle. The works of Aristotle. (1990). In M.J. Adler (Ed.), *Great books of the western world.* (Vol. 1).Chicago: Encyclopaedia Britannica, Inc.

Bechhofer, S., & Horrocks, I. (2003). The WonderWeb: Ontology language layer. IST Project 2001-33052. WonderWeb: Ontology Infrastructure for the Semantic Web. Retrieved April 15, 2006, from http://wonderweb.semanticweb.org.deliverables/documents/D1/pdf

Berners-Lee, T., Hendler, J., Lassila, O. (2001). The Semantic Web. *Scientific American, 284*(5), 34-43.

Berners-Lee, T., & Fischetti, M. (1999). *Weaving the web: The original design and ultimate destiny of the WorldWide Web by its inventor.* San Francisco: Harper.

Building the Europe of Knowledge. (2006). Proposal for a Decision of the European Parliament and the Council Commission of the European Communities. Retrieved April 20, 2006, from http://ica.cordis.lu/documents/documentslibrary/2461 EN.pdf

Chandrasekaran, B., Josephson, R., & Benjamins, V. (1999). What are ontologies, and why do we need them? *IEEE Intelligent Systems*, January/February, 20-26.

Chen, P. (1976). The entity-relationship model – toward a unified view of data. *ACM Transactions on Database Systems*, March 1(1) 9-36.

Chen, P., Thalheim, B., & Wong, L. (1998). Future directions of conceptual modeling. In P. Chen, et al. (Eds.), Conceptual Modeling: Current Issues and Future Directions. *Lecturing Notes in Computer Sciences*, N1565 (pp. 294-308). Berlin: Springer-Verlag.

CORDIS/FP6. (2006). Community Research and Development Information Service for the 6th Framework Programme. Retrieved April 30, 2006, from http://www.cordis.lu/fp6

CORDIS/FP7. (2006). Community Research and Development Information Service for the 7th Framework Programme. Retrieved April 30, 2006, from http://www.cordis.lu/fp7

Davis R., Shrobe, H., & Szolovits P. (1993). What is a knowledge representation? *AI Magazine*, *14*(1), 17-33.

DIP, data, information, and process integration with semantic web services. (2006). Retrieved April 22, 2006, from http://www.dip.deri.org/

E-Forums: SUO, ONTAC, Semantic Web. (2006). Retrieved December 3, 2006, from http://wwww.eis.com.cy/E-forums.pdf

Encyclopedic web intelligence and the world directory of things. (2005). Retrieved April 25, 2006, from http://www.encyclopedic-intelligence.com

ESSI, European semantic systems initiative. (2006). Retrieved April 25, 2006, from http://www.essi-cluster.org/

Extensible Markup Language (XML). (2004). Retrieved April 25, 2006, from http://www.w3.org/XML/

Fensel, D., & Bussler, C. (2006). The web service modeling framework WSMF. Retrieved April 20, 2006, from www.wsmo.org/papers/publications/wsmf.paper.pdf

Friedland, N., & Allen, P. (2004). *The Halo pilot: Towards a Digital Aristotle.* Retrieved April 25, 2005, from http://www.projecthalo.com

Friedland, N., Allen, P., & Matthews, G., et al. (2004). *Project Halo: Toward a Digital Aristotle.* Retrieved April 25, 2005, from http://www.projecthalo.com

Friedland, N., Allen, P., & Witbrock, M., et al. (2004). *Towards a quantitative, platform-independent analysis of knowledge systems.* Retrieved April 25, 2005, from http://www.projecthalo.com

Global information framework and knowledge management. (2005). Retrieved April 17, 2006, from http://www.bcngroup.org/area1/2005beads/GIF/RoadMap.htm

Gomez-Perez A., & Corcho, O. (2002). Ontology languages for the Semantic Web, *The IEEE Intelligent Systems*, January/February, 54-60.

IEEE P 1600.1 Standard upper ontology working group. (2003). Retrieved November 25, 2006, from http://suo.ieee.org/

IEEE-SA Standard project authorization request (PAR) form. (2000). Retrieved October 20, 2005, from http://suo.ieee.org/

IFF (2003). *The information flow framework.* Retrieved November 25, 2006, from http://suo.ieee.org/IFF/

Horrocks, I. (2003). *SIG on Ontology language standards*. Retrieved April 25, 2006, from http://www.cs.man.ac.uk/~horrocks/OntoWeb/SIG/

Horrocks, I., & Patel-Schneider, van Harmelen. *From SHIQ and RDF to OWL: The making of a web ontology language*. Retrieved February 5, 2007, from http://www.cs.man.ac.uk/~horrocks/Publications/download/2003/HoPH03a.pdf

Knowledge in depth (1994). Macropaedia. (Vols. 13-29). *The new encyclopaedia Britannica*. Chicago: Encyclopaedia Britannica, Inc.

Knowledge web FP-6 507482. (2006). Retrieved April, 21, 2006, from http://knowledgeweb.semanticweb.org/

Lattice of theories (2003). Retrieved 25 November, 2006, from http://suo.ieee.org/SUO/LatticeOfTheories/index.html

Lisbon European Council. (2000). *Conclusions of the presidency*. Retrieved April 25, 2006, from http://www.europarl.eu.int/summits/lis1_en.htm

Mizoguchi, R. (1998). A step towards ontological engineering. In *Proceedings of the 12th National Conference on AI of JSAI*, pp. 24-31. Retrieved April 25, 2006, from www.ei.sanken.osaka-u.ac.jp/english/step-onteng.html

MSO (2003). *The multi-source ontology of webKB-2*. Retrieved November 25, 2006, from http://www.webkb.org/doc/MSO.html

Novik, I., & Abdoullaev, A. (1991). *Introduction into information world*. Moscow: the USSR Academy of Sciences, Nauka (in Russian).

OpenCYC (2003). *Formalized common knowledge*. Retrieved November 25, 2006, from http://www.opencyc.org/

Outline of knowledge and guide to the Britannica, Propaedia. (1994). *The new encyclopaedia Britannica*. Chicago: Encyclopaedia Britannica, Inc.

OWL web ontology language, Overview. (2004). Retrieved April 10, 2006, from http://www.w3.org/TR/2004/REC-owl-features-20040210/

OWL web ontology language, Use Cases and Requirements, W3C Recommendation 10 February 2004. Retrieved 22 April, 2006, from http://www.w3.org/TR/2004/REC-webont-req-20040210/

OWL-S 1.1 Release. (2006). Retrieved April 22, 2006, from http://www.daml.org/services/owl-s/1.1/

Patel-Schneider, P. (2005). *A revised architecture for Semantic Web reasoning*. Retrieved May 22, 2004, from www-db.research.bell-labs.com/ user/pfps/publications/architecture.pdf

Poli, R. (1996). Ontology for knowledge organization. In R. Green (Ed.), *Knowledge organization and change* (pp. 313-319). Frankfurt: Indeks.

Project Halo. (2003). Retrieved April 25, 2006, from http://projecthalo.com/

Resource Description Framework. (2004). Retrieved April 22, 2006, from http://www.w3.org/RDF/

Semantic enabled knowledge technologies. (2006). Retrieved April 24, 2006, from http://www.sekt-project.com

Semantic web tools, applications and demos. (2006). Retrieved December 1, 2006, from http://www.csc.liv.ac.uk/~blacoe/SWtools.html

Semantic web topic hierarchy. (2007). Retrieved September 12, 2007, from http://wiki.ontoworld.org/index.php/Semantic_Web_Topic_Hierarchy

Shadbolt, N., Hall, W., Berners-Lee, T. (2006). The Semantic Web revisited. *IEEE Intelligent Systems*, May/June (pp. 96-101). The IEEE Computing Society Press. Retrieved December 2, 2006, from http://eprints.ecs.soton.ac.uk/12614/01/Semantic_Web_Revisted.pdf

Smith, B. (2004). Beyond concepts: Ontology as reality representation. In A. Varzi and L. Vieu (Eds.), *Formal ontology and information systems. Proceedings of the Third International Conference* (pp. 73–84), Amsterdam: IOS Press.

Sowa, J. (1997). *Principles of ontology.* Retrieved April 22, 2006, from http://www-ksl.stanford.edu/onto-std/mailarchive/0136.html

Sowa, J. (2000). *Knowledge representation – Logical, philosophical and computational foundations*. Pacific Grove, CA: Brooks Cole Publishing Co.

Sowa, J. (2000). Ontology, metadata, and semiotics. In B. Ganter & G. Mineau (Eds.), *Conceptual structures: Logical, linguistic, and computational issues*, Lecture Notes in Artificial Intelligence # 1867 (pp. 55-81). Springer-Verlag.

Sowa, J. (2002). Architecture for intelligent systems, *IBM Systems Journal*, September, 1-35.

SUMO (2003). *Suggested upper merged ontology*. Retrieved November 25, 2006, from http://suo.ieee.org/SUO/SUMO/index.html

Thalheim, B. (2000). *Entity-relationship modeling: Foundation of database technology*. Berlin: Springer-Verlag.

Updegrove, A. (2005). The Semantic Web: An interview with Tim Berners-Lee. *Consortium Standards Bulletin*. Vol. IV, No. 6. Retrieved December 3, 2006, from http://www.consortiuminfo.org/bulletins/semanticweb.php

USECS (Universal standard entity classification system). (2005). Retrieved April 22, 2006, from http://www.encyclopedic-intelligence.com.cy

West, M., Sullivan, J., Teijgeler, H. (2003). *ISO/FDIS 15926-2 - Lifecycle integration of process plant data including oil and gas production facilities: Data model: EXPRESS and EXPRESS-G listing*. Retrieved November 25, 2006, from http://www.tc184-sc4.org/wg3ndocs/wg3n1328/lifecycle_integration_schema.html

W3C world wide web consortium. (2004). Retrieved April 22, 2006, from http://www.w3.org/

W3C web ontology (2004). Retrieved April 26, 2006, from http//www.w3.org/2001/sw/WebOnt/

(The New International) Webster's comprehensive dictionary of the English language, encyclopedic edition. Naples, Florida, USA: Trident Press International.

WordNet, WordNet 2.0. (2004). Retrieved April 10, 2006, from http://wordnet.princeton.edu/

WSIS, Building knowledge societies - from vision to action. (2003). UNESCO High-Level Symposium. Geneva. Retrieved April 20, 2006, from http://www.unesco.org/wsis/symposium

WSIS, Building the information society: a global challenge in the new Millennium. World Summit on the Information Society (2003). *First phase*. Geneva. Retrieved April 20, 2006, from http://www.itu.int/wsis/index.html

WSIS, Agenda for the information society. World Summit on the Information Society. (2005). *Second phase*. Tunis. Retrieved April 20, 2006, from http://www.itu.int/wsis/index.html

Chapter II

Ways to View the World: A Standard Ontology as the Reality Framework and the World Code

Introduction

As a general science of the universe dealing with the most universal truths about all existence, an account of **what there is in the world** and the study of reality with its content, ontology has long had a critical import only for a small set of professionals. For the majority of researchers, not mentioning the non-committed people, it has been the realm of the occult, an esoteric domain of discourse, the region of most abstract reasoning, or the sphere of philosophical opinion. Today the field is rapidly increasing its audience and practical value. Maybe only a few people will call into question the importance of world models for such advanced knowledge domains as information sciences and computing technology. Increasingly, the universal science is identified with the life and soul of knowledge technologies and intelligent systems. The search for the **world description standard** as an exhaustive theoretical account and model of generic entities is becoming a research activity promising unprecedented advancements in the new cross area of science and technology. Particularly, this is important for creating knowledge systems of extraordinary performance, such as the emerging Semantic Web, implying the entire gamut of novel applications: knowledge management, intelligent databases, conceptual/semantic search and retrieval, software agents, speech and natural language understanding, e-commerce, and ubiquitous computing (Semantic Web Topic Hierarchy, 2007; Semantic Web Technology, 2007). Considering these design purposes, there are several technical requirements for building an ontology standard: expressivity, efficiency, scalability, compatibility, extensibility, and relative simplicity. But the most significant prerequisite appears to be its consistency with existent commonsense models of the world and scientific learning.

From a commonsense position, the whole world usually is viewed as consisting of several sorts of things:

i. Those that exist (objects)
ii. Those that take place or done (events or actions)
iii. Those that organize all the content (relationships).

In the way of the classification, it is commonly suggested: (i) that objects are in space as events are in time; (ii) that physical objects possess properties occupying different states over time; (iii) and that they are in a relationship with other objects undergoing changes through their properties and states, whereby being involved in some processes. Also, it generally is thought that events have the power, force, or tendency to be connected with other events as their effects, consequences, outcomes, or results. Following this general belief, modern professional researches are leaning to describe the concrete things, or named factual items, as objects (existing actually or mentally); parts, kinds and structures of objects; properties of objects; states of objects; changes of states of objects; processes occurring in or with objects; or complex systems of interrelated objects or processes (Bunge, 1977; Bunge, 2006).

To adhere to the established tradition, information sciences and computing ontology researchers tend to define their subject matter as the study of the ways things are: the kinds and structures of objects, properties, events, processes and relations providing a definitive and exhaustive classification of all entities in all parts of being, in every realm of reality (Smith, 2003).

In the way of methodology, there is another common trend widely practiced in designing a standard ontology: a cumulative approach marked by adding one model to another with a view that merging, aligning, and fusion of different conceptual models somehow will end up with an accumulative effect (Sowa, 1997; SUO, 2004).

In fact, to incorporate and integrate a set of generic and domain ontologies into a single and consistent whole as a comprehensive ontology library, a more fundamental approach is required. This is nothing less but formulating a universal theory of reality capable to give clear resolutions for such cardinal issues for any knowledge system:

- What is the World (Reality, Universe) of things?
- What are its basic features and contours or which ways are its things?
- What key categories, classes and rules determine the world's content and structure and behavior?
- How are all the things in the world ordered by such ontological classes and rules?
- What should the intelligent software system know about the world?
- What are the ultimate categories and meanings of reality and how should they be represented by knowledge systems?

For only after resolving these principal questions of human knowledge, we will be prized by a standard ontology of things as a single reality framework yielding anything in the way of Universal Entity Code: a united system of classification, where each object, person, event, or link maybe dog-tagged by some universal resource identifier assigning the specific item to a proper class, kind, or category. To construct such a conceptual model of the world, there is one unavoidable condition to be fulfilled: we need to renovate foundational theories of reality accommodating them with the modern scientific results using the mathematical and semantic theories as conceptual tools, as to avoid a intellectual temptation of sinking into the depths of metaphysical addiction.

It also must be noted that in the information sciences and engineering, a usual doing is to divide the conceptual model into two complementary parts:

a. Intensional (intrinsic, underlying, innermost, inherent, essential, implicit, and constitutional)
b. Extensional (extrinsic, external, extraneous, or accidental)

The former implies the generic things and relationships of the world and so inherently referred to ontology and embodied as a knowledge base schema. The latter involves particular states and cases of the world, and it deals with all sorts of specific assertions about specific instances of classes and relationships determined by the ontology, and presented as a knowledge base instance. So, the ontological work consists in providing the general classes and principles and underlying meanings, which form the base for any particular domain of special classes and individual things.

Alongside with the above methodological rules, the constituent elements and underlying principles of a standard ontology will be further established, viewed as a single coherent reality framework. It will be suggested an integrative conceptual model of entity classes and relationships, with their properties, value types, and instances, uniformly describing heterogeneous types and levels of reality. When the model of reality is designed as the unifying context of various knowledge domains and information representation formalisms, it can be applied to major computer science and technology areas: databases modeling, information and software engineering, artificial intelligence, and semantic web.

Top-Level Ontologies and Languages: The State of the Art

The activity of constructing a comprehensive conceptual schema is performed in two general directions:

- Designing formal ontology languages basing on formal logics
- Constructing hierarchical entity types as covering anything in the universe, all the basal things and relationships

Of the former, the most noteworthy looks to be the OWL, belonging to a family of the Web-based ontology languages, intended to provide advanced semantic tools, syntactic formats and specifications for knowledge contents. As a part of the World Wide Web Consortium's Semantic Activity, it is purposed to be a semantic markup language for all sorts of Web-based content and documents. The Web Ontology Working Group released a number of documents describing the final status of its research: Web Ontology Language Overview, Guide, Reference, Semantics and Abstract Syntax (OWL, Overview, 2004). This report is the production of many people: the members of the Working Group delegated from well-known universities, government laboratories, and transnational companies, not mentioning those who sent their comments online. With its three sublanguages (Lite, DL, and Full), the language is presented to the wide public as the semantic language constructing on XML, RDF, and RDF Schema (2004) but more versatile in its capacity of representing machine readable content on the Web.

Now, what is pronounced as a W3C recommendation? Realizing that ontology refers to the science of describing the kinds of entities and their relationships in the world, it eventually is presented as "a set of definitions of classes and properties'." Following a logical practice to use Property together with Class and Relation as the underlying triad of any universe of discourse (Simon & Kaplan, 1989), the OWL authors also view classes, properties, and their relationships as the basic elements. Both the syntax and semantics of such a language are constructed as the syntax and the semantics of any formal language or formal system, where the key syntactic concepts are axioms, rules of inference, and theorems, while semantics is interpreted as a branch of set theory dealing with the relations of classes (The History and Kinds of Logic).

More relevant to the purpose of a standard scheme of things appear to be the research works committed to generating comprehensive lexical taxonomy such as the WordNet Lexical Ontology (WordNet, 2004), the "starting candidates" of the IEEE Standard Upper Ontology (IEEE SUO, PAR, 2004). And extensive computing project purposed to be unified within IFF (Information Flow Framework, Kent, 2003): SUMO (suggested upper merged ontology), the CYC ontology, and the Shell's Generic Entity Framework developed as the EPISTLE Core Model (SUO 4D Ontology, 2004) as well as DOLCE (descriptive ontology for language and cognitive engineering) (Borgo, et al., 2002). Besides, there are many organizations and institutions funding research for building ontological standards in specific domains of interest, such as manufacturing, engineering, and business processes, like the PSL (Process Specification Language) of National Institute of Standards and Technology (Gruniger and Menzel, 2003). A common characteristic of the most projects, both upper level and domain specific, is that they regard all things and entities as particulars or individuals organized in a hierarchy (or hierarchies) of types by a set inclusion relation, following Russell's set theory of types. As a result, each type there has several subordinates at the next lower level and one superordinate at the next higher level. Additionally to a strict adherence to formal systems, such as set theory, formal logic, category theory, or mereology, there also is a strong appeal to the traditional classifications: dichotomous (C/C) or trichotomous (C/C/C) divisions of things in the world.

Being of special interest and value to the issue of all-comprehensive scheme of things, the WordNet project is initially designed as a lexical multi-inheritance system modeling human lexical memory, but now vastly regarded as an online lexical ontology (WordNet 2.0). A set of ground entity types is here represented as the network of distinct "semantic primes," "primitive semantic components," "unique beginners," "synsets (sets of synonyms)," or lexical concepts with *entity* at the top of the multiple hierarchies of all word meanings. *Entity* is defined as a physical thing having its own distinct existence, which can be perceived, known, or inferred. Specifically, it includes *thing* as a self-contained entity, *causal agent* as cause, physical *object, substance* as matter and *location,* as well as completely irrelevant types, such as anticipation, enclosure, expanse, and sky. A separate *thing* is viewed as covering such divisions as *anything, something, nothing, content, essential, nonessential, water, part, source, unit* and *variable.* Another important category, *cause,* is divided as *person, agent, nature, first cause, power,* etc. More generally, entity is partitioned into objects [non-living things (*natural objects, substances, artifacts,* and *food*)] and organisms [living things (*persons, plants,* and *animals*)] with all their possible *groupings or collections.* The rest of physical things constitute either states (*state and condition*) or various sorts of change: *event* and *happening*; *act, action,* and *activity* (human); *phenomenon*; and *process. Attribute* and *property*; *quantity* and *amount*; *relation*; *space, location,* and *place*; *time*; *set* or *class* are interpreted as top-level *abstractions* possessing a sort of conceptual existence and being fully disjoint of physical things. Besides, most of the "semantic primitives" actually are derivative than originative and should come up at the third or lower levels of the onto-semantic hierarchies; that is, they are all redundant in the list of the top-level terms: *animal, artifact, body or corpus, cognition or knowledge, communication, feeling or emotion, food, motive, person, plant, possession, and shape.* Despite its confusing dichotomous division of entities, WordNet is quite deservedly regarded by the natural language community as a reference lexical ontology surpassing in its scope, range, and depth all the extant generic taxonomies and typologies.

Like in the WordNet taxonomy, the SUMO accepts as its top-level things a class of physical entities (objects and processes) and a class of abstract entities. As the abstractions are admitted attributes, sets and classes, propositions, quantities, spatiotemporal relations, and relations, such as functions, predicates, and lists. As concrete individuals, there are taken two sorts of things: physical objects having spatial parts, and processes as the class of things having both spatial and temporal parts. The former includes as its subordinates self-connected objects, collections, agents, and regions, while the latter covers internal change, intentional change, dual object change, shape change, and motion (SUMO, 2004).

The CYC upper-level ontology is a more complicated, because it allows some hybrid thing named "a partially intangible" thing, a hybrid of abstract, notional, intangible and concrete, physical, tangible. As the instances of individuals, along with the physical objects named spatial and temporal things, we meet here the so-called intangible individuals that encompass time intervals, attribute values, and mathematical objects as relations, as well as situations and events. Only sets and collections are admitted as purely mathematical things (Open Cyc Vocabulary, 2002).

The Shell's data model of things, traded as Lifecycle Integration Schema, starts out its up-bottom schema from individual things carved up as abstract objects (abstraction) and really possible individual things, particular *existents* encompassing both actuals and possibles. Unlike the above representations, the particular things include not only concrete objects but also entities that can be thought of as the individual objects of thought: all sorts of things remembered, imaginary, and conceivable, whatever (whenever and wherever) has the possibility of being. Another distinction is a four dimensional (4D) relativity view of the world, as the space-time system of things where individuals, defined as spatial-temporal extents, are nothing but some pieces of the space-time continuum having one temporal and 3 spatial dimensions. Additionally to physical objects, other top categories are states and changes or activities or events or processes, where the former class specified as the temporal parts of objects, the latter as the temporal limits of states instead of being the transitions between them (SUO 4D Ontology, 2004).

Another typical exemplification of a particularistic tradition is DOLCE. With some slight modifications, it follows the trope theory division of particulars into tropes (modes, individual qualities), objects (abstract, mental and social objects and concrete objects, as masses and organisms, an aggregate of qualities) and processes. In all, DOLCE intends to lay out all sorts of particulars under five or six headings: quality and quality region, aggregate, object, feature, and occurrence. Making the most fundamental category, the particular qualities cover all manner of quality types, as well as spatio-temporal relations, while all equivalence classes of specific qualities are presented as three distinct substantial entities: aggregates (as mass and agglomeration), wholes, subdivided as physical objects and abstract things, and boundaries and features of physical objects, seemingly, like the natural features of the earth. Physical changes or processes are substituted by occurrences, a heap of states, events, processes, happenings, and so forth. (Borgo et al., 2002).

Summing up, each of the above projects revolves around the idea that the question of the standard ontology is just the matter of growing different entity data types as a library of particular modules into one general reference structure through mapping, merging, and integration. Despite seeming diversity of the above-mentioned contents, they all also share another common point: being bound up and biased by the ontology of particulars. There the classification scheme usually is built according to the dualistic presumption: "every object is either a thing (*res extensa, things proper*) or a construct (*res cogitans, ideas*), no object is neither and none is both" (Bunge, 1977). Then all *particulars* are to be split into two disjoint classes or sets: *a class of constructs* C and *a class of concrete objects* O, so that to achieve the mutually disjoint division of the whole body of individuals, that is, the physical world. The class of formal objects includes all sorts of mental constructs, mathematical abstractions and structures. The class of physical objects consists of physical individuals, concrete things (masses and organisms), and processes (occurrences, events, states, etc). The construct, in turn, subsumes concepts, propositions, and an abstract body of concepts and propositions; where the subdivisions of concept are individuals, sets, and relations, nonfunctional or functional. The functional relations are further parted as propositional (logical or attributive) relations, which are predicates, or non-propositional relations, as numerical relations. The proposition can be presented as a specific proposition or a propositional schema. At last, there is a complex construct, which as conceptual body divided into three descriptions: formal contexts, classes of formulas, and abstract theories (Bunge, 1977).

As one can see from the sketchy account, each approach surely can boast its special utility and virtue. Nevertheless, they all are the sorts of taxonomical models devoid of structural and unifying foundations, standing in want of ontological kinds, properties, and relations. To integrate the particular models, we need to work out the category theory of thing capable to organize all the specific knowledge about objects, quantities, changes, times, spaces, parts of objects, movements, substances, materials, doings, transformations, agents, groups, persons, and so forth. And the type of ontology relevant to the role of such reference structure is the one allowing the existence of generic classes along with their contingent particulars, objects, qualities, and processes. In the category theory, everything whatever exists and happens, or it is conceived as existing and occurring, can be named an entity, a thing, or a being taken in its most general sense. All beings, things, or entities are to be hierarchically divided into ontological kinds, supra-physical, abstract, general, and individuals, material or formal. That is, a few fundamental kinds of entity considered apart from the concrete and specific embodiments should make up the top-level ontological classes of being, while all sorts and manner of particular objects, qualities, events, and relationships make up their instances, cases, and exemplifications.

Ontological Fundamentals

In pursuit of the large order of creating a reference scheme of things in the world, we ought always to keep in mind that ontology is the oldest science developed by the brightest minds. There is no need to give new interpretations, which would be doomed to be ever worse than its classical and traditional definition as the science or study of being (entity or thing) taken in the most general sense, with all its major classes and relationships. We deal here not only with some science, which is knowledge of the world of nature, but rather with a fundamental science dealing with the nature of the world, with knowledge of the world in the most general sense. So the study of entity is not just a science, knowledge of the world of nature, but rather a general study dealing with the topmost ontological classes, categories and kinds; that is, it is all about being, thing, or entity *in the unqualified sense*. In other words, *ontology is a universal science of entity/relationship classes making up the world or reality as the largest universe of discourse; so it suggests a body of ultimate meanings (categories) of reality and a system of rudimentary truths, axioms, assertions, assumptions, definitions, postulates, and rules.*

Noteworthy that such a universal science of things during its evolution has been passed through three distinct successive historical research stages (see Figure 1): normative or prescriptive; descriptive or taxonomical; scientific, mathematical, and systematic. Usually in the form of a discourse and abstract deduction, laying down general directions and rules, *a normative ontology* prescribes its metaphysical theories concerning basic features of reality and realities as final standards of the world outlook, like as idealism (absolute), realism, materialism (dialectical), or nominalism. Its offshoots are prescript ethics, esthetics, politics, logics, and grammar directing the standards and rules of conduct, of value, of government, of thinking, and of language usage. Unlike this, involved with the classification of the material, *a descriptive ontology* constructs a general taxonomy of real entities and limits itself with empirical description of classes of things. Its offshoots are all sorts of descriptive sciences, the classification theory, and the biological classification of living things. All kinds of metadata models, applied as taxonomical schemes of things for information systems as computational ontologies, make a new line of research within the framework of the taxonomical ontology. At last, characterized by precise formulation, mathematical analysis, and scientific knowledge, *a systematic ontology* heavily rests on scientific methods constructing the most general theories about reality with the aid of the factual sciences, semantics, and mathematics. Its offshoots are exact scientific ontologies, universal mathematics with set theory and order theory and category theory, as well as the factual sciences themselves as physical sciences, particularly, theoretical physics. In the course of the development from

Figure 1. The knowledge sources of UFO

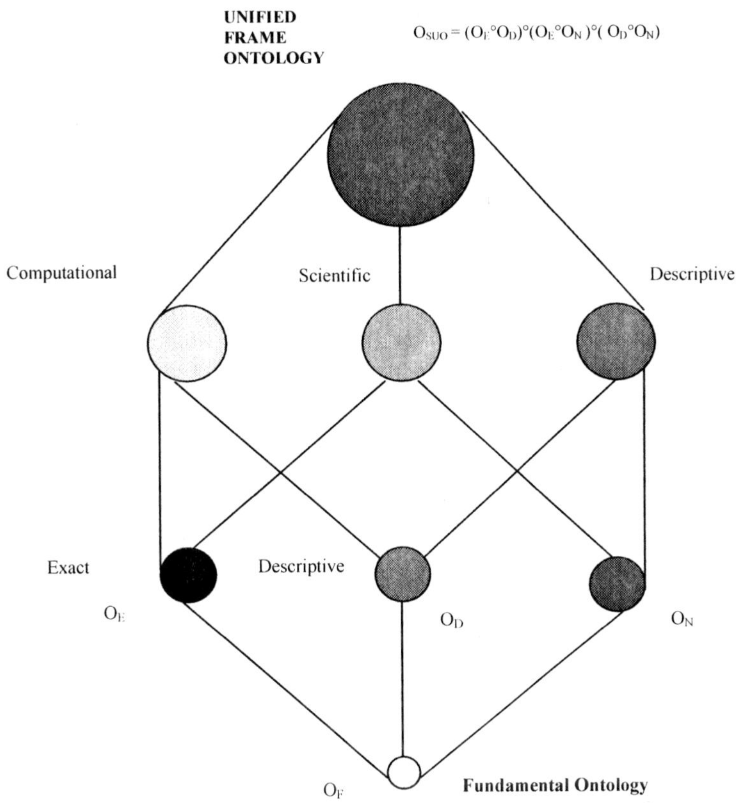

its incipient classical forms, universal ontology as the general theory of the world has been suffered several modifications through the intermediate varieties, normative, taxonomical and mathematical.

At present, mostly because of the requirements of oncoming ontology-based information technologies, like the Semantic World Wide Web and Artificial Intelligence, there is an insistent urgency to develop a single and all-comprehensive ontology establishing standards of behavior for computing machines, semantic commensurability, or communication interoperability. Evidently, if such monumental effort is possible, a standard universal ontology (SUO), it is to be constructed as the best of three breeds, **exact ontologies, descriptive ontologies,** and **normative metaphysics,** that is, as a Unified Framework Ontology or Universal Formal Ontology, designated further as the UFO or the **EIS SUO** or the **EIS UFO**. This plainly suggests that to construct an all-comprehensive reference scheme of things, one needs to investigate the scope and limits, meaning and principles of universal ontology as the general account of entity in the abstract, so that to discover a common foundation, axiomatic grounds, upon which its basic parts or modules can be united as a whole body of knowledge. For as the highest science, fundamental ontology deals with the topmost categories of thing, while its particular manifestations can be treated as fielding the special classes and kinds of being: nature, mind, society, technology, culture, knowledge, or engineering.

For a right start, we have first to formulate the substance (core) of the universe of discourse, which is nothing else but *entity* in its major meanings of *being, reality, existence, the world,* or *the universe* with its essential classes and kinds lying behind everything of the sorts of particulars, or concrete things. There are several senses in which entity can be used. Primarily and most properly, the term is employed as an abstract noun designating the abstract conception of existence or being (*ens* or *entia* from *esse* be). Another meaning is the essence (substantial nature) of something existent. But mostly popular is the sense of individual substance, something that exists or is supposed to have actual being or reality, as a distinct and self-contained entity, "that which is perceived or known or inferred to have its own distinct existence (living or nonliving)" (WordNet, 2004, Entity). Thus, from one part, the term is linked up with the concept of being. From other part, it is interpreted as a concrete individual since its meanings range from the class of all things, the most inclusive category possible in the world, to any thing, any concrete individual, an object, a person, a situation, an event, a time, a place, etc. Bearing this inherent semantic ambiguity in our mind, *entity, being* and *thing* will be used interchangeably as in the formulas, *entity is being and being is entity, as an entity is a being and a being is an entity,* the RDF devotees can here apply *resource* or *the set of resources* or *universe of discourse.* All senses in which things can be said to *be* are comprised by the term *entity*: anything, something or nothing are entities or things; individual (material) objects, bodies, or substances belong among entities or beings; minds are entities or beings; states, properties, or qualities are beings; changes, actions, acts, events, processes, and activities are beings; relations as having being are also a sort of entities.

As the highest science, fundamental ontology deals with the topmost categories of being, while the particular sciences can be treated as the applied ontologies fielding the special classes or kinds of being: nature, mind, society, technology, culture, knowledge, or engineering. Therefore, to avoid any ambiguity and equivocation, ontology should be considered at three distinct levels of abstraction: fundamental and universal; upper-level; applied- and domain-level ontologies.

The first in number and in priority is about the class of all classes, the universal class, named Being, Entity, or Thing, the class of which everything is a member inclusive the null class, known as Nonbeing or Nonentity, having no species at all, while being the ultimate unit of Being.

The second concerns the world of particulars, the infinite exemplifications of a few ground entity classes or ontological categories existing potentially in physical individuals and actually in minds, human or computational.

The third is about particular domains or realms of being, specific classes of entities. Of all the significant questions about *being*, mostly decisive appears the issue of its essential meanings, of its essence, nature, or key properties as well as the question of possible modes, kinds, divisions, distinctions, determinations, differences, accidents, or forms with the possible ways of their ordering and combinations. Specifically, the fundamental knowledge of being, entity, reality, or existence consists in finding the responses to the following challenging questions:

i. What is entity, being, thing, reality, the world, or the universe? Or, what are the essential meanings of being? And what is not to be, nonbeing, nothing, or nonentity?

ii. What are the primary kinds, classes, genera, and categories of reality? How do the primaries exist? What are their number and character?

iii. How should we divide, partition, and classify each kind or genus? What are we to assume as its species or partitions? And what is the full number of contrary divisions of a class?

iv. What are the order and interrelationships of the kinds of entities?

v. What is their meaning and how should they be formally represented?

More detailed and specified questions about entities, things, beings, realities, or existences, such as what is a particular being or thing, how it exists, what is the difference between universals and particulars, and how to systematize and classify particulars, are directly predicated on the solution of these fundamental issues.

Concerning the first issue, there is a prevalent tendency to consider being (entity, thing) as the topmost abstraction and the most general of predicates, as the ground source of particular beings, something common to all kinds of things from which they begin and to which they owe their reality and existence. The terms *being*, *reality*, *entity*, and *existence* are often conceived as identical, interchangeable, or convertible with each other. Apart this, it generally is conceded that *Being, Entity, or Thing taken together with its primary genera, kinds, species and individuals comprise the world or the universe as the maximal thing or the maximum being or the supreme entity, the whole totality of things*. For, explicitly or implicitly, it is widely accepted that being is neither unity nor plurality, but a diversified unity, a plurality in unity, the all-in-one, that is, the universal real super-class embracing everything, gathering the total of things, or all the existents and beings. For something *to be* first of all means *to be an entity, to have being, to have reality, to exist*, whether potentially or actually, in nature or in mind; because to a certain degree and extent being or reality resides in all particular things.

Regarding no less critical issue of the kinds of entities, here we also can benefit by some established and significant facts of a general methodological character. It should be first pointed out that there are in reality only few primitive classes or highest genera of being, which are different but coordinate. The unique status of a primary class should meet the necessary prerequisites:

- It cannot be referred or analyzed to other genus
- Nothing, but Entity or Nonentity, Being or Nonbeing, can be predicated of the primary class
- It is a generic classification and first-principle; it is a unity, not a complexity
- It is that which underlies differentiae, or specific differences
- The genus exists as a whole in the things whose essence it is
- The class together with the other primaries generate the particulars.

Finding a solution to the third fundamental topic, how to class things or an object of domains accurately and adequately, comes up as a highly contentious issue for the AI and Web descriptive ontologies. Not strictly holding the methodological rules confuses one of the key topics of engineering ontology: how to erect a normative classification of entity data types of things. Shortly, the underlying methodological rules here can be reduced to a number of the general propositions:

- A class is divided into simultaneous subordinate classes, or species
- A difference is an essential property distinguishing a species
- Subordinate classes are distinguished by the same way of division
- The class division is effected by the difference of the difference
- The difference must belong to a different coordinate genus
- The kinds of a genus is equal to the number of the differences
- The genera are prior to the species, as the species to the individuals
- Individuals of the same species differ in degree, but not in kind.

As to the issue of the ordering of genera and their kinds, there are at least two general established truths:

- Whatever is of a class, is of that class's class
- The invariant interrelations of the kinds of entities produce the invariable laws and rules of reality, as causal laws and mechanisms.

The last fundamental question refers to the semantics of ontological vocabulary and to the matter of formal representation, the major subject of computing ontologies.

As one already has noticed, it is of great benefit and advantage first to refer to the classical ontological treatises where the intelligent reader can lay open profound learning and thus letting us shun scratching from the zero and by-pass the repeated conceptual traps and faults. Any artificial classification not mapping the composition of reality and the order of its things will entail many conceptual troubles: entangled partitions, global inconsistency, terminological incompleteness, disjointed contextuality, and what not. For instance, the concepts may suffer from missing and wrong links; some concepts may have a few subclasses, while others on the same level a disproportionately large number of subclasses. Also, one should add here a mess of encountering concepts; high redundancy of constructs; a poor causal reasoning, etc. The root of all these concerns is evidently a lack of a strict theory of entity capable of giving exact definitions of all primary concepts complete with adequate classifications and typologies of entities done by their essential properties instead of dichotomy, formal attributes, accidental signs, or somebody's practical interests and purposes.

It emerges that the most crucial thing in the matter of constructing a standard universal ontology seems to be the right criteria of choice of primary genera from which as from the roots of all things an all-comprehensive classification of entities can be erected. It is not surprising that here lies all the difficulty: since as all the rich content of reality—with its entities, levels, and grades—is determined by a set of primary classes. Then positing the wrong genera as the first principles, causes, and elements is a fundamental error, which will ineluctably result in ambiguous, equivocal, or even preposterously absurd arrangements of things. Adding a multitude of knowledge items to a shaken background will end up with a monstrous assortment of irreconcilable facts, assertions, hypotheses, theories, and laws, instead of consistent knowledge base.

It is worthwhile to remember in our ontological prayers that a classical example of the primary classes was first given by the progenitor of scientific ontology and formal logic, Aristotle, who laid down ten categories as the primitive classes of being and the standard classes of predicates: Substance, Quantity, Quality, Relation, Time, Place, State, Position, and Activity or Passivity. Sometimes they were reduced to eight categories: Being, Quality, Place, Time, Relation, Quantity, and Activity and Passivity (or Agent and Patient as productive Cause and Effect), where State and Position considered as redundant (Aristotle, Categories, Physics). Some of the classes admit of contraries and change of extent, or variation of degree, some only partly, but some not. In general, the distinguishing characteristics and properties of these ontological classes of entities, the basic objects of thought, are determined as follows:

- **Substance**, as a twofold entity, signifies the individual, one single thing or the concrete whole, a compound of the matter with the form. At the same time, in its fundamental sense, it also denotes the species, the genus, the class, or the universal. In the primary sense, *substance* is the subject, the substratum, or the entity underlying everything else, including the genus. In the secondary sense, as the genus, it is a class of things qualitatively distinguished: "species and genus determine the quality with reference to a substance" (Aristotle, Categories, 1990, pp.7-8). Besides, substance remaining its identity has no opposite and does not admit any modification of degree, but admits of contrary qualities, the ground for change. It is the most distinguishing mark of substance, both the individual and the genus *substance*, to be subject to a process of change, accidentally, partially, or essentially.
- **Quantity**, like substance, has no contrary, no variation in degree and comes up as discrete (*number, multitude*) or continuous *quantum* (the magnitude of geometry and physical quantities—time, place, etc.). Equality and inequality are distinguishing characteristics of quantity.
- **Quality**, in opposition to substance, has a contrary and varies in degree. Due to possession of qualities, substances change or changed. In sum, it is taken to be of four different sorts: *the human qualities* (habits: knowledge, virtue; disposition: beauty and health); *the inherent capacities* (the powers of sensitivity, rationality, reactivity, etc.); *the affective or sensible qualities* (colors, odors, textures, tastes, heat, cold, etc.); *the qualities of things* (figure, size, weight, mass, and shape).
- **Relation** as *being of* or *related to something else* has a correlative; some of relatives have contraries and vary in degree, and it is held to have the least reality among all the accidents of substance. Time, being a continuous quantity, is *the measure of motion* and, as a whole, composed of past, present, and future. Place also is a continuous *quantity* and held to be the boundary of a body.

- **Action** and **Passion** (*activity and passivity,* undergoing, experience, suffering, acting or being acted upon) have their contraries and admit of variation in degree (e.g., heating/cooling, being glad/being angry).
- **Change** or **Motion** is thought to be subordinate to substance, quantity, quality, or place, hence not directly included in the elite selection of the fundamental modes of being, as Aristotle's invariant standards of the world's knowledge. According to the type of the subject of change, there are substantial change, generation and corruption, and the three types of motion: with respect to quantity, increase and decrease, to quality, alteration, and with respect to place, locomotion and rest, all sorts of motion imply time, place and void. The necessary conditions of change are determined by the five factors: (1) that which causes changes, (2) that which is under way, in the process of change, (3) that in which change takes place (i.e., in time and place), (4) that from which it passes, (5) that to which it passes. On account of this classification of entities, change is considered present in and predicable of substance, while substance is only present in change. As a result, substance is held to be the primary subject of all processes of change.

Aristotle' category theory thus held that there is no basic classes of predicates outside of substance, quantity, quality, relation, place, time, position, state, action, or affection. Of all the multiplicity of the ensuing typologies, taxonomies, and classifications of entities, the most noteworthy and valuable happen to be those that had been inspired by this definitive model of the ontological classes of things. Some of the distinctive typologies are the conceptual schemes based on the following entity categories:

- Being, motion (change), stability, difference, identity (Plotinus)
- Quantity (unity, plurality, totality); quality (reality, negation, limitation); relation (inherence/subsistence, causality/dependence, community/reciprocity); modality (possibility/impossibility, existence/nonexistence, necessity/contingence) (Kant)
- Things (matter, body, spirits, nature, operations, relations), actions (human practice, happiness), signs (mind, knowledge, representations, sounds, ideas, and words) (Locke)
- Quality (firstness), relation (secondness) and representation (thirdness) or signs (icons, indexes, symbols) (Peirce)
- Thing, the same or different, kinds, minds, bodies, time, space, subject and attribute, causal influences, the fancied, the real (James)
- Objects, qualities, and relations (causality, space-time relations, resemblance, logical relations (Whitehead and Russell).

A sufficiently rich compendium of the underlying ontological constructs can be found in the "Synopsis of the Great Ideas" (Adler, 1990) numbering a hundred of common conceptions making the basic terms of any general universe of discourse. When properly sorted, the basic ideas can be organized as a hierarchy of categories with their subordinates:

- **The categories of reality:** Being, Cause, Chance, Change, Element, Eternity, Form, Infinity, Matter, Mind, Nature, Necessity and Contingency, One and Many, Opposition, Quality, Quantity, Relation, Same and Other, Space, Time, Universal and Particular, World
- **The forms of nature:** Animal, Experience, Evolution, Fate, History, Immortality, Law, Life and Death, Man, Progress.
- **The species of knowledge:** Art, Astronomy and Cosmology, Dialectic, Education, Logic, History, Mathematics, Mechanics, Medicine, Metaphysics, Philosophy, Physics, Poetry, Religion, Rhetoric, Science
- **The elements or instruments of knowledge:** Definition, Induction, Logic, Opinion, Principle, Sign and Symbol, Truth.
- **The species and states of mind:** Angel, Desire, Emotion, Experience, Idea, Judgment, Love, Memory and Imagination, Pleasure and Pain, Reasoning, Sense, Soul, and Will.
- **The species of value:** Beauty, Duty, Truth, Good and Evil, Honor, Prudence, Courage, Temperance, Virtue and Vice.
- **The forms of human practice:** Aristocracy, Citizen, Constitution, Custom and Convention, Democracy, Family, Government, Habit, Happiness, Justice, Labor, Language, Law, Liberty, Monarchy, Oligarchy, Punishment, Revolution, Sin, Slavery, State, Tyranny and Despotism, War and Peace, and Wealth.

It is plain that the categories by virtue of which the mind describes reality make the very foundation of the great ideas. This means that the task of constructing a standard ontology is doomed to failure if it remains without being grounded on the classical theories of reality, primarily, that presented by "real" Aristotle, who profoundly investigated and explained almost all the general categories of being.

Crucially, in any class of being, in any subject matter or science, in any domain of objects or universe of discourse, these terms come up as basic elements for constructing mostly profound propositions about things, recognized as the basic truths or axioms and definitions revealing the essential nature of things. For, as the basic elements of composite expressions, they are necessary for formulating universal axioms and principles, such as, "the genera are the principles and elements of things," "all causes are beginnings," "every being is a distinct whole," "the whole holds the parts," "all relatives have correlatives," "place is the limit of space," "things are one in substance, species, genus, or by analogy," "motion is the change of place in space," "the one is the measure of all things," "the beginning is a limit," and so forth.

Besides, such propositions are inescapable for reasoning about particular classes of reality, since they serve as the root principles from which scientific knowledge of the whole body of facts can be successively derived: the definitions of key terms added with the full gamut of their meanings and together with the rules of demonstration, inference, and reasoning about them, they form the heart of any applied ontology.

Therefore, to erect a standard global schema, one needs to rest on the ontological fundamentals and primary truths as the skeleton of all things, adequately fleshed out by modern knowledge about the world.

The Elements and Principles of Reality

Built as a fundamental account of reality, the EIS SUO drew great benefits from reconciling the classical world models, such as developed in Aristotle's ontology, with modern formal, descriptive, and scientific schemas. The main points of Aristotle's theory of reality are its major principles of division of Being or Entity summarized as follows: (a) being, unity or one is the most universal of all universals or predicates; (b) categories of being are other in genus; (c) genera (or ontological classes) are universal elements; (d) there is a finite number of absolutely diverse highest genera and a non-infinite variety of species; (e) there is no intermediate species between the contrary species of a genus; (f) a class (species and genus) is determined by the quality with reference to the particular members; (g) what is predicated of the genus (the predicate) will be predicated both of the species and of the individual (the subject).

To complete Aristotle's classification theory, it is necessary to revert to Kant's theory of entity classes highlighting **the principles of the homogeneity (unity), the specification (variety) and the continuity (affinity) of forms.** The unity principle concludes that *there are no severally divided highest genera, but they all are simply determinations, divisions of one universal class or genus.* **The principle of variety** maintains that *there is an infinite variety of species without ultimately reaching the inferior species: each genus requires species, which requires subspecies and ad infinitum.* The principle of continuity grants the possibility of intermediate species of the contrary, diverse species of the same genus, like the kingdom Protista whose organisms are partly plants and animals.

It appears that if we find a way to reconcile Aristotle's classification and Kant's theory of classes, then we could give a sufficiently accurate and adequate account allowing of a definite and all-embracing typology of entities, beings, or things in the world.

At first sight, both theories appear as opposite to each other. Nevertheless, we take pains to propose the principles of division and hierarchy capable to harmonize these only seemingly contradictories. It is mostly obvious to any unprejudiced mind that there must be one *summum genus*, a universal class of things, a single all-embracing supreme genus, under which all other things could be ultimately arranged. For all things, things of every kind, type, sort, or form, come to be alike in being, inherently possessing some mode of reality or existence. On the other side, the total of all these modes, forms, differences, divisions, and modifications, again constitutes the totality of beings, or Being as the maximal Thing, carrying the connotation of Entity, the Universe, Reality, or the World.

Current views of the world differ over the issue what categories regard as prime and fundamental: individual substances (objects and their particular properties, tropes, or modes) or universals (kinds, or substantial universals, properties and relations). In most of top-level hierarchies of entities they had been used to consider as two completely distinct contraries of one common super-genus, entity, thing, or being, having the same status or power. It may be Entity or Thing equally partitioned as Contingent and Necessary (Chrisholm, 1996), as Universals and Particulars (Lowe, 2001), Abstract and Concrete (Sowa, 2000), Individual and Mathematical or Computational Thing (OpenCyc Vocabulary, 2002), or as Physical Object and Conceptual Entity (Bunge, 1977). There also are positions where particulars are made of universals, properties and relations, that is, particulars (tropes) are bundles or carriers of universals, and the former is exemplified or instantiated by the latter (Armstrong, 1997).

But mostly close to the real states of affairs stands the hierarchy of entities admitting the fundamental category of kinds, the fundamental category of attributes, the subordinate category of individual objects, and the subordinate category of particular properties, modes, where kinds characterized by attributes and instantiated by objects, as property-universals instantiated by particular properties, or modes (Lowe, 2001). For, in fact, these fundamental pairs of categories are opposed not as absolutely diverse contraries of equal in status, but rather as superordinates and subordinates identified by supertype/subtype or generalization/specialization relations.

In the order of the English language, for example, such a reference can be expressed by the prepositions "in" or "of." We can say, for instance, that *the whole is in the part* and vice versa or *the species in the genus* and vice versa, as in the definition. Or, *the matter is in the form* and *the form of the matter*. Or, *the necessary principles of contingent things*, *identity in diversity, the universal in the particular, the essence of appearances, the indivisibility of continuity, the one in the many, the unity of plurality*, and so forth.

Thus, the most true and adequate representation of reality that we are capable of conceiving is that one where the pairs of opposites are related as instances, the individuals, inheriting the specific differences or essential qualities from their higher class, the universal.

By embracing the abstract and the concrete as the different but relative and subordinate kinds, Being has both the emptiest and the richest senses by extending over the ultimate empty class and the entire reality, the universal class of all classes of which everything is a part and member. For its just complement is only Nonbeing, Nonentity, or Nothing, the empty class devoid of members at all, and included into entity as well; for the null thing is part of any thing, or the class of entities that are not identical with themselves, for example, the class of honest politicians or the class of real knowledge systems.

From all this evidence, one may with full certitude conclude that **being,** with its gamut of synonyms, **reality, existence, entity, experience, the world,** or **the universe**, is *that which is primarily*, as well as the primary object of any knower's mind, either natural or artificial intelligences. So, neither a thing nor an entity nor an object, and what not, but being in the sense of reality, the universe, or the world must be regarded the topmost class, of which everything is a member.

Another crucial dilemma is the number and character of the highest genera, and whether they are isolated, distinct, separated, and non-reducible to each other, like the Grand Ten Categories, or rather diverse but interdependent kinds of the same super-genus of being. The point at issue, whether change, relation, quantity, quality, activity, and so forth, are diverse from substance, dependent on substance, produced by substance, or reducible to substance, requires appealing to the class of state in the first place. It should be specially observed here that the category of "state" has been completely undervalued, and it looks to be the weakest point in Aristotle's ontology, as well as excluding the category of change, motion, or movements from the list of the primaries.

In contemporary science, the account of anything is based on the knowledge of the state of a thing, defined as a complex of its properties, essential, temporal, or relative; for the class of states embraces all the results of the measurements or observations of all properties of a system. Any substance, object, or system occupies a condition called a state; a state value, a state function, a state equation, a state equilibrium, along with the concepts of system, change, and law are clearly among the basic concepts of science. The notion of state is getting crucial in the exact analysis of realities, substances, and changes or motions. It appears, that the fact that the class of state must be a primitive genus, does not need a long exposition, since the claim is strongly supported by modern science, where all natural and social phenomena are described in terms of state of a thing (system) and the changes of state caused by other things (systems).

State as an indefinite number of properties not only engulfs the notions of property, quality, attribute, and quantity, but also has the power to replace the category of form no longer so much decisive in the consideration of the kinds of being. Instead of forms of existence, we can speak of states of existence. As to quality, it is an essential property of a being or thing, while quantity is a measurable or numerable property. Etymologically, the former is the condition of *being of such and such*, the latter is the condition of *being much*; accordingly, some states may be qualities, and some quantities. The *attribute as formal property or predicate* furthermore suggests the qualities of conceptual, formal, and hypothetical objects, or ideal beings. The property taken in its broadest meaning is partly coextensive with *state* by representing the distinguishing characteristics of a thing that change, as well as the typical modes of change, movement, action, or behavior. Not surprisingly, that in some theories of categories (Chisholm, 1996), things, both universal and contingent, are partitioned only as states or non-states (necessary substance and attributes or contingent individuals, substances and boundaries) or the whole world as the totality of everything is viewed as a universe of states of affairs (Armstrong, 1997). Nor it is surprising that in the largest lexical taxonomy the term "state" covers as its kinds even "being" and "nonbeing," "action" and "inaction," "motion" and "rest" (WordNet, WordNet 2.0, 2004).

Thus, the notion of "state" comes up not only as more general and inclusive, but also more definite, formal and technical term than property, as well as attribute, quality, and quantity; for it points out all the characteristics of a being, thing, or substance, as well as the modifications of change, movement action, or behavior. We speak of the state of being, the state of action, the state of relation, the state of matter, mental states, human conditions, physical states, and so forth. Thereupon, the category of state comes up as a necessary element of the ground ontological framework. Given this assumption as true, all things will become much more organized; for property and attribute, quantity and quality, eventually fall under one unifying heading of state. With that the list of the major genera will be reduced to Substance, State, Action and Passion (or Activity and Passivity), and Relation. Furthermore, the genera "activity" and "passivity" are really the species of the genus "change," and, again, this hardly will raise a strong objection. Since Action with all its varieties as acting, doing, affecting, operation, or activity is a process of exerting power to produce a change by entering into other states. The change as meaning both "making different" and "becoming different" covers here both acting on, action, and being acted upon, or passion.

As a result, Substance is coordinate with State, Change, and Relation, where **state** comes about as the intermediate between **substance** and **change**; and **change** - between **state** and **relation**. It looks that the second dilemma will find its resolution partly to Kant's favor: "there are not primitive and highest genera, which stand isolated . . . from each other, but all of the various genera are mere divisions and limitations of one highest and universal genus" (Kant, 1990, p.197). Which is nothing else but Being, and this goes to Aristotle's credit. Here the super-class of being loaded with the meanings of **reality, entity,** or **thing** can be usefully equated with Kant's single supreme and all-embracing class or genus, Entity or Thing.

With this the list of the foremost principles about reality may be reduced to the following ontological commandments:

- Entity, Being, or Thing is something common to all kinds of entities
- Nonentity as the null thing is part of every thing
- There are four primary classes ranged under being: substance, state, change, and relation
- Entity and Nonentity, along with the primary classes, their kinds, species, varieties, and individuals, constitute the world as the whole universe of discourse
- Each individual (or particular) as a distinct entity is a substance, a state, a change, or a relation.

So, our theory collects all the totality of things into one universal class and to this we subordinate the four primary genera, which have the ability and resources to produce the world, all the rest of the existents. Besides, the world is a hierarchically ordered totality of things, beings, or entities marked with the universal top and bottom elements, the whole of reality and the least ontic species of individuals. The body of the ontological principles, such as the laws of the ultimate unity, diversity, variety (multiplicity, or heterogeneity) and continuity as well as the principle of subsumption (transitivity, composition, and predication), constitute the key axioms of the theory of genera, when complete with the methodological rules.

As it has been pointed out the object of universal ontology is the entity classes, while the scope of special ontology is particular classes of being or entity. Accordingly, the business of the former is to study the universal and necessary relationships of the diverse genera of the same universal class, while the task of a regional ontology is to construct the special theories of the different species of the same genus. Thus, in its truest and primary sense, *ontology comes to be the science of being, of its forms, modes, divisions, classes, of their relationships, ordering and laws.*

As for Relation, it is another fundamental dimension on a par with Substance, State, and Change or Action. It is among the basic categories and principles of reality as the principle of ordering, laws, and rules constraining the world behavior. So, first of all, relationships are the most essential components of the world, and they are all about relations between classes, kinds, or universals. Therefore, we qualify the fundamental laws as universal and necessary relationships governing all that exists and happens. What is also important here, relation in the four-dimensional continuum of substance-state-change-relation has the same degree of reality as substance, as far as it holds among real things. When among substances, then it will be a substantive relation. When among states or properties, we shall have stative relations, like qualitative and quantitative. If among changes, it will be change-related or changeful relations, like cause and effect relationships. At last, among relations, it is something clumsy sounding as relational relations, like proportions, and analogy or likeness. One of the functions of relation is to determine identity and diversity, or sameness and otherness, similarity and difference of one thing in respect to something else, in substance, in state, in change, or in relation, as relational sameness or diversity, thereof establishing the ontological principles of identity and diversity. Furthermore, since the principal characteristic of relation is considered to be order, the order of the universe will be correlative to a relational hierarchy of beings, entities, or things.

Such are, then, the number and nature of the major classes of being, which can be defined both as the essence of all things and a world of real classes. Reality or Being exists or rather subsists in kinds, each of which has infinitely many individuals. For genera as largest ontological classes are infinitely prolific of numerous sorts of particulars, objects, states, facts, events, and processes produced by Substance, State, Change and Relation, where each separate genus is not a class of everything but only of species each of which is a substance, a state, a change, or a relation.

Again, here we touched the hottest issue of all inquiry of reality: whether the world model of things should be patterned as a universe of singulars, a universe of kinds, or rather as an ordered hierarchy of the both worlds. There are two solid arguments underpinning the latter position. First, even proceeding from a particular of any kind or sort, distributing it progressively in line with the property sharable with more and more things, we eventually will reach the universal status of Being or Entity. Second, the final object of any science is to reveal the universal properties of things in a domain of interest, to obtain the true representation of objective patterns and laws. For scientific knowledge is ultimately about the substance, the cause, the essence, the whole, the class, the one, the same, the common, the universal, but not about the object, the effect, the individual, the part, the member, the many, the other, and the unique.

Thus, it emerges that the world is nothing else but Entity taken together with the primary classes complete with their kinds, species, and individual types; every thing belongs to Being, Thing, or Entity. Next, whatever comes into being is either a substance or a state or a change or a relation. Substance, in general, is a highest class of all kinds and sorts of substances, masses, bodies, and objects. State, in general, is a highest class of all types of properties, attributes, qualities, quantities, conditions, and facts. Change, in general, is a highest class of all manners of events, occurrences, outcomes, acts, and activities. Relation, in general, is a highest class of all sorts of connections and links, comparisons and contrasts, bounds and ties, functions and bonds, processes and interactions. At last, substances are no primary entities than are states, changes, and relations; without the former there is not the latter, and vice versa.

Carving Reality at its Joints, or the Ways to Classify Things, Beings, Entities, or Resources

From our analysis, it follows that the world is not merely an aggregation of substances, an aggregate of objects, a multitude of things, a plurality of states of affairs, a collection of facts, a class of events, an association of relations, or a total of objects and events. It is rather the totality of beings, things, or entities distributed in the four-dimensional framework of substance-state-change-relation, existing in the material world of particulars as the largest continuum of objects, states, and events standing in the cause and effect relationships. The figure illustrates that we lay down four primary classes by making a fourfold division into substances (substrates or objects), states (forms or properties), changes (actions or processes), and relations (order and connections), finding in theses classes a common being, and thus including everything in one universal class. The diagram in the form of a semi-lattice (Figure 2) indicates that a classification of things proceeds from Entity split into its four major classes: substance or substrate, state or property, change or action, relation or order. Each of which is, then, further subdivided into their generic kinds and specific parts: matter, stuff, material, or body, spatial or nonspatial, material or nonmaterial, physical or conceptual; quality or attribute and quantity or magnitude; acts, activities, events, occurrences, processes; and all sorts of concrete and formal relations. And so on down, until it is reached the lowest species of individuals, nonentity, nonbeing, or nothing, the null thing.

In general, the lattice of reality is composed of four distinct modules of successive subdivisions, proceeding from the four top nodes: substance, state, change, and relation. With proper differentiation and subdivision, the four primary classes, complete with all their subordinates as tightly intertwined with each other will constitute the primary cuts of reality, the core of an exhaustive classification of the world (see Figure 2); like the primary colors of red, yellow, green, and blue through proper mixing produce an infinite multitude of different hues.

An important point represented in Figure 2 is the diversity of prime and secondary beings (substances, states, changes, and relations) presented as the positive difference between ontological classes and individual entities: the classes as essential wholes and their objects, elements, units, parts, members, or instances. Such diversity in the order of natural language is given by mass nouns for the entity types, and by count nouns for a great number of entity instances, individual parts, units, objects, or separate wholes of the classes, mediated by collective names in the singular and plural senses.

The method of division of the root classes is no less important than the character of differences of the primaries with their subordinates. It consists in that each subordinate class, a variety of a genus, is composed of the genus plus a differential. This clearly suggests that we cannot have beings of beings, substances of substances, states of states, changes of changes, and relations of relations; for being (substance) cannot differ from being (substance) by being (substance), nor relation from relation by relation. An entity cannot be a distinction (or differentia) of the same kind of entity. Differences should come from other genus (or entity types) to which the differentiated class does not belong: just such differences produce all the range and varieties of subordinate classes and individuals.

So Substance added to Being produces *a being substantial* (Being + Substance = a being substantial), while Being added to Substance produces *a substantial being* (Substance + Being = a substantial being); thus, substance becomes a property (attribute) of being, and being a property (attribute) of substance. Being can be used either in the primary sense or just only by participation in some entity. We can predicate substance of being, so that to have the proposition, "being is a substance," where substance comprises being as its essential part. Or we can attribute the power of being to substance to have the dual proposition "substance is a being," where being comprises substance as its species. The same is true for other primitive classes; such is the case with the state, change, relation, and their subordinate classes. For instance, when mutually differentiating being and change, there is *a mutable being* as a changing thing, as well as *a being mutable*, which is nothing but becoming or changing (Plotinus). Relationship in turn consists in *being related*, relativity, and *related being*, relative. Altogether, it emerges that Being is the source of the existence of the genera,

Figure 2. The universal classification of things in reality

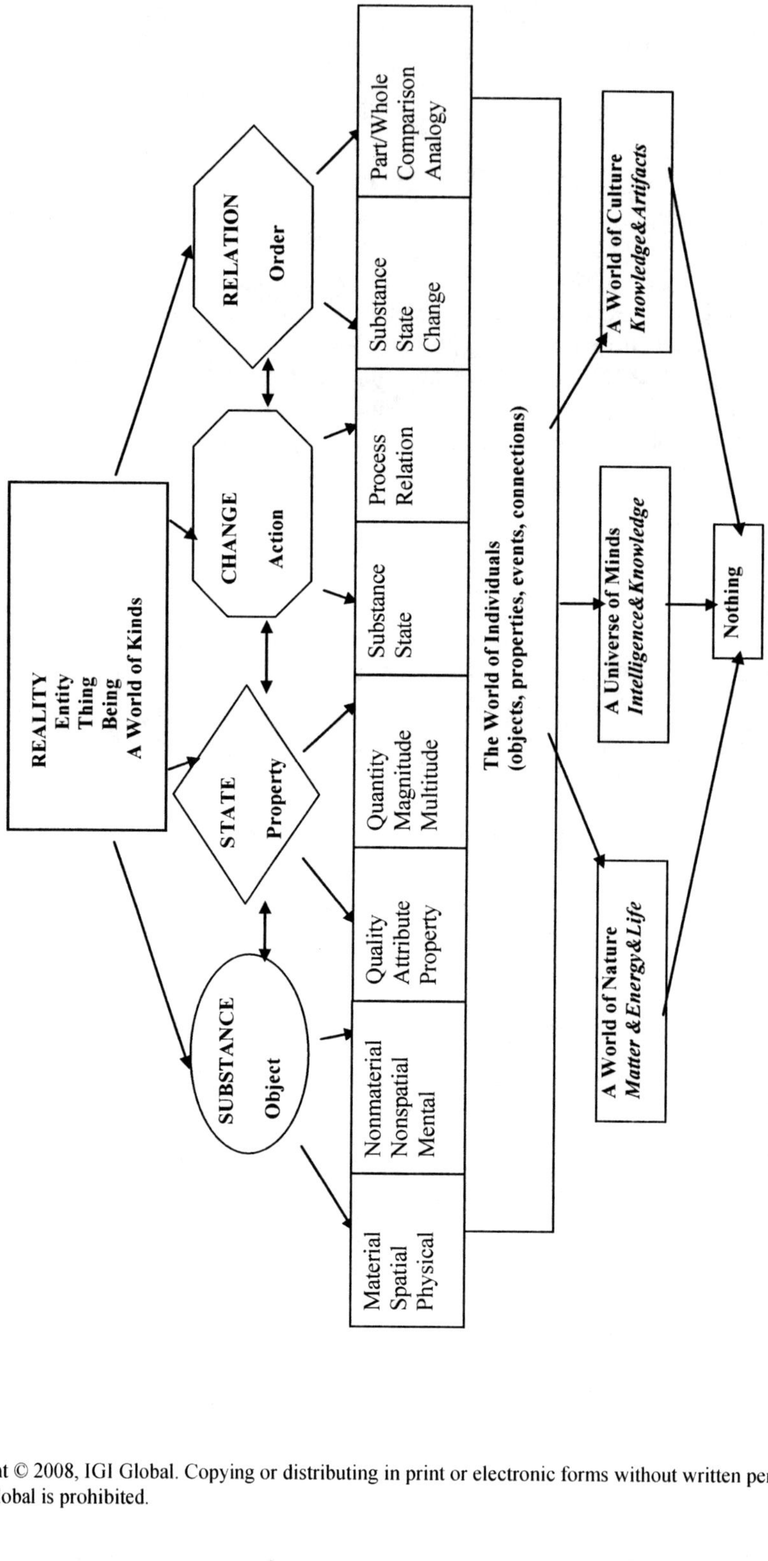

the largest kinds after Entity, while by proper mixing the general classes will generates all their divisions, species, and individuals.

This implies that the same means of differentiation can be applied to the primaries and their subordinates. So we can class the substance with all its varieties and forms on the basis of states (quality or quantity), changes (motions or causes), or relations (similarities and differences). As a result, we can posit a species of substance such as matter (material, stuff, or body) or mind; at least, three different species of state, substantial states (quantity) and changeful states (quality), and relative states; three species of change, substantive, stative, and relational; and, again, three species of relation, substantive relations, stative relations, and change connecting relations.

Thus, when we divide the genus *substance*, we need, in a sense, fresh generic blood so that not to fall into a sort of intra-generic incest: we have to take differences belonging to the classes of state, change or relation, but not of the thing differentiated. Guided by such constructive principle, the primary genera can produce all subordinate classes and particulars. That is, substance produces matter when properly qualified and quantified, and matter is conceived as the substantial aspect of reality having states, subject to changes, and standing in relationships to other substances. Dividing in such way the species of matter, we can arrive at a particular physical substance, as a single body or mass having a quale and a quantum.

With regard to *state*, after proper division, it becomes a genus of species each of which is a kind of state: properties, qualities, quantities, forms, shapes, and figures, all grouped under the class of states.

In differentiating the primary genus of change, we have to refer to the differences taken from the classes of substance, state, and relation. Then we can obtain the following distinct kinds of change: a change of substantial nature (generation, creation or destruction, decay, and disintegration), a change of state (all the relational differences between states, act and experience, alteration, modification, increase, decrease, development, and motion), and a change of relational character, causal processes. A change of change or change rate is possible only as when the process of changing alters or modifies its manner of action.

As for **relation**, it also is divided by reference to other primary classes; the kinds of relations produced by the differences of substance, state, and change. The relation between substances covers all the relations of their species and genus; between states, qualitative and quantitative relations, as spatial and temporal relations; between changes, causal relationships, or processes formed as a relation of changes. So the fundamental relation of substances involves the relations of **inclusion, diversity,** and **identity**, the relation of a substance to itself, and the relation of **part** and **whole**. The quantitative or functional relations consist in equality and inequality. The relation of qualities consists in being like or unlike, or in **similarity** and **difference**. At last, the relation of relations is *analogy* or *proportion*. Thus, **causation, inclusion, parthood, wholehood, identity, diversity, sameness, otherness, resemblance, difference, equality,** and **inequality** as well as **spatiality,** and **temporality, priority, posteriority,** and **simultaneity,** can be qualified as the species and examples of relationships and ordering.

From the above, it follows logically that relative things are those that cannot exist without reference to other things. In a sense, all the primary genera are relative, since each of them is analyzed by reference to other class. Because of such a universal relativity, everything is relative, for everything bears some relationship to something else. That is explained by the fundamental fact that the ultimate categories of reality are interrelated to each other: there are substances, there are states, there are changes and there are relationships. Any can be determined in terms of the other.

It often occurs that a thing can be classed under several genera, say, under the category of *relation* and that of *state*, as place, time, position, and situation. But having a consistent classification scheme of things makes us sure whether this or that thing exists as a whole being or a partial being, whether it is a class or a particular, or whether it essentially a substance, a state, a change, or a relative and associative entity.

Now the next important question will be how to distinguish the universe of particular entities, how properly to divide its members into classes, categories, or groups? Individual things, composites of the *summa genera*, occupy in our scheme the ground level of entities as being the closest to our sense perception and cognizance. The sorts of the particular entities are much more different than their genera. The matter here appears to be very complicated and sometimes confused basically because of a faulty selection of "the primary colors" of being suggesting the psychological primaries (black, red, yellow, green, blue, and white) to which any color can be reduced.

Moreover, there are some AI ontologies starting just from this level of reality that falling short of any primary categorical heads, that are rooted into the notion of thing conceived as a heterogeneous collection of separate individuals, that is, as the collective plural instead of the collective singular referring to a collection of things regarded as a whole. In such schemes, which are beheaded of any primary classes, Thing as the universal set and the collection of everything is usually partitioned into two sorts of contraries. It can either be split as disjoint classes of substantial individuals (substances, states, changes, space, and time) and conceptual objects inclusive all sorts of constructs (concepts, propositions, and conceptual bodies). Or, the totality of entities can be divided into substantial individuals (temporal and spatial), logical things (mathematical or computational), plus their intermediates, partially intangible things, serving all as the ground for the whole body of entities without the head and proper ordering.

In fact, in this order of reality, we deal with all the possible modifications of the primary classes. Among the realities of our earthly world are matters, parts, particles, bodies, agents, or organisms, and minds; qualities, quantities, faculties, and properties; places, times, positions, and situations; causes, acts, actions, motions, movements, events, and occurrences; connections, sequences, binds, interactions, and processes. Despite of apparent diversity and variety, on the whole, these particulars existing as instances of their primary classes can be eventually subsumed under four supreme heads: substances, states, changes, and relations.

But still there is a wide diversity in the ways of existence of particular beings. Generally, it is accepted that the particular things or individuals exist first and foremost in nature, then in mind, and in society. Hence, we can divide the totality of particulars into several broad categories: natural phenomena, living beings, mental things, and cultural entities.

Those things existing outside the mind are thought to be objective, sensible, temporal, spatial, material, palpable, physical, tangible, or visible. Nature, as the world of physical and material existences, marked off by matter, energy, time, space, body, life, and cause and effect relationships; where substance as matter is subject to exact measurement in terms of physical magnitudes, mass, force, radiation, temperature, duration, extension, and energy. With that, matter and energy, the substrate of life, constitute the key distinctions of the material universe encompassing all the species and states of matter, matter in all its modifications and sizes, from elementary particles to the whole cosmos, as well as all the forms of energy, together with their transformations and interrelations. For the time being, the bottom level of the physical universe is considered to be *the elementary physical particles*, six quarks, six leptons plus their corresponding antiparticles, and *the fundamental forces* of the universe carried by *the gauge bosons*, such as *gravity (gravitons)*, *weak nuclear force (weak bosons)*, *electromagnetism (photons)*, and *strong nuclear force (gluons)*. *The strong nuclear force*, having the short range is thought to act on quarks and gluons, causing nuclear reactions as substantial changes in matter and so forming *hadrons* and *nuclei*. *The weak nuclear force*, having the shortest range, acts almost on all particles causing neutron beta decay but without making up any stable natural systems. *The gravity*, having the most long-distance range, acts on all forms of matter and energy causing all the physical objects to attract each other and thus building any physical systems.

The last but not least, *the electromagnetic force*, having also the long-distance range, acts on all charged particles causing all types of chemical reactions and so forming *atoms* and *molecules*, all *chemical compounds* and *biochemical components*, as nucleic acids, proteins, carbohydrates, lipids, vitamins, and hormones. Hence, electromagnetism builds up not only major forms or states of matter such as *gases*, *liquids*, *solids*, and *plasmas*, but also the *organic compounds* composing the highly ordered and organized material systems such as biological, living objects (*cells, microorganisms, plants, animals, and human beings*), see Figure 3.

The world of natural particulars can be further divided according to the presence or absence of the processes of life, reproduction, growth, and development, photosynthesis, metabolism, sensory reception, and adaptation. Among them there are inorganic or physical inanimate beings as substances, states, events and processes, and organic or biological living beings, again, together with their specific substances, states, changes, and processes. In the world of nature, a sum total of living things amounts to the totality of life, the Biosphere, in turn culminating in the world of intelligent kinds, the totality of minds, natural and artificial, or the Nousphere. Thus, the totality of organic beings makes up the biosphere. This universe of living things may be divided further, according to the beings' differences of composition or structure, qualities or quantities, forms or functions, actions, behavior, and relationships, into five or six kingdoms, as microorganisms (monerans, viruses, protists, and fungi), plants, and animals. What, in turn, is distributed into sequentially narrowing taxonomic categories, phylum, class, order, family, genus, species, variety, sub-variety, or races, where mankind falls under the domain of animal beings as human species each member of which possesses intangible mind, rationality, and intelligent power. In that this specific difference (attribute, feature, distinction, or contrariety) has its reality only in minds, natural or artificial, the mind is intangible, intelligible, and nonmaterial, but still real; and distributed among mental objects, states, events, and processes. In the way of the mental objects, materials, or stuff, there are all sorts of mental signs and representations of things: images and perceptions, ideas and concepts; fantasies, opinions, beliefs, and values, memories and thoughts, intuitions, experiences, and learning, and so forth. For instance, in efficient realistic minds as the prototypes of machine consciousness and conscious artifacts (Aleksander, 2000), among mental stuff should be included the constructs and abstractions of ontological, mathematical, and logical natures, as the basic categories, truths, principles, sets, functions, and other

Figure 3. The hierarchy of natural entities

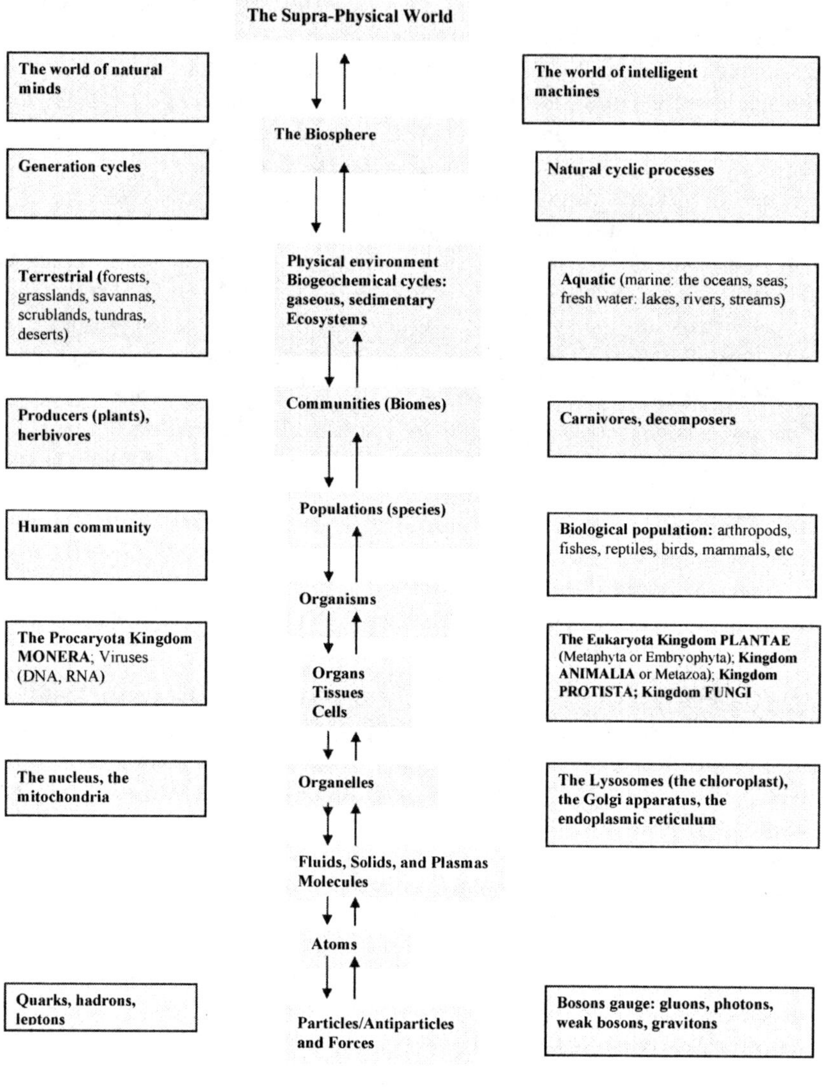

mathematical structures. Such mental objects are the subject matter and objects of the state, act, or process of knowing, understanding, thinking, or experiencing, or, speaking generally, of *minding*. On the other side, mental faculties and powers as well as all the states of consciousness, acts of memory, imagination, and willing, experiences of perception and emotion, processes of knowing and thinking, and all types of minding fall under the mental states, changes or processes. Hence, the total of the entities of mental nature comprises a world of minds where intelligence and knowledge function as specific differences, the most distinguishing characteristics.

When knowledge is objectified and embodied as social objects and structures, artifacts, languages, literature, social organization and relationships, institutions, laws, morals, science, technology, art, customs, codes, rituals, public ideas, and beliefs, we arrive at the world of culture created by human minds. Such a complex of social entities can be defined as an objective meaningful continuum of human beings and cultural objects, states, events, and processes, being in the cause and effect interrelations. Specifically, to the class of cultural objects belong all sorts of artifacts, tools, machines, books, writings, structures, buildings, and works of art, which are the composites of knowledge and material substrates. The state of acculturation of individuals and social groups suggests the condition and way of living socially: savagery, barbarism, and civilization, *such* as primitive cultures, ancient civilized societies, modern industrial civilization, mass society, emerging information culture, and imminent cybernetic society. Cultural evolution, political, scientific, and technological revolutions as well as other cultural events, acts, and actions, all come under **cultural change**. Finally the processes of production, artistic, economic, scientific, and technological, as well as education, socialization, administration, learning, passage rites, and religious practice are to be subsumed under cultural processes; whereas, such things as kinship and family, inheritance and succession, social organization, and communication fall under the class of cultural relationships.

Such, we suggest, is the way how the World or Reality as the maximal Being or Entity or Thing can be exhaustively broken up into a set of few ontological primitives, producing a great varieties of species and infinity of parts and particulars and their numberless combinations, as mixtures, compounds, complexities, organizations, systems, and so forth. Organized by the supreme ontological classes of entities, by this way reality can produce an immense world of species and particulars by taking the differences outside of the thing differentiated. The above considerations show that Substance, State, Change, and Relation are the only primary classes of the universal class of Being comprising all things, whereas the things of nature are individuals or particular beings: particular substances, particular states, particular changes, or particular relations. Further, in a more careful and detailed examination, we will undertake to demonstrate that these five genera may constitute the hardcore of standard ontology for human beings and computing machines.

Before proceeding with a formal investigation, it would be of use to reiterate some common but significant points. *When erecting a general taxonomy of things, the positing of the primary classes as the kinds of being of the first rank and importance is everything. Since the wrong positing is not an easily correctable mistake, but a fundamental error making invalid all the subsequent classification,* one should not develop a large taxonomy of things without comprehending what the topmost ontology should be. Otherwise, instead of the whole body of entities, we will have a monstrous agglomeration of things, such as individuals, times, places, quantities, parts, objects, doings, transformations, states, changes, substances, agents, organizations, actors, roles, machines, transportation, emotions, mathematical, geographic, chemical, biological objects, and so forth. (Open Cyc, 2003). Thus, a consistent set of top-level ontological categories is a necessary condition for building any viable information application, databases (domains), knowledge bases (types), logical systems (sorts), or object-oriented systems (classes); for "Any incompleteness, distortions, or restrictions on the framework of categories must inevitably limit the flexibility and generality of every program and database that uses those categories" (Sowa, 1995).

The quality of fundamental ontology is the capacity to see a deep contrast among two types of reality: ontological entities and individual things. In the first place, one needs to discern the level of the prime genera and categories plus their subordinate classes, such as matter, quale and quantum, process, ordering, and then the order of individual or physical things, as a body, a quale, a quantum, a fact, an event, a connection, which are the domain of regional ontologies, empirical sciences and technology. The ontological (theoretical) classes are at the grade of abstraction where they exist beyond specific matter, time, space, and contingency. Hence, fielding the realities of the generic status in terms of ontic categories what any true general world model is about. For the ontological classes as the supreme categories of things and knowledge and meanings together with the fundamental rules and principles of unvarying validity constitute the substance or bare bones of the world. This makes the key subject of study for a standard ontology as the single reality framework or the world codification: the code of standards and invariants, ultimate categories, principles, truths, rules and laws, governing the entire universe of things. A systematic research of the matter of the standard ontology as the world code requires a formal, rigorous description of reality, especially, how the whole world, with its major classes, kinds and levels, can be represented and formalized by unambiguous mathematical symbolism.

Conclusion

Universal standard ontology standing as a comprehensive science of reality, a fundamental form of world representation, is providing an underlying typology of things as well as the broadest classificatory schemes of different domains of objects. The underlying principles of standard ontology are formulated as follows:

- All things (beings or entities) are classified into few primary ontological classes: substances, states, changes, and relations
- The world or reality or universe or being is the universal hierarchy of classes with Entity at the top and Nonentity at the bottom, or all the things in the world form the universal hierarchy of natural kinds (forms) limited by Thing and Nothing
- Entity or Being or Thing is the class of everything, the universal thing
- Nonentity is the class of nothing, the null thing
- The primary classes make the first divisions or ultimate categories of reality.

References

Adler, M.J. (1990). The syntopicon: An index to the Great Ideas. *Great books of the western world* (Vol. 1). Chicago: Encyclopaedia Britannica, Inc.

Aleksander, I. (2001). *How to build a mind: toward machines with imagination.* NY: Columbia University Press.

Algebra: Lattice theory, Categories, Macropaedia. (1994). Knowledge in depth (Vol.13), *The new encyclopaedia Britannica.* Chicago: Encyclopaedia Britannica, Inc.

Aristotle, Logic (Organon): Categories. (1990). In M.J. Adler (Ed.), *Great books of the western world* (Vol. 1). Chicago: Encyclopaedia Britannica, Inc.

Aristotle, Physical Treatises: Physics (Physica). (1990). In M.J. Adler (Ed.), *Great books of the western world* (Vol. 1). Chicago: Encyclopaedia Britannica, Inc.

Aristotle, Metaphysics. (1990). In M.J. Adler (Ed.), *Great books of the western world* (Vol. 1). Chicago: Encyclopaedia Britannica, Inc.

Armstrong, D. (1997). *A world of states of affairs.* Cambridge, UK: Cambridge University Press.

Borgo, S., Gangemi, A., Guarino, N., Masolo, C., & Oltramari, A. (2002). *Wonder web deliverable D15: Ontology roadmap.* Retrieved April 25, 2006, from http://wonderweb.semanticweb.org.deliverables/documents/D15/pdf

Bunge, M. (1974). The relation of logic and semantics to ontology. *Journal of Philosophical Logic, 3*(3), 195-209.

Bunge, M. (1977). *Ontology I: The furniture of the world.* Boston: D. Reidel Publishing Co.

Bunge, M. (2006). *Chasing reality: Strife over realism.* Toronto: University of Toronto Press.

Chrisholm, R. (1996). *A realistic theory of categories: An essay on ontology.* Cambridge, UK: Cambridge University Press.

Gruninger, M., & Menzel, Ch. (2003). *The process specification language (PSL): Theory and applications, AI Magazine, 24*(3), 63-74.

(The) history and kinds of logic: Logic systems. (1994). Metalogic, knowledge in depth, macropedia (Vol. 23). *The new encyclopedia Britannica.* Chicago: Encyclopedia Britannica, Inc.

IEEE P 1600.1 standard upper ontology working group (SUO WG). (2004). Retrieved April 25, 2005, from http://suo.ieee.org/

IEEE-SA standard project authorization request (PAR) form. (2000). Retrieved April 25, 2005, from http://suo.ieee.org/

James, W. (1990). *Pragmatism.* In M.J Adler (Ed.), *Great books of the western world* (Vol. 36). Chicago: Encyclopaedia Britannica, Inc.

Kant, I., The critique of pure reason. (1990). In M.J. Adler (Ed.), *Great books of the western world* (Vol. 39). Chicago: Encyclopaedia Britannica, Inc.

Kent, R., Semantic integration in the IFF. Retrieved April 25, 2006, from http://www.ontologos.org/Papers/SI2003/SI2003.pdf

Locke, J., An essay concerning human understanding, In M.J. Adler (Ed.), *Great books of the western world* (Vol. 33.). Chicago: Encyclopaedia Britannica, Inc.

Lowe, E. (2001). *Recent advances in metaphysics.* Retrieved April 22, 2006, from http://www.fois-2001/keynote

Novik, I., & Abdoullaev, A. (1991). *Introduction into information world.* Moscow: the USSR Academy of Sciences, Nauka (in Russian).

Ontology, semantic technology, and knowledge society. (2005). Retrieved April 22, 2006, from http://www.eis.com.cy

Open Cyc. (2003). Retrieved April 22, 2006, from http://www.opencyc.org

OpenCyc Selected Vocabulary and Upper Ontology. (2002). Retrieved May 22, 2004, from http://www.cyc.com/cycdoc/vocab/vocab-toc.html

OWL Web Ontology Language, Overview. (2004). Retrieved April 22, 2006, from http://www.w3.org/TR/2004/REC-owl-features-20040210/

Peirce, C. (1945). *The collected papers of Charles Sanders Peirce.* Cambridge, MA: Harvard University Press.

Plotinus. On the kinds of being, Sixth Ennead. (1990). In M.J. Adler (Ed.), *Great books of the western world* (Vol. 11). Chicago: Encyclopaedia Britannica, Inc.

Quine, W. (1960). *Word and object.* Cambridge, MA: MIT Press.

Quine, W. (1963). *Set theory and its logic.* Cambridge, MA: Harvard University Press.

RDF vocabulary description language 1.0: RDF schema. (2004). Retrieved April 30, 2006, from http://www.w3.org/TR/2004/REC-rdf-schema-20040210/

Russell, B., & Whitehead, A. (1910; 1962). *Principia Mathematica.* Cambridge: Cambridge University Press.

Semantic web topic hierarchy (2007). Retrieved September 12, 2007, from http://wiki.ontoworld.org/index.php/Semantic_Web_Topic_Hierarchy

Semantic technology primer (2007). 2007 Semantic Technology Conference. Retrieved September 8, 2007, from http://www.semantic-conference.com/primer.html

Simon, H., & Kaplan, C. (1989). Foundations of cognitive science. In M. Posner (Ed.), *Foundations of cognitive science.* Cambridge, MA; London, UK: MIT Press.

Smith, B. (2003). Ontology. In *Blackwell guide into philosophy of computing and information.* Oxford: Blackwell.

Sowa, J. (1995). Top-level ontological categories. *Int J Human-Computer Studies 43*, 669-685.

Sowa, J. (1997). Principles of ontology. Retrieved April 22, 2006, from http://www-ksl.stanford.edu/onto-std/mailarchive/0136.html

Sowa, J. (2000). *Knowledge representation – Logical, philosophical and computational foundations.* Pacific Grove, CA: Brooks Cole Publishing Co.

SUMO Ontology. (2004). Retrieved April 22, 2006, from http://ontology.teknowledge.com

SUO 4D Ontology. (2005). Retrieved April 22, 2006, from http://www.tc184-sc4.org/wg3n-docs/wg3n1328/lifecycle_integration_schema.html

SUO information flow framework. (2004). Retrieved April 22, 2006, from http://suo.ieee.org/IFF/

WordNet, WordNet 2.0. (2004). Retrieved April 22, 2006, from http://www.cog.sci.princeton.edu/wn/

Chapter III

The World Code: Mathematical Ontology as the Real Road to Reality

Introduction

Formalizing the world in rigorous mathematical terms is no less significant than its fundamental understanding and modeling in terms of ontological constructs. Like black and white, opposite sexes or polarity signs, ontology and mathematics stand complementary to each other, making up the unique and unequaled knowledge domain or knowledge base, which involves two parts:

- **Ontological (real) mathematics**, which defines the real significance for the mathematical entities, so studying the real status of mathematical objects, functions, and relationships in terms of ontological categories and rules.
- **Mathematical (formal) ontology**, which defines the mathematical structures of the real world features, so concerned with a meaningful representation of the universe in terms of mathematical language.

The combination of ontology and mathematics and substantial knowledge of sciences is likely the only one true road to reality understanding, modeling and representation. Ontology on its own can't specify the fabric, design, architecture, and the laws of the universe. Nor theoretical physics with its conceptual tools and models: general relativity, quantum physics, Lagrangians, Hamiltonians, conservation laws, symmetry groups, quantum field theory, string and M theory, twistor theory, loop quantum gravity, the big bang, the standard model, or theory of everything material. Nor mathematics alone with its abstract tools, complex number calculus, differential calculus, differential geometry, analytical continuation, higher algebras, Fourier series and hyperfunctions is the real path to reality (Penrose, 2005).

The richest conceptual representations of the world structures and changes come from the intimate unification of fundamental ontology and mathematics.

For, providing the realistic foundations for mathematical quantities and relationships, ontology is concerned with an abstract modeling of the real world and its basic features. Mathematics in turn is ultimately concerned with the quantitative models of reality, so furnishing the quantitative forms for the real world content, designing the pattern of a full-dress conceptual uniform tailored to fit the world body.

It is rather symptomatic that computing ontology, as formal specification for AI systems, agent communication and knowledge reuse started as a sort of mathematical ontology, built on abstract algebra and measurement theory (Gruber & Olsen, 1994). But such a reciprocal connection opens up much broader avenues to human and machine understanding of things, rather than just formalizing mathematical modeling in engineering and establishing the conceptual grounding for some variable quantities, physical dimensions, or metrics.

Complete with their interrelations, the totality of ontological classes constitutes the world general framework as an all-defining system of things. What is mostly remarkable about the general framework is that its highest universal class and the base kinds are susceptible of symbolism and computation. Or, the ontological entities are subject to quantity and accurate representations in terms of their subordinate abstract structures: sets, classes, functions, categories, relations, numbers, and other magnitudes.

As it happens in all contemporary science, the possibility to be quantified and formulated may sharply reduce the ambiguity of the propositions and speculative reasoning upon reality, revealing its true constitution, structure, composition, and order. However, in order to correctly apply mathematical formulations and models, several considerations are necessary to bear in our minds, so that to avoid the congenital defects of set-theoretical and logical constructions of the world structure.

And the issue of issues is: what kind of reality makes the subject matter of **real mathematics** and **mathematical ontology**, assuming that there are several types of existence or being or reality:

- Ontological existence, as entity kinds, rules, and patterns of relationships at the axiomatic, absolute, ultimate, and truthful level of reality.
- Physical existence, as concrete things and processes at the factual, natural, scientific, experimental, commonsense, or objective level of reality.
- Mental existence, as abstract constructs and images at the phenomenological, perceptual,cognitive, subjective, or experiential level of reality.
- Cultural or social existence, as the artificially created level of hyperreality
- Virtual existence, at the simulated level of reality.

Crucially, there is the underlying world per se, the inherent reality, the implicit universe, the fundamental existence, independent of any conceptions, worldviews, or mathematical formulations on it. And there are then the modeled worlds, constructed existence, or represented realities, as enhanced representational (hyper)reality, imitation, simulation or simulacrum appearance of reality of artificial life forms or simulated entities in the computer systems [Reality, http://en.wikipedia.org/wiki/Reality; Hyperreality, http://en.wikipedia.org/wiki/Hyperreality]. The fundamental level of reality, its categories and rules and principles, is the subject of fundamental ontology, while its formal description involves the formal sciences as mathematics and logics.

If the empirical sciences deal with the universe of material entities (substantial systems, states, processes and dynamic relationships), the reference class of formal systems and languages is a conceptual universe populated with its abstract objects and structures possessing mental existence, while the theoretical scientific systems refer to the generic classes and kinds constituting the essence of existence. In the latter case, the universe of discourse of theoretical models is co-extensive to the reference classes of ontological theories, which are the world, or all reality as the maximal self-contained whole entity with its parts. For example, the logical model of possible worlds is a conceptual model about some formal entities, while the ontological model of possible worlds is the model representing really possible worlds.

Thus, we should use the universal language of mathematics, but with some strict qualifications and restraints. Its major constructs (objects, functions and relations) must not be employed as representing the formal properties (attributes, functions, and relations) of formal objects. Rather it should be regarded as conceptualizing the real properties of real entities, that is, the world substances, states, changes, and relations. For instance, if we use the central mathematical notion of a function, it will be used as a total state function for a certain entity kind determined by a collection of related properties. Here each component of the function represents a single property and the functional relationships correspond to the law statements about the generic class and its members.

Or, if we employ a formal concept of category, it must be applied as an abstract structure of objects (domains and co-domains) and morphisms, where the objects refer to real object-entities, the parts of the world, states, actions, and so forth, and morphisms represent real relationships among real things, as a relation of cause and effect or spatio-temporal relations. Strictly speaking, we cannot interpret the objects of the category as ontologies, which are themselves conceptual models or theories as well as the morphisms, the interrelationships of such theories. For any ontology is a theory of a certain class of being, and what should be modeled is not concepts, ideas, theories, constructs, but rather their objective meanings, the reference classes, that sorts of classes of entities having real, ontological status.

So, of a large set of mathematical objects, mostly significant and meaningful are those concepts that able to imply real significance and true meanings, which can represent, model, describe, specify, define, or conceptualize really possible beings as substantial entities and properties with their relationships.

This all means that we first of all have to construct the functional schemas of realities, the categories of the world, the lattice of the classes of entities, the tree of universal properties, the equation of things. And only secondarily there come such formal constructs as the lattice of theories, the categories of concepts, or the tree of axioms. In all, we talk here of a new sort of **real mathematics** *seeking real significance (existence) of mathematical entities*. We also talk about its complementary part, **mathematical ontology**, *a formal theory about the real world capable to establish the fundamental (causal) laws of the universe as the quan-*

titative (functional) relationships of entity class variables. And the overlap, intersection, or product of the both makes the UFO, universal formal ontology, a common ground between two universal theoretical systems.

The Standard Model of Reality

All formal theories are expressing their specific entities by systems of mathematical signs and symbols arranged into three separate groups:

- The signs of objects
- The signs of mathematical processes or operations
- The symbols of relationships

This in turn is divided as the individual signs of definite objects, operations, and relations, and the general signs of variable or unknown objects, operations, or relations. So, the most significant issue in mathematical symbolism is the question of entity variable, its definition or reference, typology or classification. A variable is commonly defined as a thing that is liable to change, as a quantity changing in value or magnitude under different conditions, and a symbol representing a set of things in a group of entity types. Any variable referring to some kind of things is represented by variable quantity and embodied by variable symbol. On the whole, there are only a few general categories of things as the key entity classes: the category of substance, the category of state, the category of change, and the category of relation.

These four basic kinds of things represented by signs, marks, characters, symbols, or equations make up the set of principal variables having a primary importance for describing all reality. United as the essential parts and elements of one universal symbol structure, the primary cuts of reality, substances (objects) **O**, states (properties) **S**, changes (actions) **C** complete with their relationships **R**, will constitute the whole symbolism of the world, as the reality codification:

$$\mathbf{W} = \langle \mathbf{O}, \mathbf{S}, \mathbf{C}, \mathbf{R} \rangle \qquad (1)$$

Here the symbol **W** stands for the universe itself, **the totality of possible worlds**;

O is the sum total of all classes of objects of any nature, **the substance space of reality**;

S is the union of the state spaces of all kinds; **the state space of reality**;

C is the sum total of all kinds of modifications and variations, **the change space of reality**;

R indicates the universal class including all types of relations, **the universal relation**.

Then, alternatively, the world with its components can be described in terms of the entity class variables:

- The world variables W_I, having as its value sets the collection of possible worlds
- The substance variables O_j, instantiated by to possible classes of substances, objects, bodies, and agents
- The state variables S_k, specified by possible classes of states, conditions, properties, qualities, and quantities
- The change variables C_l, involving possible classes of actions, processes, events, and occurrences, and activities
- The relation variables R_m, exemplified by possible types of connections, links, associations, and relationships

Here I is a set of indexes, and i, j, k, l, m are the sets of the natural numbers.

Then, in terms of the kind (class) variables, the world structure generally can be expressed by the sign structure as the world formal code:

$$\{\mathbf{W}, W_I\} = <\{\mathbf{O}, O_j\}, \{\mathbf{S}, S_k\}, \{\mathbf{C}, C_l\} \{\mathbf{R}, R_m\}> \quad (2)$$

The universal class of Being or Entity of which everything is part and member is then fully liable to mathematical analysis and formulations. Being unlimited in its scope and reach, the world codification embraces a scientific account of the physical world made in terms of object variables, property variables, and spatio-temporal variables. Besides, it covers purely formal conceptualizations of the mathematical world in terms of classes (objects), functions, and relations (predicates) variables, used for intelligent machines (Nilsson, 1991). Formally, reality is then to be described as a mathematical structure consisting of four types of variables having real reference classes: substances, states, actions, and interrelationships. Modifying Quine's logical metaphor (Quine, 1960, 1963) of what does it mean to exist, "to be, is to be the value of a quantified variable," it is rather: to be, is to be an entity variable exemplified by an instance (or value) set.

In the general sense, "being" can be described as the universal class of variables, where all other types of quantities, as object, property, and spatial-temporal variables, fall under the world entity variables. Then the problem of machine understanding of the meaning of symbolical expressions may be reduced to the matter of computing the value of appropriate quantity, as well as to locating the value (place) of the represented thing in the formal system of the variable realities.

It appears that to be rigorous, mathematic and scientific, ontology [studying the Being of Everything that exists, changes, and interrelates] matches up not only with the fundamental sciences, but also with pure mathematics, so constituting a real foundation and semantic framework for mathematical categories, axioms, and systems. To see how mathematics is correlated with *universal ontology as a comprehensive science of the world entity classes (as variables or variable quantities), their ordering or relationships*, consider the existent definitions of mathematics. In different authoritative reference sources, it is generally distinguished as:

- The science of quantitative relations and spatial forms of the real world
- The science of order, structure, and relation
- The study of numbers, forms, arrangements, and sets, and of their relationships and properties
- The study of quantity, form, arrangement, and magnitude, of the properties and relations of quantities and magnitudes
- The study of quantity and shape and arrangement

Considering that magnitude (form, figure, and shape) and multitude (or number) are two key modes and principal species of the generic kind of quantity, *mathematics is nothing else but a general formal science of quantities, of their ordering and interrelationships*. What is employed either in the abstract sense (pure mathematics) or in the concrete sense (applied mathematics). The branches of applied mathematics are probability theory, statistics, and linear mathematics. Among the branches of pure math are geometry, trigonometry and topology, calculus and analysis, algebra and arithmetic, group theory, lattice theory, and set theory, dealing with the properties and relations of sets.

And it seems quite logical that the abstract theories concerned with the initially ontological concepts of the universe, set, class, entity, object, structure and order, come to be a substantive core of abstract mathematics and mathematical logic, which is set theory with its generalization, lattice theory. It is remarkable that the relation of different parts of the same whole (class, set, aggregate) is the subject matter of lattice theory (Algebra, Lattice Theory), while the analysis of the relation of diverse species (or kinds) of the same genus (or class) is among the main pursuits of ontology. It also is noteworthy that the set theory as the general abstract theory of classes (or sets) takes full benefit from Aristotle's theory of categories, genera, or classes of entities. It is useful to mention that the conception of a secondary substance as "a class with a certain qualification" (Aristotle, Categories, 1990, p.7-8) accounts for Cantor's set-theoretic principle of abstraction. It closely resembles the meaning of a set as a single entity of individuals of any kind, a collection of distinct objects forming a whole entity by sharing common property. No wonder that the originator of set theory in his foundational work (Cantor, 1883) considered his creation as an extension of a classical ontology. Due to this close affinity, there is an opportunity to give a formal and strict account of the world (as the universe, cosmos, being, reality, and all existence): to describe its structure and order by using the notions of set theory and lattice theory, complete with algebra (higher algebras, Boolean algebra, relative algebras) and theory of categories.

As it is known, lattice theory with set theory involve such effective abstractions as *the universe, class, set, property, object, relation, partial orderings, order preserving or isotonicity, universal bounds, duality, partially ordered sets, and lattice.* Boolean algebra introduces the notions of *isomorphism or bijection, Cartesian (direct) product, Boolean polynomial*; relation algebras—*binary relation and composition*; the category theory—the notions of *objects, morphisms (or maps or transformations), composition of morphisms and the duality principle* (Algebra, Categories).

Ascribing a real significance to lattice theory, it is necessary to emphasize that its main subject is an abstract structure L of families of subsets X, Y, Z, . . .of a given set, class, or aggregate (the universe) U. Ordered by an inclusion relation $\leq$, it is marked by having a largest element, denoted by the intersection of subsets $X \wedge Y$, and a smallest element, denoted by the set-union operation $X \vee Y$: $L = < U, \wedge, \vee, \leq >$. The inclusion relation of holding $\leq$ is characterized by the laws of reflexivity $X \leq X$, transitivity, if $X \leq Y$ and $Y \leq Z$, then $X \leq Z$, and anti-symmetricity, if $X \leq Y$ and $Y \leq X$, then $X = Y$. All lattices share the properties of idempotency, associativity, commutativity, and absorption. What is more important, these algebraic structures are subject to dualization: if the union $\vee$ is replaced by the intersection $\wedge$ and vice versa and the inclusion relation $\leq$ is interchanged with the opposite relation $\geq$, the lattice will keep its original structure, although turning upside-down.

In a general form, the lattice structure is an ordered quadruple of the universal aggregate U, a binary operation of relation °, the least or bottom element 0, the null thing, and the last, universal, top element, that is, the world W itself:

$$L = < U, ^\circ, 0, W > \qquad (3)$$

The models (realizations and specifications) of the abstract structure (as a sign or symbol system with certain semantic rules and assumptions attributing formal/real meanings to its elements) may be as diverse as the types of relationship:

- The partial orderings
- The implication relation
- The subsumption relation
- The part-whole relation
- The cause and effect relation

Basing on a selected type of relations, one can realize the models as diverse as:

- Partially ordered sets
- Deductive logical systems
- Classificatory taxonomies
- Mereology
- Real systems, natural, mental or social

By bringing meaning, semantics, and content into the formal structure through specific interpretations, the lattice structure may serve for whatever model of reality one would like to prefer: the totality of things, facts, objects and events, the totality of properties, and so forth. For instance, the universal class U may be represented as a collection of set variables, where its variables X, Y, . . . , range both over sets and individuals. The structure of reality is to be modeled as a lattice produced by substantial individuals ordered by the part-whole relation. This structure will culminate in a world individual, bottomed with a least ontic individual (Bunge, 1977), as some fundamental constituent of all reality.

As a result, a *least unit* or null element in such the lattice of individual entities is supposed to form *the bottom level* or the ultimate reality of the ontological structure. The smallest and most granular element in such a model of the world ontology of facts is supposed to be:

The individual, the indivisible, the simple, the particular, the atom, the point, the element, the fact, the item, the single, or the particle.

By the agency of the binary operations as juxtaposition or superposition, the multitude of individuals forms *the second grade or scale of reality*. It includes all kinds of wholes or complex unities:

The composition, the aggregation, the aggregate, the assemblage, the group, the association, the set, the number, the collection, the mass, the body, the substance, the matter, the quantity, the class, the compound, the mixture, or the system.

The entities are here ordered by the part-whole relation corresponding in set theory to the class membership relation $\in$ of an individual and its collection.

The third grade of the physical world is made up of different bodies, masses, complexities, compositions, aggregates, aggregations, collections, systems, and wholes composed of individual parts. This level is composed of a number of large collections of collections ordered by the class inclusion relationship $\subseteq$. At the apex, the universal element is the totality of all individuals presented as the class of all classes of individuals, Thing. It is the world presented as the *many* rather than the *one*, as the total sum of particulars, as a multitude of objects, facts, events, and processes constrained by particular causation mechanisms.

In a more profound version of reality, the structure of being is generated by species and kinds rather than by individuals and particulars. That is, *the least bottom element* of reality is *the lowest species* instead of the individual entity, the increase of which number does not increase the number of species. Consequently, the variables in the ontology of universals range only over classes or sets.

The second scale of reality is occupied not by classes, sets, or aggregates of individuals, but by *essential wholes, natural unities, undivided wholes, or natural kinds of species*. The unity of which is determined by fundamental causal laws or generic causation mechanisms. The higher order of reality is composed of highest genera: the sub-worlds made up of the union of the topmost kinds: substance, state, change, and relation. Like the fusion of space and time generates a four-dimensional continuum, the interbreeding of the genera of substance, state, change and relation constitutes the world's framework. *A universal entity here is the whole being, the totality of all kinds, the union of all sub-worlds, the unity of genera or the whole body of real classes.*

In terms of ontological mathematics, reality can be represented as the largest distributed lattice with a finite number of kinds-elements X, Y . . . Z having the universal bounds: the bottom element $0 = X \wedge \ldots \wedge Z$, Nothing, and the top element $W=U = X \vee \ldots \vee Z$, Thing, Entity, or Being. Or, specifically, it can be modeled as the power set of all subsets of the universal class U, known as the power set lattice (or Boolean algebra) [P(W), $\subset$], ordered by the set inclusion relation. The descent from the universal top bound, which is Being, to the universal bottom bound, which is the null ultimate being or Nonbeing, will take a finite number of steps (see Figure 4). If the general lattice of entity types representing the world has n irreducible kinds, it will possess n steps, n+1 grades or levels, and its power set will be isomorphic to $2^n = P(W)$.

A use case of the lattice of realities is afforded by the simple case of a lattice having two join-irreducible kinds, natural or physical, *res extensa*, and constructs and representations, or thoughts and ideas, *res cogitans*. Then the universal top element, constructed by the set-union of natural kinds and mental kinds $X_N \vee Y_M$, results in an intelligent kind, natural, or artificial; whereas the bottom bound, constructed by the set-intersection $X_N \wedge Y_M$, makes the lowest species, an ultimate species of beings, for example, a monadic entity.

In the order of particulars, it can be reduced to the distributive or modular sub-lattices of concrete objects (individuals, properties, states, events, and processes).

In the order of natural language, we shall have sub-lattices of linguistic categories; terms, proper and class names, collective and mass nouns; expressions, phrases and sentences, paragraphs, and text.

In the order of knowledge, its structure will be represented by distributed or modular sub-lattices of subordinate representations: schemas, frames, or models; sketches, graphs, or diagrams; specific theories, theoretical models, general theories, and frameworks. In the order of abstract objects, the lattice of concepts will look as a hierarchy of concepts organized by the inclusion relation, with the unit concept at the bottom and the class concept at the top.

Still, all these grades and levels of reality are not separate structures. As basic kinds of reality, they are constitutive parts or sub-ordinate modules of the grand and all-embracing lattice of realities reflected by a widely distributed lattice of constructs: categories (terms, sets,

Figure 4. The lattice of reality

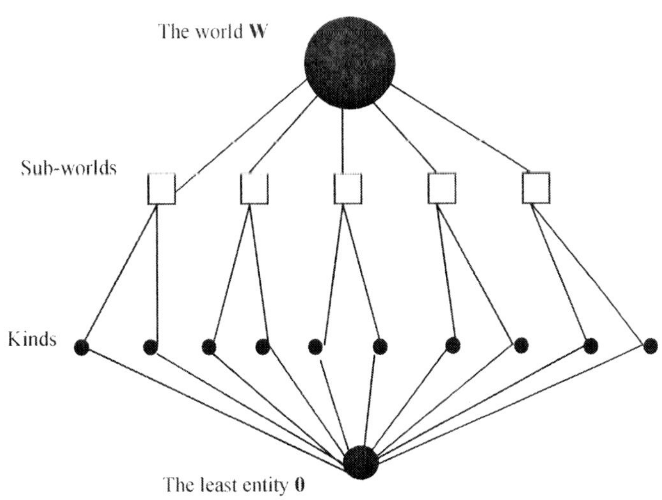

relations); propositions, contexts, theories. When completed with all the facts and specifics of particular classes of being, the universal lattice of constructs eventually will be converted into a single, consistent and complete conceptualization of the world.

It appears that foundational mathematics and universal ontology are similar in many principal dimensions. Either science is an axiomatic system and a priori study, both concerned with the notions of class and entity, structure, order, and relation. Nonetheless, there is one material distinction to be taken into account to achieve our purpose, creating universal formal ontology.

Engaged in the study of the whole being, ontology has a natural priority before all sorts of sciences, formal, and material, as the genera are prior to the species, as the cause prior to the effect, the being of the fact to the truth of the factual proposition, and as truth preceding goodness and beauty. So, the thing that differentiates the foundational ontology from the computational schemes of entities is its reality standpoint. Traditionally, computational ontologies, designed as the models or concrete realizations of lattice theory, set theory, category theory, symbolic logic, or other conceptual systems, are constructed by proceeding *from abstraction to actual domains of things through model*. As a matter of fact, we have to proceed *from reality to abstraction through model.* And the very archetype or prototype to which all the classes of models should be conformed is the world itself with its real structure, real order and universal bounds, the universal entity, Being, and the null entity, Nothing, as it is indicated in Table 1.

As the table shows, the principal difference of the ontological structure and the logico-mathematical systems is that the former is without the factual referent classes. They are all

Table 1. The World structure and mathematical representations

Classes of Entity (Being, Thing)	**Order Interpretation**	**Class Interpretation**	**Propositional Interpretation**	**Numerical Interpretation**
The totality of entities W	Abstract set U	A class of sets X, Y. . .	The set of all propositions P	The set of all real numbers R
Relation $\Rightarrow$	Partial order $\leq$	Set inclusion $\subseteq$	Implication $\rightarrow$	Inequality $\leq$
Interrelation $\Leftrightarrow$	Full order $\leq \geq$	Isomorphism $\leftrightarrow$	Equivalency $\equiv$	Identity $=$
The bottom entity 0	Greatest lower bound $\wedge$	The null class Λ	Conjunction &	The null element 0
The top entity W	Least upper bound $\vee$	The universal class V	Disjunction $\vee$	The universal element I

about abstract forms—sets, functions, relations, propositions, numbers; after conceptual structures or constructs having neither real content nor sense, nor meaning or reference. We now know that the generality of the world ontology is determined by its domain, the world at large, where all the entities are arranged into one of four broadest categories of Thing: substances, states, changes, and relations, interposed between two universal bounds: the ultimate class (Nothing) and the universal class of all classes of things (Thing): <W(orld), R(elation), N(othing), T(hing)>

In set theory (as a formal logic of classes), the universal abstract structure is formally realized as the union of all classes in the hierarchy of classes (the universe) arranged from the top level (the universal class of sets, V) to the bottom level (the null set, Λ) with respect to the class inclusion relation $\subseteq$: $< U(niverse), \subseteq, \Lambda, V >$.

But, unlike a general ontology of classes, formal systems neither explain existents, nor represent and describe any realities, nor predict changes or processes. Therefore direct application, without previous substantiation of mathematical and logical systems for building knowledge representation and reasoning systems, can result in inadequate conceptualizations of the world using ontology languages poor in substance, or real-valued contents.

As a matter of fact, the world, being, reality, or the universe as the super-class of all things should come at the top of the class hierarchy of entities. Each natural kind or species is its descendant. The major variables in the description of the world are:

- Substantial entities exemplified by objects, masses, bodies, and systems
- State entities, by properties, qualities, quantities, and conditions
- Change entities, by events, occurrences, and happenings
- Relational entities, by causation, space-time relations, part-whole relations, identity, diversity, functional relations, rules, and laws.

So, entity taken in the general meaning of the World, Reality, Existence, or Being comes up as the most universal of all real categories. When defining the world, its meaning range can run as far as from the universe and all that exists to the physical universe to the earth, its inhabitants, its countries and people. Or, from natural grand classification of earthly things to a community of people with common activities and interests to human affairs and active life, or to what affects all nations. Sometimes it relates to the state or time of human existence, to the public, public life and actions, sometimes to a total of the things, conditions, circumstances, and events affecting one person. The sense of the world as the universe, reality, existence, or the totality of all things, things of every types and kinds, turns out to be the richest, the highest as well as the most technical concept for machine ontology. The richest, because it comprises:

- **The subworlds of nature**, physical entities, and biological systems
- **The sphere of minds**, the realm of all intelligible objects
- **The world of culture**, the works of science, technology and arts

And the highest, because the category of the world as the totality of entities tops all other fundamental categories of reality: Thing, Substance, Property, State, Change, Process, and Relationship.

The category of world, its structure and order, is best analyzed from a standpoint of ontological (real or causal) mathematics. The world can be formally identified with the universal set or the power set lattice having the universal ontological bounds, the null entity **0** and the world itself **W** arranged by a partial ordering relationship:

$$\mathbf{W} = < \mathbf{P(W)}, \cap, \cup, \subseteq, ', \mathbf{0}, \mathbf{W} > \quad (4)$$

Unlike the binary operations of intersection $\cup$, union $\cap$, and the relation of a partial ordering $\subseteq$, the symbol ' denotes a unary operation, either inversion # or complementation $\bar{}$. The power set world **P(W)** is isomorphic with 2^{W}, and is equivalent to the power set of all subsets of the universal set **W** divided into three grandest sub-worlds:

- The natural universe W_N, embracing the realms of the physical W_P, the chemical W_{Ch} and the biological W_B
- The mental universe W_M, identical to the personal realities of cognition, feeling, and willing
- The cultural or knowledge world W_C, identical to the social reality of objectified knowledge

As a consequence, there are 8 parts-sub-worlds:

$$2^{\mathrm{W}} = \{W_N\}, \{W_M\}, \{W_C\}, \{W_N, W_M\}, \{W_N, W_C\}, \{W_M, W_C\}, \mathbf{0}, \{W_N, W_M, W_C\} \quad (5)$$

The algebraic manipulations of the sub-worlds are regulated by the laws of idempotency, commutativity, associativity, absorption, and distributivity complete with the laws of universal bounds:

$$W_N \cap \mathbf{0} = \mathbf{0}\ ;\ W_N \cup \mathbf{0} = W_N \quad (6)$$

$$W_N \cap \mathbf{W} = W_N\ ;\ W_N \cup \mathbf{W} = \mathbf{W}$$

and the inverse complementation laws:

$$W_N \cap W'_N = \mathbf{0}\ ;\ W_N \cup W'_N = \mathbf{W} \quad (7)$$

What is much important is that, any concept except reality itself has a restricted universality, since, however general, it divides reality into things of a certain class; for instance, all things of a natural kind and all other non-natural, artificial or incorporeal objects. Whereas the category of the world with its interchangeable terms of reality, existence, and being, apply to all things of any type or kind, incorporating also Nothing or Nonbeing, i.e., the null ontic element **0**.

Noteworthy, a fundamental attribute of any complemented distributive lattice is its self-duality under the interchange of the binary operations $\cap$ by $\cup$ and the replacement of the relation of partial ordering, $\subseteq$, by its reverse counterpart, $\supseteq$. Such attribute is inherited by the mathematical models with underlying lattice structures, like set theory, projective geometry, category theory, and symbolic logic, all allowing dual theorems, statements or formulations. But, unlike mathematical abstract systems, reality is a dynamic structure, an infinite system of objects, changes, and relations, the order and structure of which also is regulated by ordering principles, like the cause and effect relationships. So, in the order of reality, there are many kinds of regulatory mechanisms involved.

Yet the most general and underlying law regulating all real processes is likely to be the real counterpart of the duality principle laying the ontological foundation of the mathematical duality law. This general mechanism is nothing but *the principle of self-dual or reversible causation*, defined as the **reversibility principle** (Novik & Abdoullaev, 1991), with the following formulations:

- Any real process of change is convertible and inverse
- The order and sequence of causal relatives can be reversed
- Whatever comes into being as a process happens just as well as the inverse process.

The essential point of these equivalent definitions is that:

In any real and fundamental processes composed of distinct changes the sequence of their being is to be eventually reversed, converted, inversed, or turned other way round.

The immediate remarkable result of this fundamental fact of being is that *the reversibility principle is the principle of ordering of reality*. That can be illustrated by many examples from fundamental physical interactions, to physical effects, to chemical reactions and natural cycling of matter and energy, such as geomorphic and geochemical cycles or physical, chemical, and biological cyclic processes.

Then the hierarchy of the world is not simply the arrangement of the kinds of things in hierarchies of superior wholes and subordinate elements and parts, where a real whole rests as orderly arrangement and combination made up of or caused by the constituent parts. *The whole structure of a system acts upon its components and parts, as well as the parts shape the whole structure.* This universal phenomenon sometimes is described as the action of upward causation and downward causation. In fact, it is a manifestation of the underlying principle of causal order between different levels of being.

Thus the world of physical entities, the world of mental processes, and the world of culture are not merely a hierarchy of categories or the grades of being, rather, *the universe is one all-embracing whole of causally related worlds, or the universal class of causally ordered dynamic sub-worlds.*

The Categories of the World

A formal account of causal interrelationships of worlds can be best given in terms and notions of a mathematical category theory, such as objects, domains, co-domains, the Cartesian product, morphisms, compositions, functors (mappings between categories), and so forth.

Formally embodying the ontological theory of categories, the universe here can be presented as the universal class of the ordered world sets, or as the category of pre-ordered world sets **K (W, F, Dom, Cod)** consisting of a class of objects **W** and a class of transformations **F** from a domain of objects, **Dom**, to a range of objects, **Cod**. There the world of objects **W** embraces the following pre-ordered sets:

- The universe of physical matter and energy W_P, including the sub-world of chemical elements, compounds and reactions W_{Ch} ; the sub-world of biological objects and processes W_B; the sub-world of all natural objects and processes W_N
- The world of mental objects and processes, or the subjective, personal, individual reality W_M
- The social world of culture, or the objective knowledge of social reality W_C, including the sub-world of artifacts; works of science, technology and the arts W_A

The class of transformations **F** includes functions f, g, h, p, called also maps or morphisms, is accountable for designating of all the causal relationships between the world objects. So notation f_C: $W_P \rightarrow W_C$ indicates that f_C is a function from the domain of physical objects to the codomain of chemical entities. Then the world as the universal class of all sets of things is characterized by the identity function:

$$\mathbf{F}: \mathbf{W}\ \{W_P, W_M, W_C\} \rightarrow \mathbf{W}\{W_P, W_M, W_C\} \tag{8}$$

That is, each occurrence of a world state W is associated with its identity function: I: W → W. Besides, each ordered triple of world objects is associated with a composition law:

$$F(W_P, W_C) \times F(W_C, W_B) \rightarrow F(W_P, W_B) \tag{9}$$

The notion of the Cartesian product × is here no less important for the ontology of reality than for algebraic systems. The generalized Cartesian product for the whole families of the world states $\Pi_{i \in I} W_i$ can be obtained through using the inductive rule based on commutativity and associativity laws:

$$\text{if } W \times W = W^2,\ (W \times W)\ W = W^3, \ldots, \text{ then } W^m \times W^n = W^{m+n} = W^I \tag{10}$$

Among the axiomatic properties of the complete world product, there is the existence of the universal projection morphism (function, transformation, operator, or mapping):

$$\pi: W^{I} \rightarrow W \quad (11)$$

So, the body of axioms, defining the ontological structures of reality and its basic parts, consists of the following laws necessary for a mathematical account of the world:

- The axiom of existence of a world identity
- The law of composition
- The law of existence of an inverse order of processes
- The axiom of the universal projection or order-preserving function

For instance, the order-preserving mapping function describes the basic effects of the world's mapping upon its sub-worlds. There is an immediate high value consequence of that axiom. The Cartesian product W_C x W_B of chemical and biological events together with its projection functions π_C: W_C x $W_B \rightarrow W_C$ and π_B: W_C x $W_B \rightarrow W_B$ has a remarkable property. Namely, if there are functions f_C: $W_P \rightarrow W_C$ and f_B : $W_P \rightarrow W_B$, then there is a unique function f: $W_B \rightarrow W_C$ x W_B such that the following equations hold: $\pi_C f = f_C$, $\pi_B f = f_B$. And what is more important, due to the reversibility principle, the morphisms are to be invertible, that is, If f: $W_B \rightarrow W_P$ Then there is an inverse morphism g: $W_P \rightarrow W_B$. Accordingly, there is the opposite or dual world category with the same subworld objects but with the domains and co-domains of the morphisms being interchanged **K' (W, F, Cod, Dom)**. This implies that the world can be described as a causally integrated whole of the physical, the chemical, the biological, the mental, the artificial, and the cultural, as illustrated in Figure 5.

Like computer design and architecture is largely dependent on the mathematical theory of lattices, the construction of the UFO, as the interface of all fundamental theories about the universe, also involves the lattice theory but with some essential reservations. Abstract mathematical theories as lattice theory, set theory, and category theory belong to the group of abstract languages with minimal semantics, where only formal structures provide formal axioms for formal constructs, like objects, sets, functions, predicates, and relations. By contrast, the ontological theory is a fundamental system of primary categories, truths, axioms and principles of existence providing for upper-level and domain ontologies their key definitions, assumptions, and reasoning strategies. Combined with mathematical formalization, the UFO, or the EIS SUO, constitutes a general formal structure representing the world, its key parts, features, and properties. Most importantly, the universal ontology yields both mathematical and scientific analyses admitting of principal resolutions of:

- What the world is, means, or signifies
- How the world knowledge can be formally represented by computing machines (semantic systems)
- How machines interact with each other and with the world as with the largest environment.

Figure 5. The causal order of the world categories

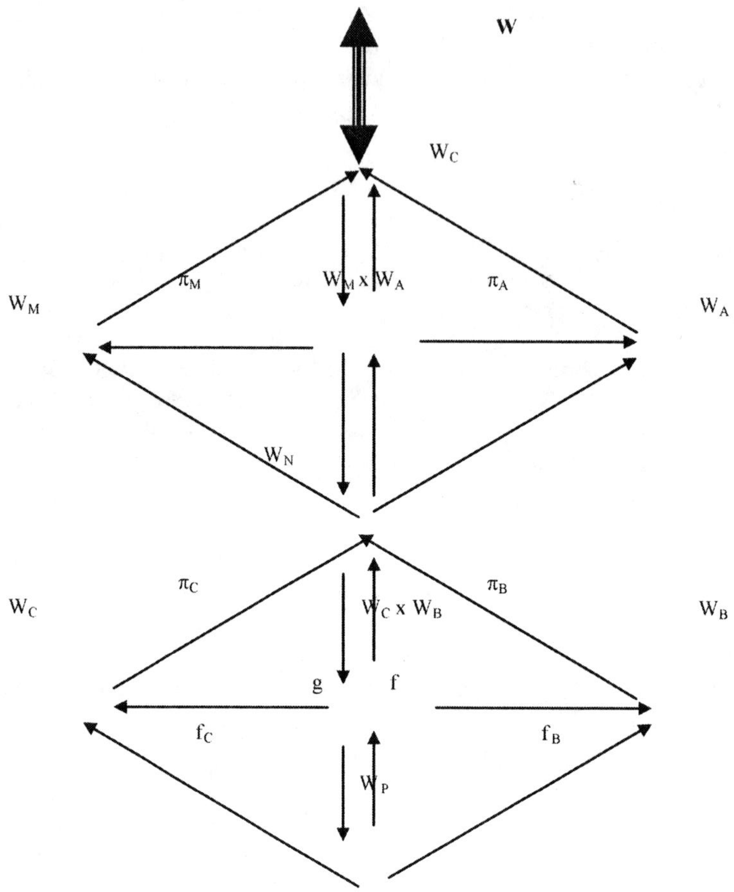

As a result, then endued with the mathematical formalism, the ontological structure is fit to adequately describe not only the world's structure, but also how machines represent the world and interact with it, or how the world's states are connected with machine states.

In all, in order to build a standard ontology for machines and people, to formulate an exhaustive conceptual framework as a hierarchical inheritance structure of entity types with the properties and relations, we need to construct a fundamental ontology formally describing all things in the world as comprised between Nothing and Being. In such a global scheme, the world entities are distributed among the primary categories (classes) of Entity or Being: substances, states, changes, or relations. Such a universal system of things determines the outlines of the world's structure by laying out its constituents and their representations as follows:

- The objects and processes of the material world under few natural genera and classes
- The word classes and expressions of natural languages under few major terms and expressions
- The contents of the mind under few concepts and categories
- The computing data or information under few metadata types

As a result, the all-entity framework unifying upper ontological schemes is to emerge not only as a universal system of the world, but also as a unifying semantic model of knowledge and data, capable of supporting the most advanced knowledge and reasoning applications.

Conclusion

Employing the system of symbols and techniques of mathematical foundational theories, a vocabulary of higher algebra, lattice theory, set theory, and category theory, we laid out a general account of the world in formal terms of entity variables. In due course, the principles of Mathematical Ontology and Real Mathematics are formulated, whereby the dynamic formal theory of complex structures and causal systems can be obtained. To truly represent reality, first of all, we have to construct the functional schemas of realities, the categories of the world, the lattice of the classes of entities, the tree of universal properties, the equation of things, and only then we can pass to such formal constructs as the lattice of theories, the categories of concepts, or the tree of axioms. In all, a new sort of universal mathematical science about the world [its levels, entities and relationships and domains (natural, mental, cultural, informational)] is suggested involving ontological principles, definitions, and axioms. Designated as the UFO, or the EIS SUO, it is emerging as the product of two complementary theoretical models about all reality: mathematical ontology and ontological mathematics.

According to the formal ontological model of reality, *the world*, the largest universe of discourse, is formally represented by the world variable W (the set of all possible world states), and its dynamics or change is represented via the transformation function F, as in a circular self-mapping equation $W = F(W)$. As its basic arguments, the world function includes the set of prime entities:

1. **Substance space** (the set of all possible substances) indicated by a substantial class variable (O)
2. **State space** (the set of all possible states) represented by a state variable (S)
3. **Change space** (the set of all possible changes) symbolized by a change variable (C)
4. **Relation space** (the set of all possible constraints and laws and rules) symbolized by a relation variable (R)

References

Algebra: Lattice theory, categories, Macropaedia. (1994). Knowledge in Depth (Vol.13). *The new encyclopaedia Britannica.* Chicago: Encyclopaedia Britannica, Inc.

Aristotle, Logic (Organon): Categories. (1990). In M.J. Adler (Ed.), *Great books of the western world* (Vol. 1). Chicago: Encyclopaedia Britannica, Inc.

Cantor, G. (1883). Foundations of a general theory of aggregates. In W. Ewald (Ed.) (1996),. *From Kant to Hilbert: A Source Book in the Foundations of Mathematics.* 2 Vols. Oxford: Oxford Uni. Press.

Bunge, M. (1977). *Ontology I: The furniture of the world.* Boston: D. Reidel Publishing Co.

Gruber & Olsen (1994). An ontology for engineering mathematics. In J. Doyle, P. Torasso, & E. Sandewall (Eds.), *Proceedings Fourth International Conference on Principles of Knowledge Representation and Reasoning*, 258-269, San Mateo, CA: Morgan Kaufmann.

Nilsson, N. (1991). Logic and artificial intelligence. *Artificial Intelligence*, 47, 31-56.

Novik, I., & Abdoullaev, A. (1991). *Introduction into information world.* Moscow: the USSR Academy of Sciences, Nauka. (in Russian).

Penrose, R. (2005). *The road to reality: A complete guide to the laws of the universe.* New York: Alfred A. Knopf Group Inc.

Quine, W. (1960). *Word and object.* Cambridge, MA: MIT Press.

Quine, W. (1963). *Set theory and its logic.* Cambridge, MA: Harvard University Press.

Chapter IV

What Makes Reality: Ontological Classes and Rules

Introduction

The ultimate purpose of standard ontology is to describe and represent all the things in the world in comprehensive and consistent ways, whereby making the fundamental knowledge explicit to the formal reason of semantic systems and cognitive agents, natural, or artificial.

To build such a formal universal framework capable of including the representation of anything, one can design a general system that includes a set-theoretic (logical) ontology constructed as a formal logical system composed of its objects-primitives (classes, individuals, and properties), logical syntax (notation techniques, formation and transformation rules), and formal semantics (model theory), as currently the Semantic Web formal languages and upper ontologies usually are constructing.

On the other hand, one can design a general system presupposing the content ontology of everything and real semiotics constructed as a unified account of things, which real meanings grounded in the world itself, kept in the mind and encoded in the cultural artifacts as the natural languages and knowledge systems. Unlike the formal ontology, the content ontology is marked by caring the real entities, underlying constraints, principles, truths, and strategic rules. Accordingly, its goal is to formulate the overall patterns and fundamental laws of the universe, while the practical role is to set the world models, rules, and reasoning algorithms, or advanced information technology or ontological semantic technology. Such technology implies the real semantic system which equals to the structure:

The Real Semantic Web = Sign (symbol) system (the SW languages, UFO, Natural Language) + axioms (ontological, mathematical, formal logical) + rules (ontological, semiotic, logical) + intelligent actions

The core of the real world semantic technology is formed by the underlying axioms about reality:

- The world or reality or universe exists
- There are entities or things or beings in the world
- Entities have properties; entities have parts
- Entities stand in relationships
- Causality shapes the world and its behavior
- There are space-time relations
- Things change over time
- Entities exist as substances, states, processes, and relationships
- Substances have masses and occupy spaces; etc.

This suggests that the crucial issues of reality modeling and representation, what classes of things compose the world and which rudimentary truths, rules, and principles constitutes its essence, are among the key topics of the master content ontology, formulated as a single categorical code for all reality, superseding the multiform logical ontologies reflecting numerous personal world views, particular mental attitudes and beliefs.

The Pillars of Reality Modeling

There are two great sciences with an unlimited scope and range of investigation, without which no knowledge domain can exist and achieve. One of them is a real science of Ontology, another one is a formal science of Logic. The object of knowledge of ontology is entire reality as the being of everything which exists, with its fundamental truths or axioms of existence, while substantial knowledge as physics and mathematics are concerned with some aspects of the world, as substance (matter), change (energy), or quantity. Unlike ontology, formal logic is concerned with the entirety of discourse about anything or everything, mostly valuable in a codified representation of human knowledge, but useless for generating new knowledge and truths about the world.

So the scopes and subject matters of Ontology and Logic, under the headings of which the computing ontology and Semantic Web activities currently are performed, should not be mixed. The real semantics or meanings of any symbolism or notations or signs are defined by ontology; for this is the only knowledge domain studying the Being of Everything which is, happens and connects. Ontology is a real science and Logic is a formal science because of the latter is concerned only with the formal parts of Discourse about Anything or Everything. Logic considers the elements (the terms, propositions, inferences or syllogisms) of the whole discourse aside from their reference to the world (or their real meanings and significance). And as a formal science, it deals with the formal patterns of discourse common to all sciences.

Unlike formal logic, Ontology deals with the matter and content, with the real components of discourse about anything, while Logic is an art, an instrument of intellectual reasoning, at least a formal science, and it has virtually little to do with reality, real significance or real meanings, a coherent representation of which is the ultimate target of Intelligent Web.

In information science and engineering and computing, formal logics of ontological discourse (a.k.a. formal ontologies) as a basic form of knowledge representation carry as the main constituents of the logical reality the following elements:

- Individuals (the base objects, instances, concrete things, abstract individuals, tokens, cases, examples, resources, values)
- Classes (concepts, conceptions, constructs, or abstract groups, sets, collections, or types of individuals)
- Attributes (features, characteristics, properties, predicates, or parameters marking the individuals)
- Relationships (the ways the entities associated or connected to one another by the taxonomic hierarchy, the subclass relationship, or the part-whole relations).

On the other hand, one can develop a formal universal ontology numbering as its primitive, Entity, with its major content classes and natural kinds, as substances, states, and so forth, along with fundamental definitions, truths, axioms, rules, and real-world meanings, as the EIS UFO scheme is building.

As it is demonstrated in the preceding qualitative and quantitative accounts, the UFO model is a reality representation and reasoning frame presuming some basic assumptions. All the natural classes form a single universal hierarchy of things modeling reality, the universe, or the world as the whole entity. And the pillars of the real world, which parts and domains and levels are sought to be represented by scientific theories and data models and programming systems, are the following prime entity classes (quantified as entity variables):

- **Substance, stuff, or substrate** (objects, material or nonmaterial, spatial or nonspatial) **O**
- **State** (properties, quantities, qualities, and attributes) **S**
- **Change** (actions, activities, events, or behaviors) **C**
- **Relation** (links, associations, connections, and bonds) **R**

In such an extensive world representation schema, all the kinds, types and instances are comprised between the null class of Nothing or Nonentity, □, and the universal class of Thing or Entity, **T**. At the level of the physical world, the basic classes (or universals and types) are instantiated by concrete instances (particulars, tokens, individuals, objects, events) in the realm of causality, time, and space. At the knowledge representation level, all they are symbolically expressed either by linguistic items or by mathematical values of variable quantities or by objects (instances) and programming classes (types) and procedures (functions).

To become a real representational language with the inbuilt ontological classes and rules, syntax, semantics, pragmatics, reasoning algorithms, and efficient computation, a general formal language must be focused on the world with its domains and levels. First of all, the language is centered on the basic construct, Thing or Entity or Being, viewed as the root source to all things existing in reality, and generalized as the class of all entity classes. The most fundamental categories and natural divisions or ontological classes of Thing are the class of Substance (Object), the class of State (Quantity and Quality), the class of Process (Change or Action) and the class of Relationship. While all the classes form the universal hierarchy, each of it is organized as a hierarchy of subordinate classes (kinds and types and forms and sorts and varieties) and individual things (instances, particulars, and concrete entities), such as individual objects, specific states, unique events and particular connections. That unambiguously implies that the formal notions of "definition," "class," "property'," "proposition," and "reasoning" should be filled up with real contents and semantics and closely tied up with the ontological predicates of entity, substance, object, quality, quantity, relationship, change, cause, effect, time, and space.

In the context of computing applications, the UFO (or the EIS SUO) can be interpreted as a common computational ontology giving an exhaustive (machine understandable) data-type system representing the semantic information about the world and its contents. Evidently, the design and implementation of such computable data structures are the necessary condition of any advanced knowledge technology. The inquiry in this much-promising field currently went in two separate ways: the Semantic Web activity, directed toward creating the W3C Web Ontology Language, and the IEEE project of standard upper ontology as high-level concepts, definitions, and relationships processed by computer (IEEE, 2003).

Recently, OWL has been declared as a "W3C Recommendation" by the World Wide Web Consortium (OWL, W3C Recommendation, 2004). The language is recommended as the representation standard for all Web information due to its capacity to formalize and define the Web content in terms of definitions, classes, individuals, and properties, enhanced with the coming reasoning power supported by rule languages. Both projects undoubtedly come under advanced undertakings deserving full attention not only by computing scientists and engineers. Nevertheless, the courses of action embarked on make the research projects a sort of blind venture, although leaving no question about the technological enormity of their final design.

To promote the great cause of a unified model of things having the crucial implications for reasoning applications, we will further show why and where the current plans of attack may be misguided. First of all, no general model or conceptual scheme of things can do without using the fundamental constructs of classes, individuals, predicates, attributes, relations, or definitions, the main constituents of any universe of discourse, however large and extensive. If a model represents the traits and features of the real world, its predicates represent real properties. In so far as the scheme describes formal objects, its predicates give description of formal properties, or attributes. Thus, there are two sorts of predicates: formal, logical attributes or real-valued, ontological predicates. Not to see such difference can result in conceptual mess and confusion. Proposing a general conceptual scheme by using the class (object)/property (relation) distinction, the general data languages like OWL risk missing the heart and soul of things. Such schematic and superficial data models can hardly adequately represent the whole gamut of the world entities with their various interrelationships.

For it will make a poor language to use the class/property distinction, where all properties are reduced to relational properties as a substitute of relationships. In fact, there are several sorts of key characteristics:

- Substantial (material, structural) properties
- Stative (qualitative, quantitative) properties
- Active (functional) properties
- Relational (causal, spatial, temporal) properties.

And all these properties of a class (genus) are to be the inherited properties of its species and individuals. Then it makes all the difference whether we just state that "a thing has attributes (properties)" or that "a thing has the properties to be a substance, to be in a state, to be acted or acting upon, and to be related to something else." The first one is a general logical proposition, while the second one is a universal ontological statement.

We then must clearly differentiate between classes, properties, definitions as logical forms and objects and entity types as ontological classes, natural kinds, and their relationships. The class and property and individual and definition of anything will always be in one of these types represented by the classes of ontological predicates and signified by the generic terms and expressions, as substance, state, quality, quantity, action, time, space, and relation. Any meaningful statements and propositions is always about one or another of these types of entity in one or another way. For example, all the propositions making the content of Web documents signify a thing's essence or its specific property, its quality or quantity or position or some other types of the predicates. It follows that, in order to have a comprehensive system of computable classes, to construct a high quality Web ontology language, we first need to formulate the universal entity framework supporting all the major classes and relations in the world hierarchy of entities. Designing such a general type system requires breaking down reality into its key components: substances or objects, states or properties, changes or processes as the classes of ontological predicates specifying diverse realms of thing, leaving for a separate analysis the real category of relationship, the adhesive of the universe.

Maintaining the perspective of the application of the reality modeling system as a universal ontology language, we will examine and formulate the distinguishing features of objects, states, and changes to formulate the common rules of reasoning about the basic entity classes, their kinds and instances. The validity of the approach is verified by the UFO's conceptual power to incorporate the content of lexical taxonomies together with upper and Semantic Web ontologies into one general world model.

The Class of Substance (Objects, Material, and Nonmaterial)

All the basic issues concerning substance (or stuff or substrate or object) still are ambiguous and confused; namely, its nature, whether it is, what it is, what its content and meaning, what the name really signifies, and how it exists.

In computer science, real world substances are modeled as objects having states (encoded in data elements and types or instance variables), exhibiting behaviors or affected by changes (actions or processes encoded in operations, procedures, functions, or methods). Objects are arranged into classes, which are hierarchically ordered by an inheritance relation; thus being divided into subclasses (contained classes, child classes, or derived classes) and superclasses (containing classes, base classes, or parent classes). Existing within a hierarchy, each subclass inherits all the states and changes of its superclass and so specifying the data elements and the operations of its instances or objects. In other words, all substance signifies both individual objects, the subject of states (properties), as well as the classes of objects predicated of individuals and ordered by an inheritance relation.

From a semantic point of view, objects are defined by means of the symbol-object correspondence (or denotation rules). Since substances are not represented but rather denoted, named by the symbols and signs and referred to by the constructs. If the denotation relation holds between terms or symbols and the objects of real kind, the reference relation holds between constructs and objects.

Therefore, objects identified with a set of referents or reference class defined as the domain of objects or universe of discourse. Identifying the nature of the reference class (universe of discourse, denotation field) of a theory (or symbolism), unfolding its ontological status, is the number one priority of ontological inquiry.

After Aristotle, all substance by nature is divided into primary substances (individual objects) and secondary substances (naturally occurring classes of substances). The *locus classicus* on substance are the passages: "all substance appears signify that which is individual"; while "a secondary substance is not an individual, but a class with a certain qualification; for it is not one and single as a primary substance is" . . . "species and genus determine the quality with reference to a substance: they signify substance qualitatively differentiated" (Aristotle, Categories, 1990, p.7-8).

Being a primitive category of reality, substance is often identified either with a substantial being (physical object, particular, concrete thing, substantial individual, and body) or with a class (aggregate) with a certain qualification acting as the underlying structure of a collection of individual objects; as the stuff, material, or matter (essence or substrate) underlying everything existing.

Such equivocation is reflected in the distributive and collective interpretations of substances, viewed either as substantial individuals, the elements of the aggregate, or as a unitary whole of its parts, which is a class (genus or species, as "man" or "animal").

The objects are, therefore, widely thought to be particular things, or specific individuals, the entities which underlie everything else, which are numerous and grammatically expressed by count nouns. Also, it is widely believed that an object is an aggregate of properties restricted to a particular space-time, and that any substance is a distinct sensible existence having mass and occupying space. Such an individual object is regarded neither predicable nor present in a subject. Generally, a particular object (substantial individual) is defined as a substance having boundaries (bounds, limits, extension); spatial boundaries, like spaces, surfaces, lines, and points, go for physical or material objects. So all material objects come up as spatial substances existing as distinct substantial entities liable to a physical sensation and instrumental measuring. The nonmaterial and nonspatial substances also should be considered as individuals, which can be exemplified by the so-called thinking substances or mental substances, such as human minds or artificial intelligences.

Following the above interpretation, natural language employs three types of substantives (or substantial nouns):

- Class names standing for a class of objects
- Mass names denoting an uncountable mass or body or mass substance
- Collective names, which meanings are shifting between the two extremities: a single entity of objects or a number (a set) of individual objects.

Under the class-names, there are subsumed all kinds of countable objects: *concrete objects, persons, animals, plants, units, and parts of a mass, body, or whole*. Under the mass names denoting all substances of uncertain quantity are included all sorts of uncountables: fluids, liquids, gases, solids, materials, languages, and so forth. The collectives as the nouns of multitude behave either as abstract singulars or concrete plurals, thus depending on the contextual meaning, standing either for a single entity or for the elements and objects composing such entity. Then, with some qualifications, concrete beings, individuals, physical objects, material existences, agents, matters, bodies, and systems come under the heading of substance as a class (*genus, aggregate, aggregation, collection, mass, total, totality, and whole* composed of matter and form) determined by a collection of lawfully related properties (states, changes, or relations).

In a realistic system of things, the secondary substances are upgraded as the primary substances. Or, a class of things, which is present in an individual subject as its nature and hence is predicable of it, is essentially prior to individual things. Being more truly entities than the individual substances, such generic substances exist in kinds, classes, groups, orders, or genera, expressed by mass nouns or some collectives, and there are many varieties or species in each class, kind, or genus. As a consequence, all substance makes up a hierarchy characterized by the rule which is equivalent to inheritance in Object Oriented Programming Languages:

Axiom 1. All the properties of a generic substance are the properties of its kinds, varieties, varietals, and individuals.

And what is predicated (predicable) about a kind is predicable of its species and individuals, or the classes of objects can be arranged and ordered by the relation of inclusion (containment, a part-whole relation).

That is, as kinds and classes, substances (or objects) with their differentiae are the classes of things not only predicable of the individual objects, but also present in them as something essential and material.

There also are some other essential consequences (theorems) of the axiom of inheritance worthwhile of mentioning:

1. The substance should fall under the same division as the kinds
2. If a kind of substance belongs to two substances, the one is included by the other
3. What contains the kind contains the genus (class) as well
4. If classes are different, their properties are different in kind

Another distinctive mark of substance, either as a qualified class or as an individual, it has no contrary, some opposed entity or exact opposition. There is no such a thing as anti-substance, and the physical example of antimatter is irrelevant here; for it is still a kind of matter, a physical substance composed of antiparticles. Also nor substance admits of any change or variation in degree; some object can not be more or less substance, somebody can not be more or less person or individual.

At the same time all real substances afford the possibility of having contrary qualities, like any real process admits its inverse counterpart. Keeping its numerical identity, one and the same substance allows of different qualities via a process of change, alteration or modification, undergoing in the substance itself. Thus, there emerges another ontological axiom:

Axiom 2. All substance is subject to a change such that the quantity of substance remains unchangeable, unvarying, permanent, constant, or invariant.

It is the paraphrasing of Kant's first analogy, principle of the permanence of substance: "*In all changes of phenomena, substance is permanent, and the quantum thereof in nature is neither increased nor diminishe*" (Kant, 1990, p. 74).

This axiom applies to physical and chemical objects as a conservation principle, when in the course of transformations (physical changes or chemical reactions) the fundamental quantities (features, properties, or functions or invariants) remain unchanged, as it is stated in the law of conservation of matter (mass) or the law of conservation of energy or the mass-energy equivalence.

The ontological fundamentals of substances also yield the principles of their classification and so simplify the task of arranging all the substantial objects, spatial and nonspatial, material and nonmaterial, into natural categories and classes, as realistic divisions in the world. That is, all substance could be arranged by the classes existing as real world features rather than imaginary concepts; since true categories are first in the world, than in the mind. Accordingly, realistic categorization, sorting, or assortment of substances sets the basics of knowledge representation and reasoning, by inferring or predicting all the basic features and behavior of individual objects incorporated in some single natural class, genus, kind, type, or species. In other words, a wrong ontological system of classification, as confused grouping, categorization, ordering, arrangement, or organization of substances, means creating reasoning applications based on the wrong assumptions about the world.

Currently the issue of substance classification and object categorization looks rather complicated. Mostly because we try to group the objects in the world according to our particular attitudes and cultures, instead of basing on their natural relationships, similarities (affinities) or differences (dissimilarities) in characteristics, composition, structure, function, or behavior.

As a result, there are many hardly reconcilable taxonomies and typologies of substances (objects) varying with each type of lexical and computational ontology, as below:

- **WordNet:** Entity, thing (group, collection), natural object, substance as matter, (plant, flora), (animal, fauna), (person, human being), (body, corpus), (artifact, food) (WordNet 2.0, 2005)
- **EuroWordNet:** 1st order entity (origin, form, composition, function) (EuroWordNet, 2004)
- **EDR:** Subject or human (person, animal, body or group, supernatural being), matter (thing or concrete object, objective subject) (EDR, 1995)
- **SUMO:** Physical object (self-connected object, collection, agent, region), abstract object (set or class, graph, proposition) (SUMO, 2004)
- **CYC:** Individual (spatial thing), mathematical or computational thing (set or collection), groups, parts of objects, composition of substances, agents, organizations, actors, materials, devices, construction, food, clothing) (OpenCYC, 2003)
- **DOLCE:** Aggregate (amount of matter, arbitrary collection), object (physical object, mental object, social object), feature (relevant part, place) (DOLCE, 2006)
- **OWL:** Objects and individuals (OWL, 2006)

Such state of affairs is caused by the widespread belief that a concrete object (as a thing) is a primary being uniting substance (bare individual, Plato's formless matter, Aristotle's primary substance) and form (having properties and standing in relations). And that the substantial individual together with its properties (states) constitutes the thing (concrete object, existent). As a result of such confusion, every set of concrete objects is partitioned into two disjoint sets, one of the physical objects, the other of the formal objects, constructs, so that "no object is neither and none is both" (Bunge, 1977).

As a point of fact, apart from the fact that substances (or objects) are no more primary entities than states, changes, and relations (no states no substances, no relations no substances, and conversely), there are three basic types of existence for substances, as:

- **What can be sensed or measured** (matter, physical stuff, and material) O_P
- **The contents of the mind**, what may be thought about, remembered, imagined, or fancied (mental or intelligible), inclusive all sorts of abstract objects O_M
- **What is produced and created through the agency of the minds**, the products of the human minds, objective knowledge, including artificial intelligences and all sorts of cultural artifacts O_C

Instead of splitting substances into a disjoint class of substantial individuals and a class of constructs, we have to classify them either as physical kinds and mental kinds and cultural kinds. All the kinds are constructed as the intersection of the greater classes, while the individuals are made of the kinds as their natures and essences.

In defining and classifying substance, it is necessary to distinguish between its two levels: a generic entity having a natural unity (as a single entity of objects of a sort), and a physical thing existing as an individual substance.

It is noteworthy that, unlike logical classes, aggregates, sets, or collections, a kind of substance is determined not only by a property or by a set of properties. Rather, it is defined by a set of lawfully related properties, such that all the kind's members share the same states (or the collection of generic properties).

It can be said that the totality of ordered classes of substances with their properties constitutes the world, or that reality is the aggregation of kinds of substances, states, changes, and their relations.

Therefore, all the substances can be split into three types: **a set of physical things, a set of constructs, and a set of cultural objects.** The union of all types of substances will constitute the universe of substances, or all substance is formally represented as the union class:

$$\mathbf{O} = \bigcup \{O_i: i \in I\} = \bigcup \{\{O_{Pi}, i \in I_X\}, \{O_{Mi}, i \in I_Y\}, \{O_{Ci}, i \in I_Z\}\} \quad (1)$$

Informally, the universe can be viewed as the largest class of objects, the aggregation of all the kinds of objects, organized in an ascending order of ever-increasing complexity from the substantial elements, concrete wholes, complexes, compositions, compounds, bodies, systems, aggregates, collections up to the totality of substances, sometimes identified with Being, Thing, or Entity. The natural order of concrete realizations and embodiments of substantial entity will schematically look like the following:

> Entity > Substance > Matter&Stuff > the Universe (the cosmos, galaxies, stars, the solar system, planets, other celestial bodies) > the Earth > Life >the Animal World > Humankind > Mind > Society > Culture (the works of art, religion, science, and technology) > Intelligent Machines >...> Null Substance

Mathematically, a universe of kinds of objects (or substances) **O** can be expressed by the power set lattice $< P(O), \leq >$. Or, as the complemented distributive lattice of the classes of objects with two universal bounds, the top element, the universal substance, the totality of substances, $\mathbf{W_O}$, and the bottom bound, the null ontic object **0**, all ordered by the part-whole relation $\leq$, as in:

$$\mathbf{W_O} = <P(O), \leq, \mathbf{0}, \mathbf{W_O}> \quad (2)$$

Out of primitive substances, more complex substances are to be formed, such as compounds, mixtures, aggregates, collections, masses, and systems while basing on the set theory axioms [the axiom of extension; the axiom of the empty set; the axiom of pairing; the axiom of separation (disjointness); the power set axiom; the union axiom; the axiom of choice (function or relation or order); and the axiom of infinity] (Set Theory, Knowledge in Depth).

Accordingly, the formal laws restricting the possible compositions of objects are:

- The idempotent law
- Commutative rule
- Associative law
- Contractive law
- Distributive law
- Complementation rule
- Universal bounds laws (Algebra, Lattice Theory)

The complex classes of objects which are possible to construct by using the set theory operations are:

- The union class
- The intersection class
- The Cartesian product class
- The complement class
- The disjoint union classes
- The complement union classes
- The universal substance, $\mathbf{W}_0$

Thus, a universe of substances, the universal class of all objects, can be represented as a complemented distributed Boolean lattice with a finite number of kinds-objects X_0, Y_0,..., Z_0 having the universal bounds: the bottom element $\mathbf{0} = X_0 \wedge Y_0 \ldots \wedge Z_0$ and the top element $\mathbf{W}_0 = X_0 \vee Y_0 \ldots \vee Z_0$.

The descent from the top to the bottom bound, occupied by the most elementary species of substances, will take a finite number of steps n, which is equal to the number of union-irreducible objects. For instance, if the given lattice of objects has three union irreducible elements, then it will have n+1=4 levels and the power set $P(W_0)$ isomorphic to $2^n = 2^3 = 8$ (Algebra, Lattice Theory).

The structure of reality is then generated by species and kinds rather than by individual objects. The null species belonging to every entity makes up the bottom level (the inferior limit) of the world. That is, *the least bottom element* of reality is *the lowest species*, which presently corresponds to the hypothetical constructions of quarks or strings.

The second scale of reality populated not by classes, sets, or aggregates, but by *essential and undivided wholes or natural kinds of objects*. The higher order of reality is composed of highest genera: the sub-worlds made up of the union of the topmost kinds of substance.

For instance, assuming the lattice formalism of objects to be an adequate representation of substances, the meaning of the conjunction and disjunction operations may be given by the simple case of a lattice of substances having two join-irreducible kinds—natural objects (physical objects) and constructs (representations or thoughts and ideas, the products of ideation processes, mental objects and content). Then the universal top element, constructed by the set-union of natural kinds and mental kinds $X \vee Y$, makes intelligent physical entities, natural or artificial, whereas the bottom bound, constructed by the kind-intersection $X \wedge Y$, makes the lowest species, the unique class the members of which are the objects common to the class of natural substances and the class of mental objects.

For the case of a particular lattice of individuals, the smallest element are defined as *the unit, the individual, the indivisible, the simple, the atom, the point, the single, the fact, the item, the particular, or the particle.* By the agency of the binary operations physically reified as juxtaposition or superposition, the multitude of individuals forms the second grade of reality. It includes all kinds of concrete wholes: *the composition, the aggregation, the aggregate, the assemblage, the group, the association, the set, the collection, the mass, the body, the substance, the matter, the class, the compound, the mixture, the composite, or the system.* The objects are here ordered by the part-whole relations formally corresponding to the class-instance membership relationship, $\in$. The third grade of reality is the total sum objects, a few large-scale collections and classes, which considered as the level of the sub-worlds of substances and described as the classes of species of individuals ordered by the class inclusion, $\subseteq$. At the top, the universal element is described as the totality of all individuals presented as the class of all classes of individuals. It is the world presented as the many rather than the one, as the total sum of particulars, as a multitude of objects tied by the part-whole relation.

To sum up the topic, the system of ontological axioms of substantial entities together with lattice theory of abstract entities may constitute the base for general machine reasoning about substantial beings (things, entities). Briefly, the primary principles of substantial ontology can be summarized as follows:

- The category of substance is one of the first divisions of entity
- The category of substance is a single entity of objects with different classes, kinds and species
- All substances form a semi-lattice hierarchy of kinds of objects with universal bounds
- All the properties of a substance are the properties of its species and individuals
- The world is the totality of generic substances standing in various relations

The Class of State (Properties, Qualities, and Quantities)

Since objects are characterized by properties (attributes, qualities, or quantities), they always have to be in some state or condition, which is nothing but one of the most critical categories of the world. We say of the states of the mind, states of knowledge, states of affairs, states of things, or states of the world. In its most general meaning, State is a state of being (existence) of an entity with respect to substantial properties, quality and quantity, or with reference to attributes for formal objects. That means that the state acting as the essential differentia of the substance amounts to the sum of the properties and modifications of substances that alter (qualities and quantities), and in virtue of which objects are said to change or do changes. Or, more generally, **state** is the way (manner, mode) an entity is, with reference to its substantial properties (qualities and quantities). The way something is, with respect to its formal attributes, refers to the states of formal objects. If property is an essential state shared by all members of a class of things, attribute is a formal property shared by all members of a class of formal objects; for attributes and predicates just represent properties, substantial or formal.

For the case of individual objects, state is defined as the union of its individual properties. Accordingly, the totality of all the individual's states constitutes its state space (or property space, quality space, or quantity space, or attribute space), within which all its behaviors and changes take place. Unlike individual states, the generic states are constructed as the aggregates of universal properties. Therefore such state types are to be classified with respect to the kinds of substances they relate. Then according as the types of substances, we can specify three types of states:

- **Physical states**, the collection of all universal material properties, like the states (forms) of matter (phase, liquidity, solidity, gas) S_P
- **Mental states** (cognitive states of the mind), the total of cognitive powers, capacities, abilities, along with conceptual states embracing all formal properties as attributes, functions, and predicates S_M
- **Cultural or social states**, the collection of all possible traits, characteristics, features, qualities of social objects and institutions S_C

Being the most crucial mode of reality, the category of states, spanning from the state of being to the state of not being (nonexistence), comes as the generalization of properties, qualities, quantities, and formal attributes, as well as of states of affairs, conditions, and situations.

Thus all the possible states of being can be split into three broad classes, *a class of physical states, a class of cognitive states, and a class of cultural states*. Then the totality or the sum total of all the state spaces, or the State of Being as the lawful state space of the world, can be constructed as the union class of three state types:

$$\mathbf{S} = \bigcup \{S_j : j \in I\} = \bigcup \{\{S_{Pj}, j \in I_X\}, \{S_{Mj}, j \in I_Y\}, \{S_{Cj}, j \in I_Z\}\} \tag{3}$$

Being generally defined as *the way a thing is with reference to its substantial properties*, the UFO generic states will embrace as species the following particular classes of the computing ontologies and taxonomies:

- **WordNet:** State (condition, situation, the state of mind, etc.), abstraction [(quantity, amount, measure, quantum and attribute (property, quality, shape, form, etc.)], phenomenon (state of matter), psychological feature (knowledge, property, attribute, dimension)

- **EuroWordNet:** 2nd order entity (static situation and situation component, condition, existence, quantity)
- **EDR:** Event/occurrence (condition/state)
- **SUMO:** Abstract object (quantity, attribute, internal)
- **CYC:** Intangible individual (static situation, attribute value)
- **DOLCE:** Quality (color, size, shape, smell), quality region (qualia), occurrence (state)

As for its child classes, there are two kinds of state types as stative properties:

- Quality as an essential property (attribute) of an entity
- Quantity (measure, amount, or quantum) as a measurable property (attribute) of things

In science, quantity is regarded to be prior to quality, since quantities are likely to be universal qualities of objects, whereas qualities are the necessary parts of experience, as the qualities of sensations, perceptions of the mind, mental powers (see Figure 6). As a collection of universal properties, the category of state subsumes as its subordinate kinds both the class of quantities and the class of qualities, as it is shown in the hierarchical scheme of state types:

Entity (Being, Existence) > State > Property > Quality > Quantity > Multitude > Magnitude > Energy >...

That is, states exist in such a hierarchy that if the union of quantity and quality constructs the higher genus of state, their intersection results in a qualitative quantity or quantitative quality, such as form, figure, or shape.

As with substances, the state types are marked by the similar rules of behavior:

- The state type should fall under the same division as its kinds
- If a state belongs to two types, then the one type is included by the other
- What contains the state subtype contains the state genus as well
- If states are different in kind, their equivalent classes of substances are different as well
- All the states of an entity are the states of its kinds, species and individuals

As to the quantitative type of state, quantity, there exist several essential axioms and assumptions. Above all, it is the rule: *quantities are wholes consisting of parts, having or not*

Figure 6. The taxonomy of states

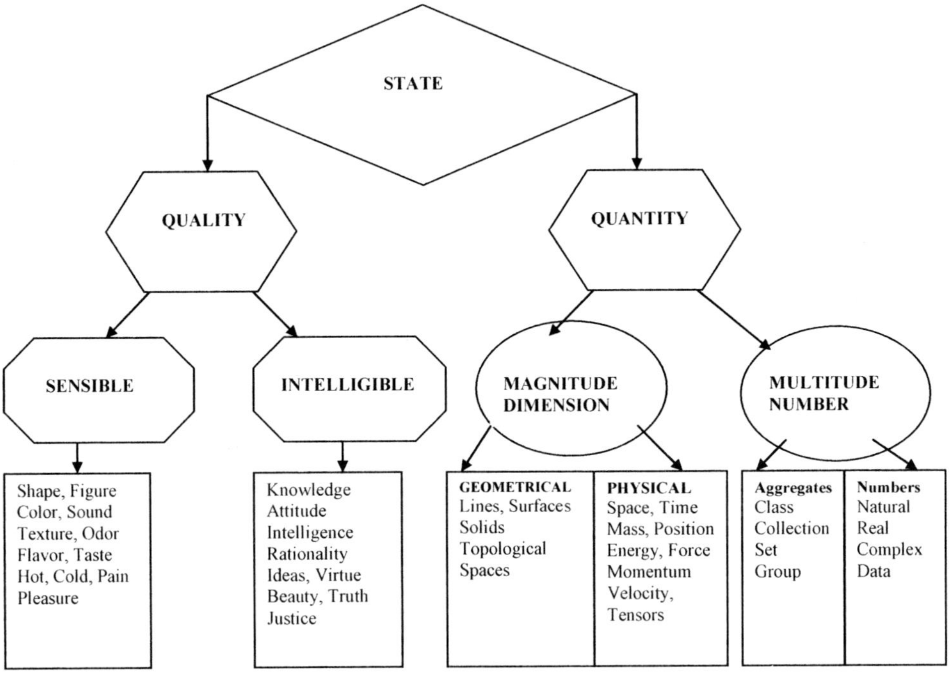

having a relative order to each other. Their most distinguishing character is to be predicated by the relations of equality or inequality. Or, the basic character of any quantity is that the relationships of equality or inequality are stated about it, unlike quality marked by likeness, similarity or difference, diversity. Like substance, quantities have no contraries, but unlike qualities, which do have contraries.

Also, quantities are thought incapable of admitting of variation of degree, unlike qualities that are liable to alteration (variation, modification) in degree or in kind.

Along with analyzing its nature and classification, the theme of quantity involves most critical issues in modern science, such as the relation of magnitudes and multitudes, dimensionality, equality, proportion, the measurements of quantities, the units of measurements, number and numbering systems, the types of numbers and their relations to each other as numerical ratios.

It should be observed that traditionally quantity (or quantum) is regarded as a fundamental ontological and scientific category, and it was classified into two different types, magnitude and plurality:

'Quantum' means that which is divisible into two or more constituent parts of which each is by nature a 'one' and a 'this.' A quantum is a plurality if it is numerable, a magnitude if it is measurable. 'Plurality' means that which is divisible potentially into non-continuous

parts, 'magnitude' that which is divisible into continuous parts; of magnitude, that which is continuous in one dimension is length; in two breadth, in three depth. Of these, limited plurality is number, limited length is a line, breadth a surface, depth a solid. (Aristotle, 1990, book V, chs. 11-14)

So, by its basic meaning, quantity is the state of being "so much," or the property of a thing marked by a magnitude or plurality (number). By its definition, quantity is an ontological term used to refer to any type of quantitative properties or attributes of things. Initially, quantity was introduced as quantum, an entity having quantity.

Of entities that pertain to quantities, some are such by their inner nature, while others are functioning as the states (properties, dimensions, attributes) of some sort of a thing, like heavy and light, long and short, broad and narrow, small and great, or much and little. Traditionally, there are discrete quantities (with no common boundary among the parts as number) and continuous quantities (with common boundary as space, time, or shape). Of which, what is divisible into continuous parts makes a magnitude (as length, breadth, and depth), while what is divisible into discrete parts make a plurality. Two basic divisions of quantity, magnitude, and multitude (or number), imply the principal distinction between continuity (continuum) and discontinuity. The continuous/discontinuous distinction of quantities also comes from Aristotle: "everything continuous is divisible into divisibles that are infinitely divisible." That means that the parts divided have their extremities as being one and the same and in mutual contact, while everything discontinuous is divisible into indivisibles being in contact with each other only as whole with whole (Aristotle, 1990, Physics book VI, ch. 1).

So, quantities exist either as continuous or discrete, without having a relative position or relative order, as any magnitude (state variable) and multitude (number), or not having any relative order of parts. So, being a whole of the parts, quantities consist of the parts that either possess a relative order and position or do not. For example, time and space and number have their parts (continuous and non-continuous) as all marked by some sort of a relative order: like past, present, and future, for time, or like having a unique position in an ordered sequence of parts, for number. But what is without any position and indivisible in quantity is a unit (one in quantity), while that which is divisible and with position is a point. Each sort of quantity has its specific type of unit of measurement, of time, of distance, of mass, of energy, and so forth. The units of measurement are all sorts of standards of measurements used in physics, computing, and economics.

All modern science is basically about quantity and measure; for quantitative knowledge is the measure of all things. Formally, quantities (numbers and magnitudes), their ratios, proportions, order and formal relationships of equality and inequality, are studied by mathematics. The essential part of mathematical quantities is made up with a collection variables each assuming a set of values and coming as scalar, vectors, or tensors, and functioning as infinitesimal, arguments, independent or dependent variables, or random and stochastic quantities. In mathematics, magnitudes and multitudes are not only two kinds of quantity, but they also are commensurable to each other. Specifically, the topics of the discrete quantities as numbers, number systems, with their kinds and relations, fall into the number theory. Geometry studies the issues of spatial magnitudes: straight lines (their length and relationships as parallels, perpendiculars, angles) and curved lines of different kinds with their relationships (tangents, secants, and asymptotes). Besides, it encompasses surfaces and solids, their transformations, measurements, and relationships.

Physics is fundamentally a quantitative science. Its progress is chiefly achieved due to rendering the abstract qualities of material entities into the primary quantities of physical things, by postulating that all material bodies marked by quantitative properties or physical dimensions, which are subject to some numerical measurements and observations. Setting the units of measurement, physics covers such fundamental quantities as space (length, breadth, and depth) and time, mass and force, temperature, energy and quantum. In thermodynamics, the notion of physical quantity is examined as extensive quantities and intensive quantities.

In logic and semantics, in respect to quantity, propositions are grouped as universal and particular, applying to the whole subject or a part of the subject to be predicated. Accordingly, there are existential and universal quantifiers. In relation to the meaning of a construct, quantity involves two semantic dimensions:

1. **Extension or extent** (determining the specific classes or individual instances indicated by the construct)
2. **Intension** (content or comprehension or definition) measuring all the implications (relationships and associations involved in a construct, its intrinsic, inherent, internal, built-in, and constitutional implicit meanings and relations).

The notion of quantity is essential in natural language as well. With respect to quantity, all entity names can be divided as the names of multitude (the discrete things that are numerable, countable, or denumerable) and the names of magnitude (the continuous things that are measurable). The former involves all cases of collective nouns: *army, fleet, flock, government, company, party, people, chorus, crowd, mess, and number*, while the latter involves all examples of non-collective nouns: *the universe, matter, mass, energy, liquid, material, animal, plant, tree. . .*

In English, number is a syntactic category, along with person and gender. The quantity is expressed by identifiers, definite and indefinite, and quantifiers, definite and indefinite, as well as by three types of nouns:

1. Count unit nouns or countables
2. Mass nouns, uncountables, referring to the indefinite, unidentified amounts
3. Nouns of multitude (collective nouns)

The word "number" belongs to a noun of multitude standing either for a single entity or for the individuals making the whole. As mentioned, an amount in general is expressed by a special class of words, identifiers and quantifiers. The amount may be expressed by: singular form and plural from, ordinal numbers before a count noun singular (first, second, third. . .), the demonstratives; definite and indefinite numbers and measurements (hundred/hundreds, million/millions), or cardinal numbers before count nouns. The set of language quantifiers covers "'a few, a great number, many, several (for count names); a bit of, a little, less, a great deal (amount) of, much (for mass names); all, plenty of, a lot of, enough, more, most, some, any, both, each, either, neither, every, no." For the complex case of unidentified amounts, the parts and examples of a mass are indicated with respect to the following: a measure of a mass ("two kilos of rice and twenty bottles of milk or ten pieces of paper"); a piece or part of a mass ("part, element, atom, item, article, drop"); or a shape of a container ("a basket, box, case, cup, bottle, vessel, jar").

The role of quantity in human society is hard to appraise. Quantities and numbers and numerals are widely used to represent information as well as to code and communicate data about persons. In our social life, to identify an individual, a range of identification numbers is employed; namely: registration number; social security number; personal identification number (PIN); bank identification number (BAN); mobile telephone number; and Internet protocol (IP) number, and so forth.

All in all, quantity is split into physical quantities (dimensions or variables), geometrical magnitudes (lines, surfaces, and solids, figures and shapes), and mathematical multitudes: non-referential, abstract state variables, and numbers (natural, rational, irrational, real, complex, and data). An ordered collection of numeric data or information, such as scientific, technical, financial or statistical data together with data-processing operations makes up numeric databases. Therefore, computational ontologies sometimes are defined as a collection of information (data types) about classes and relations of a domain of knowledge.

From an onto-mathematical perspective, a state type as a collection of properties can be described by its representative, a state function or state variable, the set of values of which corresponds to the set of properties of an entity type. The collection of all lawfully possible states constitutes a generic state space, such as material, mental, and cultural state spaces described with their specific state functions. Then the totality of all generic state spaces makes up the universal state space **S**, which is formalized as the total state function $\mathbf{f\,(E)} \subseteq \mathbf{F}$ mapping from the entity (or relationship) classes $\mathbf{E} \subseteq \mathbf{W}$ into the value (instance) sets V_I or a product of the instance (value) sets:

$$f_E: E_I\ (\times\ E_I) \rightarrow V_I\,(\times V_I) \qquad \textbf{(4)}$$

This implies that states are represented as the mathematical relations (functions) from the domain of possible worlds to the range of entity instances, individuals, particular properties, or values. For the case of particulars, the function maps an entity individual to a certain value, a relationship instance to a certain tuple of value. In the event of general things, the state of an entity variable is represented by the ordered differences in the state measured by a set of values, while a relation class, by an ordered set of value sets.

Shortly, the state functions representing properties of all kinds and sorts instance, exemplify, or illustrate generic states of all kinds and sorts. Crucially, the formal expression of states indicates that there are several sorts of state function as describing four types of generic states:

i. The states of substance
ii. The states of process
iii. The states of state
iv. The states of relationship

All is varying according to the type of state substrate. As there are several sorts of properties: (1) substantial (intrinsic, permanent, simple, or basic); (2) qualitative or quantitative; (3) active (transient, emergent); (4) relational (mutual and complex), where change-like properties belong to a group of second-order properties, and so forth.

The most traditional substrate of states, a substance, can be viewed as a state of having the property of being a substance, or a zero-order state. A state of a substance is a first-order state marked by having the property of being a state. A state of a first-order state is then a second-order state, like having of property of changing from some initial state (or phase) to some final state (or phase) and so relating or ordering the pair of differing states, measured by the pair of state instances (values). A natural conclusion follows, thence: each entity is a member of a hierarchy of state types covering infinitely many particular things.

Any ontological category can be defined in terms of state. Even the entire realty can be viewed as an enduring limitless state space defined above as the world state space and having as parts different state types, such as substances, conditions, changes, and relations. This means that canvassing the nature of state, to discover its essential features and meanings, we need always to see which sort of state is under examination:

a. States as states *of* and *in* substances (or systems of substances)
b. States as states of qualities or quantities
c. States as states (or conditions) of changes
d. States as states of relationships

For instance, all substances can be modeled by a functional schema **O** = $<O_i, f_0>$, where the substance state function through its components $f_k(O_i)$ represents all the possible properties of the substances as the Cartesian product of the possible values of the functional components, $V_1 \times V_2 \times \ldots \times V_n$. A set of relationships among the functional components as well as a number of constraints imposed on the possible values of these components will then constitute a body of laws and rules L as a core part of a general theory about the substances, T (O_i). Then such a law function of substances will be written as, L: T $\rightarrow$ L(T). If the range of the state function V_i is made of vector spaces, then the entity state function will become the state vector function. The lawful state space for objects is described as a set of the cross product of all the possible values of the components of the entity state function L $= <f_1, f_2, \ldots, f_n>: O_i \rightarrow V_1 \times V_2 \times \ldots \times V_n$ subject to some lawful restrictions. As a result, in the functional schema or models of substances, the objects represented as the object variables can be entered as the arguments of the substantial function $f_0: O_i \rightarrow N$; N is here a finite set of natural numbers, and f_0 stands for the populations (number) of objects of any kinds and sorts: atoms, molecules, organisms, or persons varying their numbers over time and space. Now any empirically verified relation among the possible values of the components of the substance state function will constitute a law relationship (or statement) of the theory.

Noteworthy that unlike the object variables, the entity state function does not denote but rather represents or measures the properties of objects such as physical quantities and dimensions, as mass, size, temperature, pressure, and so forth. Like it happens with an atom's quantum stationary states $f(o) = <n, l, m, s>$, where the members stand for the specific quantum numbers characterizing the properties of an atom. It seems that the power (capacity) of an entity is determined by the number of its states.

To generalize the concept of a functional schema, one should combine the formula of the world (**2**) and the world transformation rule (**8**) taken from the Part III, obtaining a universal functional schema for the world W_I, as in:

$$W = F(E_I) = F(O_j, S_k, C_l, R_m) \quad (5)$$

The expression indicates that the various momentary states of the world are characterized by the momentary values of all its basic variables. As much as the specific states of a machine (as a general-purpose computer) are characterized by the specific values of all the variables in the program. So the equation obtained describes the state of reality in general as well as the state of any real system of entities. The universal state function F is now clearly defined in terms of the class variables. By its role, the universal function represents or measures the state of the world determined by the total sum of results of the observations or measurements of all the properties or quantities relevant to the state measured. The changes of the state can be represented by an infinitesimal or finite change in magnitude of the property *d* or by the rate of change of the state with respect to a parameter of quantity, for example, the change parameter of time $t \in T$.

Several remarkable implications for physical entities can be deduced from the world state formula. From the equation (**5**) we can obtain the universal properties possessed by all material objects, such as mass property or energy function measuring the changeability of objects. First of all, it allows formulating one of the most universal state functions of real objects, which is physical energy E:

$$E = E(O_r, \Phi) \quad (6)$$

Here the symbol Φ stands for the class of reference frames. The meaning of the energy function becomes clear for a concrete system having a definite position at a given time: its value measures the changeability of the object with regard to a certain reference frame.

The equation (**5**) can be used for studying the general behavior of material systems exchanging energy and material with its environment. For any isolated collection of matter, characterized by a set of values of energy E, a number of species of particles N_k, and constraints R_s imposed by the environment, the universal equation will be specified as a simple mathematical function:

$$F = F(E, N_k, R_s) \quad (7)$$

It is noteworthy that this function is a principal equation used in classical thermodynamics for describing the changes of states of any physical system passing through or being in a special condition of stable equilibrium states. It is clear that any physical property can be presented as a function of the set of other properties considered as independently varied arguments.

So substituting a property F with entropy S and energy E, the function **(7)** will be properly modified to describe the relation of energy and entropy in all stable equilibrium states:

$$S=S(E, N_k, R_s); E=E(E, N_k, R_s) \quad (8)$$

It is remarkable that when the entropy change is equal to zero *dS=0,* the sequence of states (process) involved will be reversible. That is, the work-producing process executes reversibly at constant entropy, with no net work effects in any of the systems involved, like as *reversible cyclic engines*, *reversible nuclear reactions*, and *reversible chemical reactions*. A remarkable engineering advantage is that the efficiency of the reversible machines much greater than that of the irreversible ones.

If the category of generic substance as a class of objects constitutes an ontological base for a set-theoretical fundamental concept of set (or class or aggregate), a state type determines the class of objects as much as the class of substances determines the class of states. Or, the state type as a collection of properties is referred to the substance types as the class of sets of objects as much as the classes of substances assigned to the state types, as in the expression: $2^O \leftrightarrow 2^S$. So, the scope of a state type as a collection of properties is the class of substance types possessing such state or condition $2^S \rightarrow 2^O$, while inverse scope function assigns each class of sets of substances a state type (a collection of properties): $2^O \rightarrow 2^S$. As a consequence, the lattice of generic states (universal properties) and the lattice of substance types are isomorphic: $< 2^S, \cap, \cup, \subseteq, > \leftrightarrow < 2^O, \cap, \cup, \subseteq >$. Another significant result is that additionally to the set theory principle of extension, where a class is defined by its elements, members, or objects, it can be determined by some state condition making another significant axiom, the principle of abstraction for nonfinite sets defined by its properties, formalized as $\{X: S(X)\}$, and read as "the set of all entities X such that the condition or state S(X) holds true."

It emerges that the universe of discourse and the subject matter of the universal ontology is nothing else but the world state space, or the state space of the world, marked by the fundamental properties and dimensions of quantity and quality (Abdoullaev, 1999). Just as a state of an individual thing (entity) is an aggregate of its individual properties, the collection of all possible states of the thing is its state space, so the world's state space is the sum total of all possible states, or the totality of all kinds of properties of entities.

As an abstract space is characterized by dimensions without limits, and extension in all directions without bounds, so the state space of the world is defined by an infinite number of quality dimensions where transitions, variations, and relationships between states constitute changes, or processes. As the substructure of reality, the state space of the world is inherent in or underlies the cardinal levels of reality:

- **Natural reality**, or the physical space of all matter and energy with all the dimensions of physical quantities
- **Mental reality**, or the mental (power) space with cognitive, volitional, and affective dimensions
- **Social reality**, the cultural (information) space with semantic, human, and social dimensions

To conclude the lively subject, the primary principles of state ontology can be summarized as follows:

- The category of state is among the first divisions of entity
- The collection of properties in virtue of which a thing changes makes a state type
- All states form a hierarchy of state types with the state of being at the top and the state of nonbeing at the bottom
- A state is a member of a hierarchy of states where a first-order state is a state of substance
- All the states of an entity are the states of its species and individuals
- The world is the totality of generic states standing in various relationships

The Class of Change (Actions, Activities, and Events)

Being identified by states, all substances change or produce changes, either becoming different in property, or causing to change something else. In some or other respect, any thing is in a state of flux, in condition of a continuous or discrete change. For one entity, of any sort and kind, has more then one distinct state. That means that whatever exists is given or likely to or capable to change; thus *every being is changeful, alterable, variable, or changeable*. Changefulness or changeability or mutability is the constancy (identity, permanence, and fixity) of the world. If there is something without change, it will be nothing, a constant, changeless, and enduring being, since *nothing (not any entity) endures except Nothing*, and everything is in flux: *to be is to be in change, not to be is not to be in change*. Thus comes up the first self-evident truth about changes: *everything changes*, or *all things, beings, or entities make different or become different in state and so participating in a process of change.*

Being another base class of reality, along with substance, state, and relation, change is a manifold, multiform, and complex entity, too. It can be essential or accidental, superficial or deep, material or conceptual, partial or whole, natural or artificial, regular and constant or spontaneous, chanceful, and eventful. This character of change is expressed by a multitude of synonymous terms:

alteration, conversion, damage, difference, diversity, development, devolution, flux, innovation, mutation, modification, novelty, passage, qualification, regeneration, renewal, renewing, reversion, revolution, shift, transformation, transition, transmutation, turn, variation, variety, or vicissitude.

Because of its status as a pivotal class of entity, the knowledge of change is among the key area of study in modern sciences. Namely:

i. Its nature and forms of existence
ii. Kinds, forms and manner of action
iii. Its causes and effects
iv. The ways of analyzing and measuring of different sorts of motions and movements.

As to the nature of change, corresponding to three types of real entities, there are typically three forms of change: natural C_N, mental C_M, and cultural C_C.
But whatever its domain of action or operation, as a primary class of being, change operates and acts as a process and a sequential process during which entity or being becomes different either in substratum, or in state, or in action, or in relation. In the language of properties, a change in general is a change of the property of entity:

i. The property of being a substance
ii. The property of being a state
iii. The property of being acted upon or of acting in a certain way
iv. The property of being related to something else

In accordance with the 4D (substance-state-change-relation) reference system of the UFO, any specific change, natural process, mental action, or social activity, should fall under one of the following top headings:

- A change of substance
- A change of state (quality and quantity)
- A change of change (in action, variation)
- A change of relation (in correlatives)

Altogether, there are four different forms of change with distinct natures:

- Substantial changes
- Stative changes (qualitative changes and quantitative changes)
- Variable (functional) changes
- Relational changes

Under these four change types can be subsumed the six traditional forms of change: substantial processes, generation (coming-to-be and becoming) and degeneration (destruction or passing away); quantitative processes, increase, growth, and decrease, diminution in magnitude or size; qualitative processes, alteration in quality; a change of position, a relational change, motion, movement and rest, equilibrium, as well as a change of form (physical states) as a sort of stative processes.

The first sort of processes is a change of a substantial nature, coming into being from other being, genesis, a process of change during which one or more substances transform into one or more other substances, as in chemical reactions. What undergoes change here is the substrate itself, when the entity changes as a whole, from one substance to one or many other substances. Substantial changes seem to be most dramatic and fundamental changes as it is illustrated by generation, production or corruption, and damage (annihilation and extinction) of physical objects, chemical substances, living organisms, ideas, or whole civilizations. For example, any process of forming of a compound or a complex as a single substance in which the entity parts lost their individuality and identity comes under radical, deep changes. In nature, this sort of change covers all kinds of subatomic interactions and chemical reactions, as decay, fermentation, corrosion, digestion, combustion, and so forth. In human life, birth and death, creative processes, discoveries, making novel artifacts also fall under the substantial changes.

The second in the list of distinct changes is stative changes, a modification of the state of an entity as a transition of an entity from one state to another, as a difference between states, while the substance retains the same identity. For instance, a transition of an object from one state position to another is specified as motion or movement, which is exemplified by a mechanical motion. Another example is a change of physical states, sometimes called a change of form, from solid to liquid, liquid to gas, gas to liquid. Under the change of quality fall material changes resulting in producing a heterogeneous mass, body, aggregate, collection, complexity, or mixture in which different entity parts still retain their identity. Here also come some biological changes, such as genetic and chromosomal mutations, mitosis and meiosis of cells, the series of division changes.

A subclass of stative change, the change of quantity includes decreases and increases of the magnitudes of objects. In the order of physical quantities, a quantitative change is a change of energy, including all forms of energy, as well as the energy change pertaining to the transfer of energy and energy interactions of objects such as heat and work. Again, any changes of state with regard to human conditions, emotions, and desires, as want and satisfaction, belong to the class of stative changes of conditions, either of a qualitative nature or a quantitative form.

The third change type, dynamic change, is a process of change during which the state of action, operation, or agency changes its mode of acting or operation under the action of other change, what is typically referred as the changing cause of change. Examples of such difference in the manner of action may be the transitions from one distinct form of change to other, from rest to motion, destabilization, from motion to rest (negative acceleration), as well as all sorts of accelerative acts and processes increasing the rate of change. All changes in policy, practice, doing, and use are the changes of human action and activity. Here belong also the irregular cycles of generation and destruction, growth and diminishing, rotary and linear motions.

The fourth type of change, relational change, is a process during which one sort of relationship changes into other kinds (marriage/separation). As there are changes in respect of substance or essence, state or quality and quantity, change or activity, so there is change in respect of relation. When one or both correlatives alter, we have another relation; for if the husband dies, the wife becomes a widow, thus changing their family relationship.

Whatever changes also happens and occurs at a certain time and place, and whatever happens shows changes. All events are changes, but not all changes are events, as all liquids are fluids, but not all fluids are liquids. Anything that happens by chance or occurs at a certain time and place refers to an event as a result and outcome of the changes caused and actions accomplished or performed, like as act, case, chance, consequence, doing, end, episode, eventuality, fact, fortune, incident, occurrence, or outcome.

So, whatever people do or cause to happen, their actions, activities, acts, doings, proceedings, can be classified whether as radical substantial changes, qualitative changes, quantitative changes, variable changes, or relative changes.

Having being divided into four distinct types, the generic changes, then, will embrace as its subordinate species the relevant instances listed in the lexical taxonomies and formal ontologies:

- **WordNet:** Event (happening, occurrence, natural event), phenomenon (natural phenomenon, process, consequence), act, human action, human activity (action, group action, social event, activity, etc.), psychological feature [(cognition, knowledge, noesis (cognitive or mental process and operation, unconscious process), (motivation, motive, need), (feeling)]
- **EuroWordNet:** 2nd order entity (situation type, dynamic, situation component, cause, experience)
- **EDR:** Event or occurrence (phenomenon, action or deed, movement, change in state or change in condition)
- **SUMO:** Physical process (internal change, intentional change, dual object change, shape change, motion)
- **CYC:** Partially intangible (event), doing, transformations, changes of state, transfer of possession, movement, emotion, perception, waves
- **DOLCE:** Occurrence (dynamic events, processes and accomplishments)

The common characteristic of the above taxonomies is a disinclination to consider a change class as a fundamental ontological category, although generally accepting that any things that happen (particular events, occurrences, acts, activities, processes, or goings-on) imply (signal or presuppose) a change thus involving it as a necessary condition of their existence.

In accordance with the EIS SUO, change in general is the basic class of all things that take place (*occur, happen, go on, pass, pass off, fall, come about, come up, come to pass, or come to being*) in (*at, on, near, or by*) some places at a time (over a period of time). So *an event is a change that happens at a given location and time to a certain individual (or system of individuals)*. In other words, any event is a change endowed with spatial properties as spatiality and location, and temporal properties as temporal order, duration and rate. Any event is a temporally and spatially spread-out particular, one-off entity, while change is the class of entity beyond time and space. Therefore, *events as happenings go through states (stages and phases) thus having beginnings, middles, and endings, modifying from one phase to another but within the same general change*. In other words, events are nothing else but particular changes, another name for individual changes.

As an example, the EIS classification includes all the basic senses of change specified by the WordNet, namely change as alteration or modification of states (phases), as a relational difference of states, as the action of changing, and as the result of modification. Under the four key headings, substantial changes Co, stative changes of conditions, qualities and quantities, Cs, variable or fluctuating changes Cf, and relational changes Cr, there comes the following senses:

Sense 1 - (alteration, modification, an event that occurs…)

=> mutation, Cs

=> birth, nativity, nascency, nascence, Co

=> separation, breakup, detachment, Cr

=> variation, fluctuation, Cf

=> death, decease, Co

=> decrease, lessening, drop-off, Cs

=> destabilization, Cs

=> increase, Cs

=> transition, Cs

=> transformation, transmutation, shift, Cs

=> transmutation, Co

=> damage, harm, impairment, Cs

=> development, Cs

=> revolution, Co

=> mutation, genetic mutation, chromosomal mutation, Cs

Sense 2 – (a relational difference between states)

=> difference, Cr

=> gradient, Cs

Sense 3 – (the action of changing something)

=> about-face, volte-face, reversal, policy change, Cs

=> adulteration, Cs

=> move, relocation, Cs

=> reduction, simplification, Cs

=> decimalization, decimalization, Cs

=> metrification, metrication , Cs

=> variation, Cf

=> turning, Cs

=> diversification, variegation, Cs

=> flux, Cs

=> switch, switching, shift

=> substitution, exchange, commutation

=> change of state, Cs

=> alteration, modification, adjustment, Cs

=> motion, movement, move, Cs

=> movement, Cs

=> motion, movement, move, motility, Cs

=> change of direction, reorientation, Cf

=> change of magnitude, Cs

=> change of integrity, Co

=> conversion, Co

=> updating, Cs

=> change of shape, Cs

=> nationalization, nationalisation, Co

=> communization, communisation, Co

Sense 4 - (the result of alteration or modification)

=> depolarization, depolarization, Cs

The WordNet's taxonomy of changes can be used with some essential reservations. First, all the species of change must be properly reclassified according to the arguments presented above. Second, the meanings of changes ought to lie within one distinct class, which is neither relation nor quality, but change. For a change is divisible only into changes, as a magnitude into magnitudes. Third, both events and human actions are the subtypes of change as according to the order of changes:

Entity > Change > Action > Process > Phenomena > Event >Happening, Occurrence, or Outcome > Human Act, Action, or Activity > … > Nonentity

In the descending order of changes, *the event is a change happening at a given place and time, the action is a change that is done by some agents, substances, forces, chemicals, or intelligent agencies. This intends that events are merely particular changes, physical processes, changes in concrete things*. Like it is proposed by SUMO, where a process of change has a traditional dictionary definition: "the class of things that happen and have temporal parts or stages," or "anything that lasts for a time but is not an Object" (SUMO). Following such a temporal definition of change, the SUMO subtypes of change are presented as follows: the changes in an object, or internal changes (biological, chemical, physical state change, quantity change, surface change, damaging, and creation); the changes of an object (shape change, motion); the changes between two objects (dual object process, contact, attaching and detaching); and intentional changes (all sorts of human activity or action, from making to searching to social interaction to organizational processes).

To have a full and consistent classification, changes are correctly to classify only with respect to the kinds of substances they refer to or the sorts of states they relate. Accordingly, we can specify three broad groups of processes of change:

- Physical processes, the collection of all natural processes, physical, chemical, and biological, C_P
- Mental events (cognitive processes of the mind), the total of psychological processes and experiences, together with formal, logical processes, C_M
- Cultural processes, the collection of all possible human actions, activities, and acts, C_C

Thus all the particular changes can be split into three specific classes, *a class of physical changes, a class of psychological processes, and a class of cultural or social processes of change* [the intersection class ($C_P \cap C_M = C_C$)]. Then, formally, the universe of change **C** will be represented as the union class of three subclasses:

$$\mathbf{C} = \bigcup \{C_k\colon k \in I\} = \bigcup \{\{C_{Pk}, k \in I_X\}, \{C_{Mk}, k \in I_Y\}, \{C_{Ck}, k \in I_Z\}\} \tag{9}$$

We have to refer to the classical learning about change (Aristotle, 1990, Physica) when searching the right answers to the fundamental questions about change:

1. What it is to be a process of change
2. What its nature and essence, its content and meaning
3. Whether has it a being or not
4. How it appears, its forms and species, and all the lawfully possible instances, cases, and exemplifications

Such a classical reference, when properly extended and modified, tells us that everything that changes is changing not only in essence (as generation and destruction and creation) and state (quality, quantity) (as increase and decrease and alteration), but also in action and relationship (as change of place or position). Besides, the generic types are exemplified as a multitude of particular changes, each of which is "a change or becoming of something from something to something else" (Aristotle, 1990, Physica, p. 306). And that, altogether, we have five elements in any particular change:

- That which changes (quality or quantity)
- That which is in change (substratum, body or state or action)
- That in which change takes place (time or space)
- That from which it starts (initial state)
- That to which it proceeds (final state, an effect, end, purpose, consequence, or result of change)
- That what causes a change

At last, every change suggests the governing rules used for general machine knowledge and reasoning about any process:

- There are three modes of changes, essential, partial, or accidental
- There is a substrate underlying the processes of change
- There are changes in kind and in degree
- There are six sorts of changes, generation and destruction, increase and decrease, alteration and motion
- Substances may undergo all the changes at the same time, such as destruction, motion, and growth

From the onto-mathematical view, the concept of change involves the reality-centric formal representation. It is a common fact that all the differential calculus of mathematical analysis is about computing the infinitesimal changes of variables or quantities, where the central notions are a function (rule, formula, or expression) and the rate of change, plus real variables, dependent and independent (Analysis [in Mathematics]). Given two sets of state variables S_i and S_j, a function f is a well-defined rule (formula, expression, or law) that assigns to each state-element (the independent variable) S_X of S_i a unique state-element (the dependent variable) $S_Y = f(S_X)$ of S_j. Now the rate at which the dependent state variable changes in respect to the independent state variable is the difference quotient, or the ratio of the difference $\Delta = f(S_X + H) - f(S_X)$ to H, where H is any real value of change of the independent variable S_X. Such an average rate of change, a result of S_X changing by H, can be transformed to the derivative of the function f, if H tends to zero, and denoted as $f' = d\,S_Y/d\,S_X$, where the symbol d is brought in as a differential (an infinitesimal increment in a quantity). The result of mathematical function representing the instantaneous change of one state variable with respect to another also is named as a derived function, derivative, first derivative, or differential coefficient. Repeating this process, we can obtain higher derivatives, a derivative of order n denoted as f^n. A mathematical relation connecting the function S_Y of an independent variable S_X and its derivatives is an ordinary differential equation, as in:

$$f(S_X, S_Y, S'_Y, \ldots, S^n_Y) = 0 \quad (10)$$

Or, for a case of implicit relation between a solution (integral) of the differential equation and the independent variable, $F(S_X, S_Y) = 0$. In reality, such differential equations, linear or nonlinear, homogenous and heterogeneous, correspond to natural laws describing the behavior of rates of changes of (state) variables (quantities) in respect of independent variables, as time and space.

Again, the possible kinds of the changes generally are of four sorts: substantial, stative, variable, or relational. According to this division, a multitude of differential equations describing the lawful behavior of rates of changes of states can be ordered by the entity classes described above. As examples of substantial changes are the rate of decrease of the reacting substances in chemical reactions and the rate of disintegration of substances (a substantial change) in nuclear radioactive decay, or the growth of populations in a given region over a time period, or the rates of change of biological compounds in the human body; of active processes, the rate of change of velocity in a moving body; of stative (quantitative) changes, the rates of changes of physical quantities; of relational changes, the rate of change of causal connections.

Alternatively, in set theory, using a special operation of symmetrical difference, which unions the difference sets, as $(S_i - S_j)$ or $(S_j - S_i)$, the change C_{ij} between two distinct generic states S_i and S_j can be expressed by the symmetric difference Δ of the sets of states: $C_{ij} = (S_i - S_j) \cup (S_j - S_i) = S_i \Delta S_j$. In the most general mathematical description then the processes of change can be described as a partially ordered universe of state sets $\mathbf{C} \subseteq \mathbf{S} \times \mathbf{S}$. As a result, the universe of generic changes can be formalized as a category of classes of state sets and functions. Because any sort of processes of change, chain processes, converging processes, diverging processes, or circular processes, can be represented as a sequence of states, the history of each change can be presented as a trajectory in the lawful state space. In other words, the functional representation of the universe of changes can be expressed in terms of generic states, as in the triple:

$$\mathbf{C} = < \mathbf{S}, \mathbf{f}, \circ > \qquad (11)$$

This triple structure is the process of change category, where the class of objects is all the possible states of the world S, $S' \subset \mathbf{S}$; the class of morphisms (or functions), denoted as **f**, is the set of all lawful transformations f, g, h...$\subset \mathbf{f} \subseteq \mathbf{S}^{\mathbf{S}}$. Earlier this identity function $\mathbf{I} = \mathbf{f}$, having the same domain and range, was introduced as the state function, the totality of all functional transformations (maps, morphisms) of **S** into itself, representing all lawfully possible changes and processes. The operator ° signifies the composition of functions (transformations), connecting each triple of states, like the constitutive parts of the world state space, the natural state space S_P, the mental state space S_M, and the cultural state space S_C.

Now the process category can be interpreted as the category where the class of objects is made up of the state spaces, physical, mental, and cultural, $S_P, S_M, S_C \subset \mathbf{S}$. The class of morphisms includes the transformations like $J: S_C \rightarrow S_M$ and $G: S_M \rightarrow S_P$, where J, G indicates that the sets of changes are of kind J or G. To each ordered triple of object states S_P, S_M, S_C and two morphisms, a law of composition is prescribed such that $J \circ G$ is the morphism from the cultural state S_C to the physical state S_P.

Also, to each state S identity morphism $I_M: S_M \rightarrow S_M$ is associated such that the compositions $JI_M = J$ and $I_M G = G$ defined. Due to these equations, for each morphism J or G there exists the reversible transformations $J': S_M \rightarrow S_C$ and $G': S_P \rightarrow S_M$.

So, the identity function I_P for the physical state space embraces all sorts of changes and processes the material substances undergo. Namely, *a change of position*, movement or motion; *a change in the physical state of matter*, a transition of gas to liquid, liquid to solid, and solid to plasma; *a change of chemical substance*, a variety of chemical reactions, such as burning, decay, fermentation, film development, corrosion, gas evolving, digestion, synthesis, and so forth.

As to the mental world, the identity function I_M involves sensations, perceptions, images, memories, expectations, beliefs, thoughts, emotions, desires, aversions, intentions, and actions (see Figure 7).

At the same time, the function I_C describes all major groups of cultural change: demographic, technological, economic, social, and political. However, to establish connections between distinct kinds of change, such as change in the natural environment, conceptual change, and technological innovations, we need to use another primitive category of entity, the primary class of relations, which is the next important thing in our consideration of the nature of reality and its consistent formal representation.

Such is then the number, nature, and formal description of the kinds of change, or what it is to be a change. Concluding the topic, we must point out that if the essence and content of change consists in the action of changing of entities by making (or becoming) different in substance, state, action, or relation, its basic meaning and consequence consists in being a cause or effect. For every change is at the same time also a cause and/or effect. *Nothing happens without a cause-change, and nothing exists from whose being an effect-change does not follow.* In other words, every change (process) is followed and preceded by other changes (processes). Neither object, nor state, nor relation but changes constitute causes and effects, the components of causality. Human actions, wars, diseases, emotions, crimes, births, deaths, and so forth, are both changes and causes or effects. By this reason, *the knowledge of change and behavior is the knowledge of cause and effect.* Since revealing the nature, meaning, and content of change is equal to revealing the number and nature of the kinds of cause.

In all, some major principles and postulates to be used as the premises of general knowledge and fundamental reasoning about changes and processes can be summarized as follows:

- The world is the totality of processes of change
- Nothing endures but Nothing, all entities are mutable

Figure 7. Mental processes: The materials of the mind

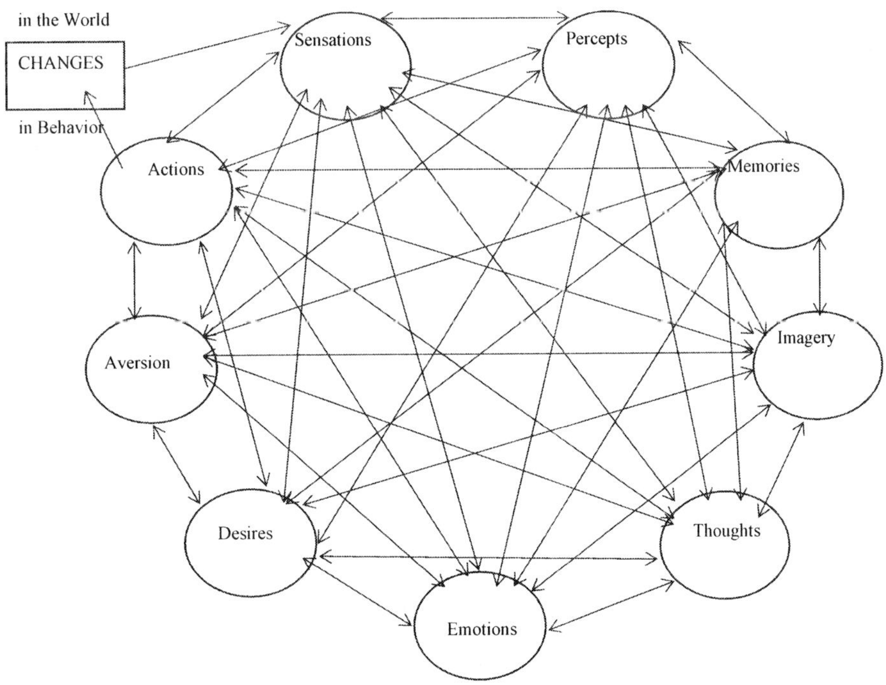

- Any change entity is a composition of two or more states, or a change is a second-order state
- Changes differ in type (decay, increase), sphere (substance, quantity), and manner
- Any process of change is a sequence of differing states of the same change
- Whatever changes is changing (turning) into a different substance, state, action, or relation
- There are substantial changes, stative changes, variable processes, and relational changes
- Every process of change is followed and preceded by other processes
- Every change is either a cause or effect; the essence of changes is to enter into causal relations, to be and to have causes and effects
- Every event, happening, or occurrence is an instance of a process of change

Conclusion

A possibility of a global schema of reality capable to incorporate scientific knowledge, upper computing ontologies, and Web ontology languages is suggested. Substances, states, and changes are exposed among the primary ontological classes, a closed set of basic entity types with different kinds and species. A formal description of the primary entity classes as the fundamental categories of reality and the template classes of the universe of discourse is given. It is proved that to have a well-developed ontology of computable classes, to construct a high quality Web ontology language, the universal entity framework supporting all the major classes in the general hierarchy of entities should be formulated. Designing such a machine-centric system requires a formal inquiry into objects, states, and changes as the classes of ontological predicates specifying the structure of any realms of the world or any knowledge domains and bases. Having in view the close perspective of creating the unified modeling language, we formulated the distinguishing features of the key entity types, their definitions, axioms, properties, classification, and rules, so that to formulate the common rules of reasoning about the world and its key features. Some of the rules are as follows:

- Reality consists in ontological classes, realities, and fundamental rules (laws)
- There are several prime classes of entity and reality: substances, states, changes, and relations
- The class (genus) falls under the same division as the kinds
- If a species belong to two classes, the one is included by the other
- What contains the species contains the genus as well
- If classes are different, their properties are different in kind
- All the properties of a class are the properties of its species and individuals

References

Abdoullaev, A. (1999). *Artificial Superintelligence.* PDC, Bowling Green State University, Ohio, USA.

Algebra: Lattice theory, Categories, Macropaedia. (1994). Knowledge in Depth, Vol.13, *The New Encyclopaedia Britannica.* Chicago: Encyclopaedia Britannica, Inc.

Analysis (in Mathematics). (1994). Macropaedia, Knowledge in Depth (Vol.13). *The new encyclopaedia Britannica.* Chicago: Encyclopaedia Britannica, Inc.

Aristotle, Logic (Organon): Categories. (1990). In M.J. Adler (Ed.), *Great books of the western world* (Vol. 1). Chicago: Encyclopaedia Britannica, Inc.

Aristotle, Physical Treatises: Physics (Physica). (1990). In M.J. Adler (Ed.), *Great books of the western world* (Vol. 1). Chicago: Encyclopaedia Britannica, Inc.

Aristotle, Metaphysics. (1990). In M.J. Adler (Ed.), *Great books of the western world* (Vol. 1). Chicago: Encyclopaedia Britannica, Inc.

Bunge, M. (1977). *Ontology I: The furniture of the world.* Boston: D. Reidel Publishing Co.

DOLCE (a descriptive ontology for linguistic and cognitive engineering). Retrieved May 1, 2006, from http://www.loa-cnr.it/DOLCE.html

EDR (electronic dictionary research) dictionary. (1995). Retrieved May 1, 2006, from http://www2.nict.go.jp/kk/e416/EDR/index.html

EuroWordNet top concepts. (2004). Retrieved April 25, 2005, from http://demo.ontotext.com

IEEE P 1600.1 standard upper ontology working group. (2004). Retrieved April 25, 2005, from http://suo.ieee.org/

Kant, I. (1990). The critique of pure reason. In M.J. Adler (Ed.), *Great books of the western world* (Vol. 39). Chicago: Encyclopaedia Britannica, Inc.

Open Cyc. (2003). Retrieved April 22, 2006, from http://www.opencyc.org

OWL web ontology language, use cases and requirements, W3C recommendation 10 February 2004. Retrieved April 22, 2006, from http://www.w3.org/TR/2004/REC-webont-req-20040210/

Set theory, Knowledge in Depth. (1994). Macropaedia (Vol. 27). *The new encyclopaedia Britannica.* Chicago: Encyclopaedia Britannica, Inc.

SUMO ontology. (2004). Retrieved April 22, 2006, from http://ontology.teknowledge.com

WordNet, WordNet 2.0. (2004). Retrieved April 22, 2006, from http://www.cog.sci.princeton.edu/wn/

Chapter V

What Orders Reality: Relationship, Relatives, and Relations

Introduction

To model reality by mapping its content to computable forms, we need to know how to represent the first-class entities of any existence, relationships, the adhesive of the world. Both human and machine understanding of the universe, its parts and realms, consists in knowing the cardinal relationships and underlying rules and making valid inferences from them.

The Semantic Web ontology is often identified with a schema defining relationships between different resources. For instance, the OWL markup language is supposed to specify the types of relationships represented in RDF language employing an XML vocabulary, with a view to determine the relationships (and hierarchies) among diverse Web data resources identified by URI. And the formal specification of relationships determines the meaning (semantics) of knowledge domains, and the universality and credibility of any ontology, its rigor, cogency, validity, and richness, come from the capacity to fully describe all the possible types of relationships in a domain of knowledge or practice. Without a systematic theory of relations, it hardly is possible to form a universal account (language or theory) describing reality, its entity classes, properties, individuals with their particular properties, on which human or machine reasoning has to take place.

In the EIS world schema, an ontological entity of relationship is among the primal cuts of reality along with substance, state, and change. In the context of computer science, a top-level ontology is a knowledge base structure sought to be implemented in computer programs. The task of such a computational ontology is to supply a set of high-level classes manipulated by general rules mapping the relationships among entities in the universe of discourse about anything. So, finding the answers to the fundamental questions of the meaning, existence, and kinds of relations (*what* relation is, *whether* it exists, and *how* it is/works) appears to be crucial both for constructing world models and data -type structures of specific realms and domains. Also, it is particularly important for building a unified [Semantic Web] ontology language to resolve the following critical issues:

- What it is to be a relationship
- What key relation types may be
- How it must be formally represented

The extant SW languages define a property as a binary relation (or n-ary relations) connecting two or more individuals. The approach is just an extension of the set theory's axiom of choice, reducing a real relationship to a mathematical function, the sets of ordered pairs, where each individual element of one disjoint set (domain) is associated at least with one element of another disjoint set (range). Such conception comes from a long-established misplacement of relations for their relative terms: "those things are called relative, which, being either said to be *of* something else or related to something else, are explained by reference to that other thing." Or, "those terms, then, are called relative, the nature of which is explained by a reference to something else. . ." (Aristotle, 1990, Categories).

Revising the extensional definition having a strong influence on the set theory of relations and, hence, SW languages, we argue: while existing really, a relation has its own status and nature quite different from the nature of its foundation, relative things, for which the relation to something else is a necessary condition of existence. Au fond, *a relation has its own essence different from its components, it has a distinct reality*, being, and existence, although not easy to perceive. Again, a relation exists really, but not according to a conception of the human minds. Something is said to be a relative only by virtue of its relationship, as much as somebody is said to be a parent only by virtue of its real relationship of parenthood, and not other way around.

In the essence, a relation is composed of the relationship, expressed by prepositions or other relative words, and a multitude of relative entities, as in:

Relation = Relationship + N-Relatives (Components, Elements, Arguments, Terms, Relata)

Here the relationship also may enter as the relative element, and vice versa, like *less* and *greater*. In other words, as much as individual entities (as objects) are defined by their classes (kinds) of things, the class of relationship is a definition of all the relative terms of a specific type. While studying the nature of relationships, their basic properties also are properly examined. Among relational traits, the basic one is the property of being reversible (inverse), and reciprocal or correlative. Since *to be inverse and convertible is a necessary and sufficient condition of existence for any relational entity*, like the meaning (or signification) relation between names (URIs) and things (Web resources,) or causal relationship, implying correlation or co-variation as a necessary condition.

In all, giving readers a clear idea of relationship, its nature and meaning, it is demonstrated that classifying the relationships should be performed both by their real properties and formal attributes. Widely practiced in data modeling methodologies the classification of relationships by degree (number of entities connected), connectivity and cardinality (the number of connected occurrences for each of the related entity types), direction (the parent entity and the child entity), mode of existence (contingent, or optional and necessary, or mandatory), or reflexivity (symmetry or transitivity) is proved to be only a part of the whole matter.

How to Define and Describe Relations

Commonly, relation is accepted as an indemonstrable and indefinable entity, which meaning and existence should be posited, assumed, or taken for granted. Although, it is intuitively plain that relation is real if it holds between real things, like temporal and spatial relations, and formal if it holds between abstract objects. And that the nature and meaning of relationships consist in establishing various connections between things in the real world. Still the issues of its nature and comprehensive systematization suffer many controversial interpretations, mainly because most top-level computing schemas and lexical taxonomies, under risk of confusion, view relations as abstract things along with classes, properties or attributes, times, spaces, quantities, and sets.

Any general ontology language in order to become such has to include a core set of relationship kinds accompanied with the essential properties and formal attributes. This is not the case with the current Semantic Web ontologies created to formally describe properties (as binary relations) and classes, their relations, cardinality, equality, and other formal characteristics (OWL Web Ontology Language, 2004). Such formal languages are lacking in substantial classes of relations, both formal and real. Among real types, poorly represented or even misrepresented, is the relation of cause and effect, an essential construct in an emerging OWL-based Web service ontology aimed to furnish "a core set of markup language constructs facilitating the automation of Web service tasks, including automated Web service discovery, execution, composition and interoperation" (OWL-S, 2005).

To build a viable Web ontology, instead of classes and relational attributes, rather we have to talk about entity classes and their relationships as two sorts of distinct but existential entities. A relationship is correctly understood when considered as a real kind of entity organizing, structuring, and ordering the things in the world. Then the definition as an association, or relationship, or link among real-world entities will most correspond to its nature. Namely, at the class level, *relationship is an entity class relating (or ordering) the kinds of things*, each having a relationship property (as its role). At the instance level, it is a set of links or connections among individuals, particular cases of generic real-world associations. Both statements imply that any good ontology should avoid a replacement of relations with relational properties, distinguishing internal and external connections as formal or existential relations. If the same type of predicate is asserted of its kinds (a quantity predicated of magnitudes), it signifies its class. Again, if one type of predicate is asserted of another type, as the relations of equality and inequality are predicated of quantity, the relations of sameness or difference, of substances, it signifies rather a property than an essence (Aristotle, 1990, Topics). For example, *being a parent* is merely a relational property, while *parentage* (or parenthood) is a relation of parent to child. Or, more generally, the relational property of *being a cause* is just a monadic reduction of *causality*, the relation of cause to effect. In short, the proper kinds and sorts of relationship cannot be mixed with its roles, or associative properties, though facilitating its further division.

Another crucial consequence of such an explicit definition is that any relation is a complex unity of a relationship and its **relata** (related terms or components, correlatives, or correlates). This means that its associative part is the entity relating (correlating, linking up, associating, tying, or pasting) the parts of the world and instantiated by relator (link, connector, connection, connective, tie, bond, function, or operator, or whatever) by which both real and formal things are interrelated. The relata as the things connected by the relationship are, in turn, presented by arguments, components, terms, roles, participants, correlatives, or correlates.

On establishing the nature and kinds of the correlatives as substances, states, changes, and relationships, a set of basic relation types can be constructed. It will comprise as follows:

- Object-to-object relations for spatial (material) substances and non-spatial (abstract) objects
- State-to-state relations for attributes or properties, qualities and quantities
- Change-to-change relations for natural, mental, and social events
- Relationship-to-relationship relations as analogies and proportions of various sorts
- Cross-entity connections (substance-to-state, substance-to-action, etc.)

The comprehensive classification comprises the OWL mathematical (logical) relations between sets (classes) making only a small part of an enormous universe of relations. Also, it covers the rich listing of the WordNet relation types running from the part-whole relations to causality to spatial relations to temporal relations to social, human and business relationships to magnitude relations to semantic, linguistic and formal relations.

The Ontology of Relations

There are few assumptions guiding the fundamental inquiry of relationships, their rudimentary truths and underlying principles. The long-standing disputes of the reality of relations, whether relations exist or not, or are they of our own creation or made by Mother-Nature, can be solved by invoking the very fact that all beings and things have a reciprocal order and mutual relation. We just need to see that there are three general kinds of connectedness: there are real relations between real things— like times and spaces, and there are relations created by the human mind—logical relations, like a hierarchy of types, and there are mixed relations—partly real, partly formal, like semantic relations.

It is more than evident that no understanding of relations, like cause and effect, no knowledge of reality and its basic truths. But deep knowing and reasoning upon relationships or interrelationships asks for the ontological postulates, first principles, basic truths, or primary premises,. Which is a set of foundational assumptions about its existence, nature, meaning, and essential properties, employed as the ontological knowledge base for further automated demonstration and machine reasoning on different sorts and manner of relations. In as much as the reality of relationships comes from the very being of relationship, there comes out the number one in priority, *the existence axiom*:

Axiom 1. Relationship exists really by (in or of) itself, over and above the things connected.

In human perception and actions, there are many instances of relations, like causal links, spatial relations, and temporal relationships, which are parts of our everyday experience as much as the properties of matter. By its nature, relation is an existential entity having full existence, actual in nature and mental in the mind.

Next self-evident truth is *the order axiom*:

Axiom 2. Relation is the principle of order that determines the ways all things are arranged or associated and organized.

The axiom states that each and every relation goes with a certain order among entity types and instances. In that each and every entity is a part of the world and a relative term at the same time, it can be postulated that all things in the world are disposed to or placed in a certain kind of order—what makes the universe an orderly and well-regulated place (or system) subject to few fundamental laws and principles.

Third, being the principle of order, relation inverts, reversing the order of things in the opposite direction; for if one entity has a relation to another, then the other will have the converse relation to the first one. Thus, if there is a relation, then we need to inquire into its capacity to be inverted, reversed, or converted; for in the relations of things the sequence of their ordering, if not the sequence of their being, can always be reversed. This assumption implies that a relation in its full entirety and completeness is an interrelation, interrelationship, or interrelatedness, a reciprocal relation, relation of mutual dependence between entities. Conversion, inversion, reciprocity, or reversibility is an intrinsic property of relation of any level, specific or general. Thus, we come to the **complementarity principle** stated as *the order reversal axiom*:

Axiom 3. Every relationship has its inverse counterpart (opposite or complement).

Or, complementarity, convertibility, reversibility, and inversion is an essential property of any relationships, external, which are unaffected by the sides they connect, and internal, which are affected by the terms they relate. From *the reversal law* it follows that no inversion, no relation and vice versa, like it is reflected in the duality principle of lattice theory, no dual statement, no true statement at all. The axiom simply postulates that a relation must be two-directional, where both directions should be determined and defined, and that a full relation is combined of the given connection along with its reciprocal relation.

It emerges that any relation by nature comes with a definite order among entity classes and kinds (or species) where the ordering may go in the opposite, contrary ways:

- From an entity class to other class and vice versa, making fundamental laws
- From a class to its kinds, making the relation of class division, classification, sub-typing, or subdividing (direct)
- From the kinds to their class, making the relation of class inclusion (reverse)

Further, seeing that the universal scheme of entities implies that particulars are instances of generic entities, individual members and elements of the primitive classes of being, we can formulate *the instantiation axiom*:

Axiom 4. A particular relation is a generic relation holding among particulars.

As far as the world holds many manifold links, connections, bonds, and ties, their nature is to be revealed not in the particular correlatives, but rather in their generic entities exemplified as cases and instances. It also follows that any relation by nature comes with a definite order among entity kinds and instances, where the ordering may go in the reverse ways:

- From a kind (species) to its individual members (direct), making instantiation relations of concretization, materialization, specialization, or realization
- From individuals to their kind, making the relation of generalization, aggregation, or composition

So, depending on their direction—forward, downward, upward, and backward—we can determine a few general types of order:

1. Forward and backward relations as constituting one-level interrelations of classes
2. The class inclusion from kinds to its next higher level, super-ordination
3. Specification from a class at a higher level to its kinds at the lower level, subordination, for example, the relation of class division
4. Exemplification or instantiation relations from a class to its objects, elements, and individuals
5. The relation of generalization from particulars or species to their class, for example, the class membership, generalization, or abstraction enabling inductive reasoning.

Additionally to the power of ordering things, there are few other essential attributes of a relationship. First of all, it is a feature of relation to compare things of any nature and kind as to be the same or different, knowledge of which is embodied in all our reasoning about sameness and difference. Both sameness and difference will differ with regard to the kind of entities. Besides, things can be the same or different in number, species, and class, that is, numerically, specifically, and generically. In the case of numerical **sameness**, we speak of absolute identity or self-sameness, the reflexive relation of one and the same entity to itself, while other forms of sameness belong to **resemblance**, likeness, or similarity, sometimes called relative identity. By analogy, in the event of numerical difference, we speak either of absolute diversity or relative difference. Altogether, in four respects, either with regard to substance and matter, state, the mode of changing, or the sorts of connection, the following sorts of comparison relation can be established among entities:

- Generic sameness, similarity, or similitude
- Generic diversity, otherness, and heterogeneity
- Specific sameness and diversity
- Particular identity and diversity

For example, we speak of individuals as distinct existences that they are different (many) in number and yet also the same (one) in substance, quality, or change, which is their generic kind or nature. Or, we can talk of the specific difference between two beings with reference to their intelligence, at the same time possessing the generic sameness with respect to animality/species; things may be similar or different as the diverse kinds of the same class.

In the order of knowledge, we can consider representation as sameness or likeness between the agent knowing and the thing known. The similar reasoning applies to the sameness and diversity in the significance of terms, symbols, and words, always accompanied by the generic difference between the representatives (signs) and representees (things). Thus we come to *the comparison axiom*:

Axiom 5. Everything is comparable as identical, like or unlike.

Namely, anything can be related to anything on the ground of anything: objects, properties (quantities, qualities, attributes), actions (changes, functions, operations, processes), and relationships. Briefly, this general feature of relation to compare things can be formulated as the following equation of **comparison** relation founded on similarities and differences:

Difference (Similarity) = **Sameness** (Diversity) + **Diversity** (Sameness)

Using this ontological rule, we can find out the following species of difference or dissimilarity or unlikeness or unsimilarity between two related things: otherness or distinctness, distinction, divergence or variance or disagreement or discrepancy, discrimination, variation or variety, inequality, differentia, and differential. The fundamental consequence from the rule is that in the world there are no things that are either absolutely diverse or absolutely identical. This rule applies to relations as well, which can be the same or similar, being diverse in class or genus at the same time. Such relational sameness can be exemplified as analogy or proportion, identity without similarity, a resemblance of relationships belonging to different classes, commensurate or incommensurate. Symbolically, it will be written as X: Y = Z : W where the formula means "as X related to (in) Y, so Z stands to (in) W." The relative relationships underpin all sorts of reasoning by analogy or from similarity, where the discovery of the similarity of two relationships is based on matching their correlatives: one term can be related to a second as a third one to a fourth. Being out of line of the nature of relations, the number of models for reasoning from the known resemblances of relations can become uncontrollably enormous (Hoffman, 1995).

In general, all this suggests *the universality of diversified sameness*, since at the same time *all entities* are one and many. Namely, some entities may be one (others) in number having one (different) substratum, one (others) in species having one (different) definition, one in class having the same (different) genus and predicate, one in analogy having the same (different) proportion. So, many different things may be one kind of entity, say, one kind of quantity as magnitude or number, one kind of matter, and so forth. In other words, all classes of things are the same as the divisions of one supreme universal class—Entity (remember Kant's principle of ultimate homogeneousness). Again, they also exist as different kinds of entity [substance, state, change, and relation] having a finite variety of species and sub-species of individuals (remember Aristotle's rule of heterogeneity of things).

As enlargement/extension to **comparison,** the relation indicating differences and likeness of things, can serve **opposition**, the relation between opposite entities, a variety of comparison relation. The relation of opposition comprises as its species dissimilarity or contrast, polarity or mutual opposition, contrary, contradiction, contrariety, gradable opposition, ungradable opposition or complementarity, contradictoriness or inconsistency, inverse, reverse, antagonism, antipode, antithesis, conflict (WordNet 2.0, 2004, Opposition). Altogether, entities are to be opposed to one another in four distinct ways: as correlatives or relative things, as contraries, as possession to privation, and as contradictories (affirmation to negation) (Aristotle, 1990, Categories). Pairs of correlatives are defined with reference to each other, as two relatives that imply and require one another, as one thing to be in comparison with other. Pairs of contraries either refer to contrary properties of a class or to contrary classes, thus dividing a class into types that differ in degree or in kind, as in existing classifications of animals with and without intermediate species. In a sense, sameness and diversity, like heat and cold, wisdom and folly, can be considered as the contrary things not existing in the same entity at the same time, but allowing the middle term to exist, while similarity or difference as sharing the property of both. A pair of contradictories belongs to the opposite terms that exclude each other as existing in the same thing at the same time and applying to everything in the world, as positive and negative terms in a logical dichotomous division of things, for example, human and non-human, being and non-being.

On a par with comparison and opposition, there is other general sort of relations applying to all sorts of things, the part-whole relation, as that which is contained to that which contains, and vice versa, the whole-part relation, a relation of a whole to its parts and elements. It should be noted that the part-whole synthetic relation together with its complement, the whole-part analytic relation as that which contains to that which is contained belong to the highest relation classes, **hyperrelations or metarelations**, always coming up as we divide entity classes or instances. As far as Entity or Thing or Reality or Being is the class of all classes, of which all things are members, we can hold that substance is an essential fragment of reality, state is a substantial piece of being, change is a crucial part of entity, and relation is a determinative part of reality. But each of the four types also is an aggregate of its elements (parts) formed into a whole entity, whole thing, or unity. Substances, states, changes, and relations are the topmost classes of thing, each of them consists of parts, either as a subdivision of its highest division, the class as a whole, or as one example of a class of thing. The former usually is defined as a specialization, subsumption, subordinate, or class-inclusion relation, while the latter as an instantiation, or class -membership relation; the inverse of both is a generalization, superordinate relation.

The number and character of specific types of the part-whole relation is to be defined by the specific nature of its opposite correlatives, a part as a part of a whole and a whole as a whole of parts. From one part, there is the general term "part" with its specific meanings as atom, amount, component, constituent, division, element, fraction, fragment, ingredient, member, particle, piece, portion, section, segment, share, and subdivision. From other part, the general term "whole" with several meanings as aggregate, aggregation, amount, assemblage, body, bulk, collection, complex, composite, composition, compound, entirety, group, heap, mass, matter, object, system, substance, sum, thing, total, and totality. The meanings of the part-hood, its possible kinds and types, may be determined by its constitutive elements, which are the different senses of the part at one side, and all the senses of the whole as a complex unity of its parts at the other side.

"Part" means that into which any whole entity as a complex unity might be divided, as a complex substance into simple substances or as a class into kinds, a quantum into a magnitude and a number, a process into its stages, a relation into its subrelations. It can measure the whole, or can be a part of the class, the matter in which the form is contained, the stuff of which some compound or mixture made, and so forth. As the general term, it includes the sets of cognate meanings: {atom, particle, piece}, {component, constituent, element, factor}, {amount, portion, share}, {division, subdivision, fraction}, {ingredient}, {fragment, section, segment}, {detail, particular, item, point}, {member, object, individual, unit}. The part's whole signifies such an entity that forms some sort of a unity from all its constituent elements, a composite entity. It also covers a wide range of meanings that can be grouped according to the nature of the elements, objects, or members: {aggregate, aggregation, assemblage}, {mixture, compound, composite, complex}, {amount, number, quantity, quantum, sum, total}, {body, mass, matter, substance), {class, set, group, collection}, {entirety, totality, all}. Moreover, the parts of the whole can be again viewed as other wholes divisible into their parts, that is, parts can have other parts as in a sequence of subdivisions or subparts:

State → quantity → magnitude → solid → surface → line → point → unit (in length, breadth, depth, speed, weight, force, etc)

Then the ultimate part of any whole is a unity, that which is the one, no more divisible in quantity (number, unity) or in quality (kind, elements), the measure of all things. Evidently that the part-whole relation inherits all essential differences of the general class of relationships: to be real, to be ordered, to be inversed, to be specified, and to be compared. In accordance with the nature of the terms entered into the part-of-relation, we can distinguish four types:

- A subordinate classificatory relation (species/genus); "a species is a member (division) of genus"
- A particular relation specifying a particular thing of a class (individual-collection)
- A constituent relation specifying a necessary element, stuff, material, characteristic, or stage of a composite whole: (stuff/object, part-process); "an atom is a part of a molecule," "a word is an element of speech," "a beginning is a part of process"
- A structural relationship (component/object, portion/mass, place/area, head/ body, piece/mass)

Of all these part-whole sub-relations, the first one is of highest value. For, possessing the formal properties of asymmetry and transitivity, the subordinate classificatory relationship superimposes a hierarchical order on the whole being or entity, its classes and subclasses, organizing them into a single inheritance system of things. Accordingly, all key entity classes, as substance, quantity, quality, action, and relation, as well as their divisions are the wholes consisting of parts. As, for example, substance in general is an aggregate, totality (e.g., matter, body, mass, compound, or mixture) of its substantial parts (components, elements, constituents, fractions, fragments, ingredients, portions, particles, or atoms); or state in general as a whole of properties (qualities, quantities, conditions); change in general as a whole of actions (occurrences, acts, and events); and relation in total as a whole of its various members (time, space, process, resemblance, causality).

Thus, the kind-of-relation must be included into the part-of-relation as its subordinate type. Here, it must be mentioned that there is an interesting classification proposed by empirical semanticists, which distinguishes seven sorts of the part-of-relations as component-object, member-collection, portion-mass, stuff-object, feature-activity, place-area, and phase-process (Winston, Chaffin, & Hermann, 1987). Despite its originality, the principal defect of such a classification is the suggestion that the kind-of-relation should be viewed as a semantic relation possessing a special status on a par with the part-whole relations. As it happens, with all other relations, the part-whole relationship possesses the property of reversibility, reciprocity, or mutuality: each constituent part is determined not only by the other parts making the whole, but also by that whole to which this part belongs. This suggests that the relation of part and whole includes an upward relation of part and whole and a downward relation of whole and part. The former means that what goes on a higher level can be reduced to the next lower level, like upward causation in physics explaining macrosystems in terms of elementary particles. The latter means that a whole may act upon its parts so that to form a structural and functional unity, like all examples of downward causation in nature (Abdoullaev, 1999).

Knowing that things stand in a certain relation to each other, to properly define or describe some sorts of things, first we need to *compare* them, to identify their sameness and likeness, or unlikeness and diversity. Next, we *oppose* or *contrast* the things putting them in opposition in order to reveal dissimilarity or unlikeness in a most general sense. Then we *differentiate* them identifying the specific differences of things most closely alike. At last, we *discriminate, distinguish, or discern* those differences, so that to acquire clear logical distinctions, descriptions, or definitions of the entities related, compared, opposed, contrasted, and differentiated. Finally, having defined a certain class and its properties, we divide it as a whole into parts, species or individuals, inquiring all possible relationships among their parts, as well as the parts and their whole.

Comparison with opposition and part-whole relations does not exhaust all the ontological, fundamental, and existential relationships, however. In conformity with the theory of entities proposed, the further division of the class of relations will be directly determined by the nature of the entities standing as connected, associated, tied in, linked up, or interrelated. As far as the number of higher-order entities is established as substances or objects, states, changes, actions, or processes, and relations, the full list of entity-entity relationships will include all the possible associations and connections between the primary kinds of things—which are the object-object, object-state, object-change, object-relation associations; the state-object, state-state, state-process, state-relation linkages; and so forth. In such the Entity-Entity Relationship, the most interesting relationships appear to be the diagonal connections between the entities of similar types, as in:

- The relationship of substances (objects, natural and abstract, and artifacts), or substantial relationships
- The relationship of states (of properties, quantities, and qualities), or stative relationships
- The relationship of changes (actions, activities, operations, and processes), or active and dynamic relationships
- The relation of relationships, relative connections or associations

For instance, a hierarchy of material substances, from elementary particles to large-scale physical aggregates, belongs to the field of substantial relations. Equality and inequality, as well as logical relations of ideas and mental relations of conscious states, fall under types of relations involving states, quantitative and qualitative. The relation of cause and effect, and temporal relationships come under the sorts of relations involving changes, actions, or processes. At last, analogy or proportion, the resemblance of relations, together with logical relations of relational universals will be subsumed under the relations of specific relationships. Depending on the nature of correlatives, we can have several distinct forms of analogical reasoning: substantial proportions, qualitative and quantitative analogies, functional analogies of change, or meta-relational proportions.

Thus, in its essence, relations constitute the universal class of ordered classes of entities, the nature of each of which is determined by the specific type of the related kinds of entities, such as substances (objects, spatial and non-spatial), states (quality, quantity), changes (action, activity, passivity), and relations (time, space, position, place). When the relational members, or relata, are substances, we deal with the first-order relationships among objects, and when the relata are states, we have the second-order relationships among states of all kinds and sorts. In case the relata are processes or activities, we will come to the third-order relationships among changes of various sorts. And when the relata are relations, we will have the fourth-order associations among different relationships. That implies that the parts of relation, a relator and a relatum, are exchangeable or replaceable with the order-changing effect: the relator may become the relatum, and vice versa, by changing the order of connections. In the first case, the relator will become a relatum but in the newly formed relation of a higher order complexity, while the relatum becomes a relator not within the same relation where it was an associative entity but already relative to the relata of a lower order complexity (Gunther, 1970). Then, even substance may be defined as a complex relation of bare matter and its inherent properties, while a state as an aggregate of properties, and a change as a relational difference of states. It emerges that each entity is a member of a hierarchy of entities, infinite for individuals but finite for types. That is why sometimes states can be viewed as arranging into a hierarchy of states, where a first-order state is a state of substance, a second-order state (event) is a state of a first-order state, a third-order state (relation) is a state of a second-order state, and so forth. (Chisholm, 1996).

A Universal Classification of Relations

A UFO model of relations opens up an opportunity to construct a single comprehensive classification comprising existent taxonomies of relations:

- **WordNet:** Abstraction (relation, time, space), state (relationship, relationship)
- **Roget's Thesaurus:** Relation, Causation, Order, Time, Space
- **EuroWordNet:** 2nd order entity (situation type, static, relation, situation component, communication)
- **EDR:** Location, locale, place, time
- **SUMO:** Abstract object (relation, proposition, attribute, relational)
- **CYC:** Mathematical object (relation, formal), temporal relations (times and dates), spatial relations
- **DOLCE:** Quality (position and place, time interval), quality region (space, time)
- **KM Component Library:** Entity (object)-to-entity (object) relations, event-to-event relations, event-to-object relations
- WebKB-2: Entity (object)-to-entity (object) relations
- **RDFS:** Properties, relationships between resources (ObjectProperty, DatatypeProperty, type, subClassOf, subPropertyOf, domain, range, isDefinedBy, subject, predicate, object)
- **OWL:** Set-theoretical properties or relations (in) equality, equivalent property, sameAs, differentFrom, AllDifferent, distinctMembers; inverseOf, TransitiveProperty, SymmetricProperty, FunctionalProperty, Inverse FunctionalProperty; complementOf, unionOf, intersectionOf, sameValuefrom).

There have been several attempts to uncover the fundamental species of relations, so that to arrange all the connections in the world into few categories.

Hume made an effort to analyze relations by dividing them into two groups: internal relations of ideas, thoughts, or concepts, and external relations of matters of facts. The first part of the dichotomy included identity, resemblance, quantity and number, quality, and contrariety, while the second part contained causality and spatiotemporal relations (Hume, 1990).

Subsuming substance under relation as a prime category of understanding, Kant distinguished as the relationship kinds: inherence, causality or dependency, reciprocity or community (Kant, 1990).

The position that there are "such entities as relations between things" was firmly supported by Russell, who considered them as the sort of classes (universals) represented by verbs and functions words. Assuming that any statement expresses a relation among things, Russell admitted as true classes of the world of being the relation of resemblance or similarity, relations of space and time, the relations of relations or analogy, as well as logical relationships (Russell, 1990).

Trying to create an exhaustive sorting, some researchers proposed to add as internal relations, the existence of which are determined by the existence of the related terms, difference, part-whole relations and aggregation or composition, classification and instantiation, generalization and specialization, grouping and membership, semiotic relations.

There also are taxonomies where as generic relationship types are numbered as the following:

- The relations involving changes (effective relations)
- Property-based relations
- Topological connections
- Semiotic associations
- Referential relationships
- Mereological relations, or external relations (Relations, 2002)

The effective relations include existential relations among things, binding connections as influences and bonds; case relations between agents, actions, and objects; transitional relations between states; scenario relationships, and causal relations between situations, states, and events. Comparison, magnitude relations (equivalence and nonequivalence), and spatiotemporal relations fall under the external or nonbinding connections, unaffected by the related terms. Among property-based connections are the inherence relation between a substance and its properties; the opposition relations as contrast, contrariety, contrary, contradiction, antagonism, opposite, and reverse; and logical relations as the class inclusion and the set-membership. Topological relations comprise attachment and inclusion; referential relationships, relevancy and dependency; and semiotic relations, syntactic connections among signs, pragmatic between users and signs, semantic or representation relations between signs and things expressed as iconic, indexical, and symbolic. At last, the part-whole relations constitute the mereological relationships.

Still, of all the existent classifications, the richest taxonomy can be found in the WordNet division of relations (Relation, WordNet 2.0, 2004):

=> social relation (communication (message, infection or contagion, message or content, linguistic, written, visual, auditory, paralinguistic or psychic communication, show, signal, sign, indication, style, vocalization, voice, document); (relations or dealings or transactions; professional relation; politics; interpersonal chemistry)

=> position, spatial relation (inclination; slope; placement; point of view; coincidence; centrality; marginality; anteriority; posteriority; outwardness; inwardness; malposition; northerness; southerness; horizontality; verticality; way; direction; angular position)

=> ownership (community; severality; property right; proprietorship; employee ownership; landholding; stockholding)

=> causality

=> relationship, human relationship (personal relation: bonding, obligation)

=> function

=> association (colligation: generalization, induction)

=> logical relation (contradictory; contrary; transitivity; reflexivity; modality or mode; implication or conditional relation)

=> mathematical relation (function: inverse, Kronecker delta, metric, transformation, mapping, map, or correspondence, operator, trigonometric, threshold, exponential; parity; transitivity; reflexivity)

=> foundation (footing, basis, or ground; grass roots)

=> connection, connexion, connectedness (series; alliance or bond; linkage, communication; concatenation; bridge; involvement, inclusion or containment; relevance (materiality, cogency, point, with reference, regard, respect, or relation to, applicability or pertinency); relatedness (bearing))

=> unconnectedness (irrelevancy, unrelatedness)

=> linguistic relation (grammatical relation, agreement, transitivity, intransitivity, coreference, conjunction, complementation, coordination, subordination, modification, mood, anaphoric relation, voice, inflection, aspect; semantic relation, subordination or hyponymy, superordination or hypernymy, synonymy, antonymy, whole to part relation or holonymy, part to whole relation or meronymy, troponymy)

=> part, portion, component part, component (linguistic or language unit; item or point; basis or base; detail, particular, or item; unit; member; substance; remainder, balance, residual, residuum, or rest; subpart)

=> affinity, kinship (rapport; sympathy)

=> kinship, family relationship, relationship (affinity or phylogenetic relation; descent, line, lineage, or filiation; affinity or kinship by marriage; consanguinity or blood kinship; parentage or birth; fatherhood or paternity; motherhood or maternity; brotherhood; sisterhood; marital relationship) ,

=> magnitude relation, quantitative relation (scale; ratio (profit margin, abundance, efficiency, frequency, productivity, quotient; rate (acceleration, fertility, mortality, flow rate, flux, frequency, growth rate, speed)

=> control

=> business relation (competition; clientage)

=> reciprocality, reciprocity (complementarity; correlation; mutuality or interdependence (commensalism, parasitism, symbiosis, sharing); mutualness; reciprocal)

=> interrelation, interrelationship, interrelatedness

=> temporal relation (antecedent, chronology, synchronism, asynchronism)

=> comparison (imaginative comparison: simile, metaphor, allegory)

=> opposition, oppositeness (antipode, antithesis, conflict, contrast, mutual opposition or polarity, ungradable opposition, conradictoriness, contradiction, contrary, contrariety, tertium quid, reverse, inverse, antagonism)

=> change (difference, gradient)

Although comparatively extensive and very instructive, the listing suffers from several principal errors: the wrong definition: relation is not an abstraction but rather a reality, an entity connecting two or more things. Besides, it calls for new reclassification where the spatial relation, the human-human relations (social, business, personal, familial, affinity relationship), with human-property relations go under substantial relationships. The interrelation, connection, and association should not be marked as separate but rather used as synonyms of relation, instead; for typically, relation is defined as an association or connection among things. The part, portion and foundation are not relations, but rather the terms or correlatives. Again, the taxonomy is missing some significant relation types that are not distinctly identified, namely the relations occurring in nature (physical interactions, chemical reactions, and biological relationships), in mind (mental, psychological relations), in society, and in history. The scheme also misrepresents the nature and role of causality and causes, closely interpreted as a particular sort of individual events, although the types of causal relations expressed by causative verbs are well specified. The temporal relation should include the range of temporal properties (temporal order, duration, and rate). The mathematical relations are missing the set-theoretical relationships. Reciprocity is the inherent property of any relations. The magnitude relation together with the quality associations will fall under the stative relations. Last but not least, the semantic (meaning) relations of signs, constructs and extra-linguistic things, like significance, denotation, designation, intension, comprehension, sense, extension, reference, and representation, are confused with the ontological relations: the whole to part, the part to whole, and the hierarchy relations, subordination and super-ordination.

So, in the EIS model of relations supported with the axiomatic system composed of the existence axiom, the order axiom, the reverse-the-order axiom, the instantiation axiom, and the relativity axiom, all relations are to be distributed into few classes according to their natures and kinds. Some of them hold for any kind of entities: the relations of part/whole and whole/part, and comparison relations as identity (similarity) and diversity (difference), and opposition relations of correlatives, contraries, contradictories, possession and privation. The other remaining four are determined by the nature of the entities related: substances, states, changes, or relations.

As for substances, they are terms of substantial relations: the spatial relations between bodies, the association of ideas, the mathematical relations of objects, and first-order predicate relations, linguistic relations of substantives, the social relationships between the classes and the masses, the relations of affinity or kinship, the business relations among legal entities, the ownership relations of human beings to property, the marriage relation.

States are terms of the stative relations of conditions or states of affairs, the relation of qualities, the magnitude relations of quantities, functional relations.

Changes are terms of the relation of cause and effect, the temporal relationships of simultaneity, contiguity, priority, and posteriority, physical interactions of events, relationships of human actions, activities or acts, as well as the higher-order logical relations.

At last, the relations themselves are the terms of relational proportions or analogies as a ratio of ratios, resemblance or differences of relations, making up the bulk of all possible relationships.

Conclusion

There are relationships in the world with their relative terms, the status, significance, and role of which are defined by the relation itself. We examined the relationship as the universal adhesive thing, formulating its definition and axioms as follows:

Def. Relationship is a prime kind of entity ordering all the things in the world [each carrying a relationship property (as its role)].

Axioms

- **The existence axiom:** Relationship exists really by (in or of) itself, over and above the things connected.
- **The order axiom:** Relationship is the principle of world order.
- **The order reversal axiom:** Every relationship has its inverse counterpart (opposite or complement).
- **The instantiation axiom:** A particular relation is a generic relationship holding among particulars.
- **The comparison axiom:** Everything is comparable as identical, like or unlike.

The definition and axioms define the major species of relation as its subordinate types, like as:

- The relationships in space and in time
- The relationships of cause and effect
- The relations of parts and the whole
- The opposite relations of entities

- The comparative relations of similarities and differences, qualitative and quantitative, direct or rhetorical.

References

Abdoullaev, A. (1999). *Artificial superintelligence*. PDC, Bowling Green State University, Ohio, USA.

Aristotle, Logic (Organon): Categories. (1990). In M.J. Adler (Ed.), *Great books of the western world* (Vol. 1). Chicago: Encyclopaedia Britannica, Inc.

Aristotle, Logic (Organon): Topics. (1990). In M.J. Adler (Ed.), *Great books of the western world* (Vol. 1). Chicago: Encyclopaedia Britannica, Inc.

Aristotle, Physical Treatises: Physics (Physica). (1990). In M.J. Adler (Ed.), *Great books of the western world* (Vol. 1). Chicago: Encyclopaedia Britannica, Inc.

Chen, P. (1976). The entity-relationship model – toward a unified view of data. *ACM Transactions on Database Systems*, March *1*(1) 9-36.

Chrisholm, R. (1996). *A realistic theory of categories: an essay on ontology*. Cambridge, UK: Cambridge University Press.

EDR electronic dictionary. (2005). Retrieved May 22, 2005, from http://www.iijnet.or.jp/edr/E_Struct.html

EuroWordNet top concepts. (2004). Retrieved April 25, 2005, from http://demo.ontotext.com

Hoffman, R. (1995). Monster analogies. *AI Magazine*, 16, No. 3, 11-35.

Hume, D., An enquiry concerning human understanding. (1990). In M.J. Adler (Ed.), *Great books of the western world* (Vol. 36). Chicago: Encyclopaedia Britannica, Inc.

Gunther, G. (1972). Cognition and volition: A contribution to the theory of subjectivity. In *Collected Works of the Biological Computer Laboratory.* Urbana, Illinois, USA: University of Illinois.

Kant, I. (1990). The critique of pure reason. In M.J. Adler (Ed.), *Great books of the western world* (Vol. 39). Chicago: Encyclopaedia Britannica, Inc.

KM component library. (2006). Retrieved April 25, 2005, from http://www.cs.utexas.edu/users/mfkb/RKF/clib.html

Open Cyc. (2003). Retrieved April 22, 2006, from http://www.opencyc.org

OWL web ontology language, overview. (2004). Retrieved April 22, 2006, from http://www.w3.org/TR/2004/REC-owl-features-20040210/

OWL web ontology language, use cases and requirements, W3C recommendation 10 February 2004. Retrieved April 22, 2006, from http://www.w3.org/TR/2004/REC-webont-req-20040210/

OWL-S 1.1 release. (2006). Retrieved April 22, 2006, from http://www.daml.org/services/owl-s/1.1/

Relations. (2002). Retrieved December 26, 2006, from http://web.cs.mun.ca/~ulf/gloss/rels.html

Roget's thesaurus. (2002). Retrieved April 22, 2006, from http://poets.notredame.ac.jp/Roget/contents.html

Russell, B. (1990). The problems of philosophy. In M.J Adler (Ed.), *Great books of the western world* (Vol. 55). Chicago: Encyclopaedia Britannica, Inc.

RDF vocabulary description language 1.0: RDF schema. (2004). Retrieved April 30, 2006, from http://www.w3.org/TR/2004/REC-rdf-schema-20040210/

SUMO ontology. (2004). Retrieved April 22, 2006, from http://ontology.teknowledge.com

SUO 4D ontology. (2005). Retrieved April 22, 2006, from http://www.tc184-sc4.org/wg3n-docs/wg3n1328/lifecycle_integration_schema.html

WebKB. (2006). Retrieved April 30, 2006, from http://www.webkb.org/

Winston, M.E., Chaffin, R., & Hermann, D. (1987). A taxonomy of part-whole relations. *Cognitive Science* 11, 417-444.

WordNet, WordNet 2.0. (2004). Retrieved April 22, 2006, from http://www.cog.sci.princeton.edu/wn/

Chapter VI

What Organizes the World: N-Relational Entities

Introduction

Relationship as a primary constituent of reality setting a mutual order of things in the world commonly involves a large number of the terms or components or arguments that it interrelates. Realistic, actual, factual, existent, or real relationships consist of a multitude of elements, or parts, or terms, thus being polyadic, N-term connections. This implies that a true modeling of N-relational entities should consider their real nature, types, and properties (ontology) as well as the meanings of their correlatives (semantics) and formal attributes (mathematics). The first considers the reality of relations, telling us:

- How they exist, intrinsically and inherently (in the very nature of things) or extrinsically and extraneously (as connections among things)
- What kinds of relational species are there
- How relationships are classified

The second one indicates the primary meanings (definition or intension) with the connotative senses (extension) of the relations, or how they draw their meanings, from the correlative terms or from the relation itself. The third expresses the relationships in terms of mathematical sets, quantities, variables, functions, arguments, and values.

As with a binary relationship, the basic meaning of an N-ary relation consists in the way of ordering of correlative things, while the extension in the number of the things it connects. As it was discussed, the SW ontology languages have a poor expressivity for an adequate representation of real objects, states, changes, and relations. The same observation goes for a recent W3C project on modeling N-ary relations (Hayes & Welty, 2006), analyzed only with respect to the number of terms they connect, disregarding both their reality and such inherent properties as cardinality, symmetry, transitivity, reflexivity. Such a trivial reduction of relations to their correlative terms is rooted in a set theoretical axiom of choice, where a relation is defined as a mathematical function, the set of ordered pairs of individuals, members of two disjoint sets (the domain and range parts). This conception underpins the property construct of the RDF Schema, defined as relationships between resources, viewed as representing everything in the universe of discourse. Instead of defining a class in terms of its properties, the class/property distinction is solved to the benefit of the properties determined in terms of the classes of resource connected as the domains and ranges of properties (RDF Schema, 2004). In other words, relations as properties are simply reduced to the sets of ordered pairs of subsets, exemplified by common attribute-value pairs.

A deeper view comes from a relational logic based on the relations as denuded of their terms, an alternative to the subject-predicate logic based instead on the terms related, and considering properties (predicates) as inhering in substances (subjects). It is as old as Kant's shift from objects (substances) to the relations of substance and accident. Russell further generalized this relational approach as that "it is neither in space nor in time, neither material nor mental; yet it is something" and that "every proposition should be regarded as expressing a relation between two or more things" (Russell, 1990, pp. 272-273). Symbolically, it is written in the from R(a, b, c,...), where the arguments or terms are unknown constants, or as the relational variables that are neither individual variables nor classes, but just objects.

In fact, relationships should be considered with respect to the nature and meanings of its components; for it is a real entity (like interactive causality and space-time relations) rather than an idea, although linguistically expressed by special relational verbs, propositions, common nouns, and relational adjectives. Not all verbs indicate just connections, but they also show *existence* ("be," "exist"), *state* ("have," "possess," "know"), *action* ("do," "change"), as well as *relationships* ("relate," "cause," "represent").

Proposed as another W3C standard, the SW N-relational modeling appears as a trivial approximation of real relationships; for the real Semantic Web grounded to complex nonlinear reality asks for the N-relational ontology of things. As a matter of fact, relations are marked not only by the number of terms they connect but rather by the nature of relationships (or ordering) and the types of the terms related. Therefore, we speak of connections of diverse ontological types: substantial relations (identity and part-whole relationships), quantitative relations (equality or inequality), qualitative relations (similarity or dissimilarity), causal connections, space-time relations; natural relationships, mental associations, logical connections, semantic relations, social links, or cultural ties, and so forth.

Now, the statements (factual sentences) actually express some types of relationship among the terms associated by the agency of verbs, which show also existence, states, and actions. It clearly is fancied wrong to represent the proposition "'man created God and angels in his own likeness"' just by the symbolic form R (a, b, c). Organized by nonlinear relationships, this ever-changing dynamic complex world is more mysterious and enigmatic than the notional world peddled by formal logicians. Besides, we again confirm the golden ontological rule (Aristotle, 1990, Topics, book I, ch. 9): *to be related to real world domains, a core set of predicable relations of "definition," "class," "property," "inheritance," "difference," "sameness," and "statement,", a base of Web ontologies, should be reified in terms of the ontological predicates: the entity classes of reality ordered by real relationships.*

To have a complete description of relationships, the associative entities ought to be distinguished as having (1) real properties as existence (whether a relation exists), nature (what the nature or meaning of it is), type (which natural kind it belongs), direction (what the direction of the connection is); and (2) formal attributes.

The Mathematics of Real Relationships

The necessity of having a comprehensive account of relations may be determined by the fact that currently many computational upper-level ontologies are restricted by the assumption that relations are formal abstractions, without any real status. Namely, they exist only formally as a mathematical relation over n entity sets or as a truth-valued function of n variables, an n-ary predicate. Such an n-ary relation is commonly represented as a subset of the Cartesian product of n entity sets, or as an n-tuple, a finite sequence of entities of length n, where the integer n is called the arity (adicity) of relation. So, for n=1, 2, 3, 4, …, n one will have unary (monadic), binary (dyadic), ternary (triadic), quaternary (quadratic), or a n-ary relation.

As it has been observed, the universality and power of ontology come from its capacity to fully describe all the cardinal types of relationships in a domain of reality, its ability to distinguish the real relations and their conceptual abstractions, logical specification, and formal modeling. All the starting participants of the IEEE SUO project refuse the reality of relations with a risk never to achieve the declared goal. For instance, SUMO allows three relationship types: predicates, functions, and lists, where the first two designate sets of ordered n-tuples, whereas the lists are particularly ordered n-tuples. Considered as abstract objects, under the heading of relation also are brought quantity relations, spatial relations, probability relations, and n-ary (n-term, n-adic) relation. In the Open CYC upper ontology, all relations are the species of mathematical and computational things; thus they are reduced to formal relations: polyadic relation of fixed arity, truth function with predicate, functional relation and scoping relation with quantifier. Shell's Life Cycle Integration Schema gives rather a tautological definition of relationship as "a class of abstract objects whose members are members of relationship." As its specific kinds, the Schema enlists the subclasses of relationship involving individuals and abstract objects: composition, connection, classification, representations of several sorts, assertion, approval, location, specialization, temporal sequence, functional mapping, individual causes, property, and so forth. (SUO 4D Ontology, 2005). Being at the scheme level of conceptual representations of things, devoid of ontological axioms and systematic formalization, as other upper ontologies, such listing is missing the high-order relations, such as comparison, causality, the part-whole relation, or containment. Altogether, all three schemes are missing the clear fact that relations as associations between real-world entities belong to the class of things having full existence, actual and mental or conceptual, over and above the things related. The view of relations as formal properties also is against data schemas widely adopted in information engineering and computing applications, such as the entity-relationship model assuming that *the real world is the totality of entities and relationships* (Chen, 1976).

It must be especially indicated here that there are promising projects of building a universal formal ontology started by the biomedical ontology research Institutes at the University of Leipzig in 1999 and dubbed as GOL, a general ontology language for knowledge domains (Degen, Heller, Herre, & Smith, 2001). The GOL is intended to build ontology as reality representation considering the set-theoretical constructions of general things as one part of the upper-level ontology. Accordingly, the studies of relationships are performed as the entities linking up the things of the real world consisting of three distinct levels—the material, the mental, and the social strata or layers. But because of the trichotomy, the C/C/C dissection of the world as consisting of (i) sets or classes or collections, (ii) universals or categories, and (iii) concrete entities or individuals, the classification system allowed many redundant or nonexistent relationships (Heller, Herre, Burek, Loebe, & Michalek, 2004).

Because reality is the totality of entities and their relationships, its formal description by natural ways requires using the notions of entity sets (classes) and relationship sets (classes), the major constructs of set theory and relation theory. In mathematics, relationship is traditionally defined in the following simple way. Given two sorts of variables X, Y representing two distinct classes of entities, a relationship set is a mathematical relation between the entity variables, written as the set $\{\{X\}, \{X, Y\}\}$. The complex set is made of the singleton $\{X\}$, whose only member is X, and the unordered pair $\{X, Y\}$, the set whose only members are X and Y, like $R_1 = \{\{X\}, \{X, Y\}\}$, where R_1 stands for a binary (two-termed) relation. Alternatively, a binary relation between two entity sets can be recorded as an ordered couple, $<X, Y>$. This is equivalent to a Cartesian product (or direct product or cross product or join product), where the relation is a subset of the product of two entity sets, $R_1 \subseteq X \times Y$, ranging form zero to full relations. In the case R consists of empty sets, it is called an empty relation R_0; while $R_1 = X \times Y$, it is called a full or complete relation. If x is a subclass of class X and y is a subclass of class Y, then the relation also can be written as $R_1(x, y)$ or as xR_1y. If we have a binary relation $R_1 \subseteq X \times Y$, then the class X is the domain of relation DomR, whereas the other part of the couple Y is the range, sometimes called the image of binary relation. Domain is defined as the number of all values that an independent variable (X) of a relation can have, while its range is the number of values that the dependent variable (Y) of the relation can take.

Then taking into account the definition of relations as composed of two parts, associative relationships, that is, the Cartesian product, and the entities connected, that is, entity sets, its adequate description will be written as in: $R_1 \subseteq <X, Y, X \times Y>$. For the case of individual entities, the relationship set is the set of all ordered pairs in which the first element is taken form an entity set X, and the second element is taken form an entity set Y, like as $X \times Y = \{x \in X \& y \in Y\}$. For arbitrary but finite number of entity classes, an n-ary relationship set R can be inductively introduced as a construct that orders an unary relation and an (n-1)-ary relation and described as a sequence of the direct product of its n terms: $R = X \times (Y \times Z \times \ldots \times U)$. Again, for the case of n entities, the Cartesian product will be here redefined as the class consisting of all ordered terms $<x, y, z\ldots w>$ in which x is an element of X, y is an element of Y, z is an element of Z, and w is an element of W. For the special case of identical classes $X=Y=Z=\ldots U$, the n-ary relation will be reduced to $R = X^n$. Again, following the definition, the n-ary relation set, R, may be represented as a mathematical relationship over n entity sets altogether with the components: $R = <X, Y, Z\ldots U, X \times (Y \times Z \times \ldots \times U)>$. Or, it may be presented as an (n+1) tuple $R = <X, Y, Z\ldots U, G(R)>$, where G (R) is a subset the cross product of the entity sets, named and identified as the graph of R and written as G (R)

$= \{<x, y, z\ldots w>;\ x \in X, y \in Y, z \in Z, w \in W\}$, where the entity sets are named domains, and <x, y, z...w> is called a tuple of entities. As we can see, relations can be described not only with reference to the nature but also in connection with the number of entity sets related, or its degree. Now, according to the number of arguments in the direct product, the possible relations may distinguished as zero relation; unary relation; binary relation; ternary relation; quaternary relation; and n-ary relation, a 5-ary relation, a 6-ary relation, and so forth.

In mathematical logic, any n-ary relation with bases X, Y, ..., U can be described by its species, a functional relation, consisting of the domain of classes, the range of classes, plus a function rule associating the values of truths to each n-tuple <x, y, z...w>. As it as is in $f(x, y, z\ldots w) = \{1,0\}$, where the n-ary function f is called a logical (the n-ary) function or the n-ary predicate P (x, y, z...w), a truth-valued function of n entity variables. More general description can be made for multi-valued n-ary logical function $f: X^n \rightarrow A$, where A represents a set of truth values (or a class of statements). Using such a function as mapping from a relationship set into a set of values V or a cross product of value sets, we can introduce the formal notion of an attribute of relationship and its value set, $f: X^n \rightarrow \times V_i$, the particular instances of the attribute obtained by experimentation or observation.

As far as the binary relations are the constitutive elements of any composite relationships, we consider the possible binary associations between entity sets. The dyadic relations may be described as a relationship between entities classes or as a relationship between two entities, individual elements or members, within a given entity class, like the membership and equality relationships. As to the sorts and kinds of binary relationships between entity sets, they are of various types. Most essential of them are equivalency, complementarity, super-ordination, subordination, connectivity and cardinality, and the order type, as in:

- To be equivalent to ($\sim$)
- To be a class of ($\supseteq$)
- To be a subclass of ($\subseteq$)
- To have the same cardinal number (size) as (n , the integer)
- To be in one-to-one correspondence (all good mathematical functions)
- To be in many-to-one correspondence (sentences/topic)
- To be in one-to-many correspondence (universal/instantiations)
- To be in many-to-many correspondence (teachers/students, words/meanings)
- To have the order type ($\geq$)
- To be a complement (reciprocal) of ($^{\#}$)

The first three set theoretical pairing relations are widely used in constructing all sorts of generalization, type hierarchies, while the connectivity and cardinality, the order type, and the complementarity of a relationship seem to be a less employed in ontological data models, regardless of their significance. Regarding the order of connection, the connectivity of relationship represents the one-to-one, one-to-many, many-to-one or many-to-many correspondences of connected entity instances (subclasses or individuals) in the stated relationship. In turn, the cardinality of a relationship measures the possible number of entity instances of one its part connected to a given instance of another part, such that the cardinal of the empty relation set is defined as zero. Then a one-to-one relationship between two entity sets occurs when only one instance of an entity set X is matched, paired, or associated with an instance of another entity set Y. The relation of one-to-many correspondence between two entity sets occurs when for one instance of an entity set X, there are two and more instances of Y. The relation of many-to-one correspondence happens when for many instances of an entity set X, there is only one instance of Y. At last, a many-to-many relationship takes place when for one instance of X, there are many instances of Y, and for one instance of Y, there many instances of X. Thus, to determine the cardinality and connectivity of any relationship, one first needs to specify as existent one entity instance taken from any side of the association, to see how many entity instances are tied to it from the opposite side, then to repeat the procedure from the opposite direction. For the cardinality value of a connection should correspond to the cardinality value of its reciprocal relationship.

Since, in accordance with the essential feature of relations of **order reversal**, any of these relations must be two directional. This implies that the both sides of a relationship should be determined with respect to which one is now acting as the beginning or originating entity as defining the nature and name of the relationship. It is essential that any of these relations is to be inverted with the "increase of being," like in a many-to-many (family or love) relation of marriage: any marriageable man may be a husband (lover) of any nubile woman, as well as any nubile woman may be a wife (lover) of any marriageable man. So, due to the fact of convertibility, we really have two distinct sorts of being here, husband and wife.

Additionally to being formally sorted by the number of arguments, cardinality, and connectivity, the binary relations may be analyzed and classed by the order type, namely by their symmetricity, reflexivity, and transitivity properties, as in:

- Symmetrical, asymmetrical, or nonsymmetrical
- Transitive, intransitive, or nontransitive
- Reflexive, quasi-reflexive, and irreflexive, or nonreflexive

Intuitively, a symmetric relation is a relationship marked by equivalence of both parts and direction, the same relation holds for both directions, like "is equal to" or "is identical to," and so forth. In the case the relationship holds only for one direction failing to hold for the opposite way, we shall have an asymmetric relation, like "is bigger than." When neither the first nor the second example applies, we will come to a nonsymmetrical relation.

A transitive relation is the relation that is transferable along a series of connected terms, so that it allows bypassing the mediate, in-between terms, leaving the relation the same, like "is less than," for example, "less than the less is less than the more," or like "is a cause of," for example, "cause of cause is cause of effect." A intransitive relation is not transferable, like "is mother of," while a nontransitive relation{s} is neither transferable nor nontransferable, like "is a first cousin of."

A reflexive relation, the relation of a thing to itself, is a relationship holding between any entity and itself, like "is of the same quality," while to be irreflexive means to never hold between any thing and itself, like "is bigger than." Keeping the same logic, to be nonreflexive is to be neither reflexive nor irreflexive.

Although each of these properties comes up as independent of each other, and one and the same relation may be nonsymmetrical, transitive, and irreflexive, such as the subsumption classificatory relation, or reflexive, symmetrical, and transitive, such as the identity relation, there are some restrictions here. Namely, if an association is transitive and irreflexive, this relation must be asymmetrical; or, if some relation is asymmetrical, it must be is irreflexive. Generally, each of these types of relation can be axiomatically introduced using the formal laws of relation: (1) reflexivity $R \supseteq \Delta$; (2) transitivity $R^2 \subseteq R$; (3) symmetry $R \subseteq R^{\#}$; (4) asymmetry $R_1 \cap R^{\#} = 0$; and antisymmetry $R \cap R^{\#} \subseteq \Delta$, where $R = X \times X$ and Δ is the identity relation consisting of all the couples (x,x). In more specific terms, the reflexive law holds when for all R, $R \geq R$ is true; the transitive law, when if $R_1 \geq R_2$ and $R_2 \geq R_3$, then $R_1 \geq R_3$; the antisymmetric law, if $R_1 \geq R_2$ and $R_2 \geq R_1$, then $R_1 = R_2$, and the symmetric law, if $R_1 \geq R_2$, then $R_2 \geq R_1$. As to the signification of the order relation, $\geq$, it can be instantiated by a variety of specific orderings: the relation of cause and effect, the inclusion relation, the subsumption relation, the relation of implication, the relation of inequality, the relation of division, the pairing relations, and so forth.

What is important, the combination of these formal properties results in the property of order types distinctive of relations, like as:

- **Quasi-order** marked with reflexivity and transitivity
- **Partial order** distinguished with reflexivity, transitivity, and antisymmetricity
- **Equivalency** characterized with reflexivity, symmetricity, and transitivity
- **Strict order** characterized with transitivity, antireflexivity, and antisymmetricity

The above order types differentiate many important relation types. For instance, the comparison relation being reflexive and transitive has the order type of the quasiorder; nets, trees, chains, or inequalities are species of the partial ordering, while equality, the distinctive mark of quantities, is a sort of the equivalency relation. The strict order can be exemplified by a singular causation or a set of natural numbers. Again, those relations that have the same order type are called isomorphic, like a total ordering and a tree, which is a partial ordering, where for each element the set of its predecessors comes to be totally ordered. It is essential that the order type of a relationship persists to be the same for its complementary, reciprocal relationship. It sometimes is suggested the above properties composing the order type are mere formal and that none of these relations may be realized externally, for instance, reflexive relations as the relations of entities to themselves, or asymmetrical relations. Actually, each of them has its ontological referent, like an asymmetrical relation of parent and child, or all unequal quantitative relations asymmetrically related. If symmetrical relations are always convertible, without "increasing of being," other relations do clearly increase the number of beings, like it does the inverse relation of child and parent.

When being considered apart from the nature and character of the terms, relations may be abstractly analyzed not only with respect to their degree, cardinality, connectivity, and order type, but also calculated by mathematical methods developed in relation mathematics constructed as a formal structure of entity sets, relationship sets (or rules) along with the operations over the relations. So relation algebra is a formal system for manipulating relationship sets (classes or aggregates), where the operands are relation sets, while the operations are the common set of theoretical operations completed with the rules distinct of only relations, like the mathematical process of composition or transitive enclosure.

For the special case of relations of the same arity (the number of components), $R = \{R_1, R_2, R_3\}$, where $R_1 \subseteq X \times Y$; $R_2 \subseteq Y \times Z$; $R_3 \subseteq Z \times U$, the basic algebraic operations over relationships are described by the following well-established set of rules and laws:

The laws of idempotency:

$R_1 \cup R_1 = R_1$

$R_1 \cap R_1 = R_1$

The distributive laws:

$R_1 (R_2 \cup R_3) = R_1R_2 \cup R_1R_3$; $R_1 (R_2 \cap R_3) = R_1R_2 \cap R_1R_3$

$(R_2 \cup R_3) R_1 = R_2 R_1 \cup R_3 R_1$; $(R_2 \cap R_3) R_1 = R_2 R_1 \cap R_3 R_1$

The commutative laws:

$R_1 \cup R_2 = R_2 \cup R_1$; $R_1 \cap R_2 = R_2 \cap R_1$; $R_1 \circ R_2 = R_1R_2 \cup R_2 R_1$

The associative laws:

$(R_1R_2) R_3 = R_1 (R_2R_3)$; $R_1 \cup (R_2 \cup R_3) = (R_1 \cup R_2) \cup R_3$; $R_1 \cap (R_2 \cap R_3) = (R_1 \cap R_2) \cap R_3$

The law of composition or transitivity:

$R_1 \circ R_2 = X (R_1R_2) Z = X R_1 Y \circ Y R_2 Z$

The rule of ternary cyclic relation:

$R_1 \cup R_2 = X \times Y \times Z$

The rule of order relation:

$(R_1 \leq R_2) \equiv R_1 = R_1 \cap R_2$ or $R_2 = R_1 \cup R_2$

The rule of order preserving (isotonicity):

$(R_1 \leq R_2)$ and $(R_3 \leq R_4) = R_1R_3 \leq R_2R_4$

The non-monotonic rule of analogy:

$X (R_1) Y \circ Z (R_3) U = R_1 R R_3$

The rule of transitive enclosure:

$\hat{R}_1 = E_X R_1 E_Y \cup E_X R_1^2 E_Y \cup ... \cup E_X R_1^m E_Y$

The existence of an inverse relation:

$Y (R_1^\#) X = X (R_1) Y$

The existence of an identity relation:

$R_1 R_1^\# = \Delta = E_X \times E_X$

The double inversion law and the double negation law:

$(R_1^\#)^\# = R_1$; $(R_1')' = R_1$; $(R_1^\#)' = (R_1')^\#$

The rules of inversion:

$(R_1 \cup R_2)^\# = (R_1^\# \cup R_2^\#)$; $(R_1 \cap R_2)^\# = (R_1^\# \cap R_2^\#)$; $(R_1 R_2)^\# = R_2^\# R_1^\#$

The laws of negation:

$(R_1 \cup R_2)' = (R_1' \cap R_2')$; $(R_1 \cap R_2)' = (R_1' \cup R_2')$; $(R_1\, R_2)' = R_1'\, R_2'$

The laws of monotony:

$(R_1 \leq R_2) \subset (\hat{R} \leq \hat{R})$

$(R_1 \leq R_2) \subset (R_1^{\#} \leq R_2^{\#})$

$(R_1 \leq R_2) \subset (R_3R_1 \leq R_3R_2)$ or $(R_1R_3 \leq R_2R_3)$

Additionally to the symbolic calculi, the relationship sets permit both graph and matrix representations. So, in terms of graphs, any R can be depicted as the full directed and symmetrical graph whose vertex-set is {1… n} and edge-set is $\{(i, j) \mid i \neq j \;\&\; a_{ij}=1\}$.

Besides, in the language of matrices, a binary relation R_{ij} of two variables X_i and X_j is equivalent to a set of elements (entries or numbers) arranged in the *i*th rows and the *j*th columns, designated as a_{ij}. Then the four characteristic sorts of relation can be specified as the empty relation $a_{ij} = 0$; the full relation $a_{ij} = 1$; the diagonal relation $a_{ij} = \delta_{ij} = 1$ when i=j, and the antidiagonal relation $a_{ij} = 1 - \delta_{ij}$.

All the typical algebraic operations over relations, union, intersection, composition, inversion, and transitive enclosure, are presentable in the language of matrices. The intersection of two relations $R = R_1 \cap R_2$ is expressed as the multiplication of two matrices, for example, with *m* rows and *n* columns. Where the elements in the *i*th row of R_1 is multiplied with the elements in the *i*th row of R_2, and the elements in the in the *j*th column are multiplied with the elements in the jth column: $r_{ij} = a_{ij}b_{ij}$. The union of two relations $R = R_1 \cup R_2$ is equivalent to the sum matrix $r_{ij} = a_{ij} + b_{ij}$.

The product or composition of two relations $R = R_1R_2$ corresponds to the multiplication of matrices where the number of columns of the first matrix is equal with the number of rows of the second matrix: $r_{ij} = \Sigma a_{ik}b_{kj} = a_{i1}b_{1j} + a_{i2}b_{2j} + \ldots + a_{im}b_{km}$, here the summing index k runs from 1 to m. Unlike the product of relations, the multiplication of matrices is not commutative, although associative and distributive over addition. Further, the unit relation corresponds to a unit matrix, a square matrix with zeros everywhere except ones on the main diagonal $r_{ij} = 1$. Expressed by nonsingular inverse matrix $R^{\#}$, the inverse relation is obtained from the original matrix R by the symmetrical reflection to its main diagonal. At last, the transitive enclosure is expressed as the unit matrix $r_{ij} = 1$, if the sets of elements, objects, states, changes, or relations, denoted as X_i and X_j, are in the given relations X_iRX_j, the zero matrix $r_{ij} = 0$ when such is not satisfied. Alternatively, the relation equation can be rewritten as the structural equations describing either direct or inverse relations:

$$X_i = r_{ij}X_j\,;\; X_j = r_{ji}X_i \tag{1}$$

In both equations, the structural coefficient acts as a functional operator transforming the independent variables into the dependent variables, as in: $r_{ij}: X_j \rightarrow X_i$ or $r_{ji}: X_i \rightarrow X_j$. Therefore, the first subscript of the coefficient corresponds to the output or dependent variable, while the second subscript relates to the input or independent variable. It is evident that the complexity of any structural equation is determined by the complexity of its structural coefficient as well as its mathematical decision. So the totality of all transformations of the state space is contained in the structural coefficient.

Summarizing our review, it can be concluded that there is a well-founded formal theory of relations, as the theory of partial and total orderings as well as the general theory of relations developed in terms of set theory and mathematical logic (Shreider, 1971; Alexandrov, 1977; Fraisse, 1986). And any good formal ontology of relations should draw its strength from the relational mathematics, using its basic rules for rigorous description of real world associations and relationships.

On the other side, the knowledge about the nature of relations should constitute the very foundations of any formal view of things, and the entity-relationship schemes claiming to be adequate to the things in world must be based on solid ontological assumptions. In other words, to construct relations as tuples, classes (of sets) consisting of ordered pairs of elements (sets), may be enough for mathematical descriptions, but it is not sufficient for any real analysis. First of all, it is because the entity of relation is one of most generic classes of thing, entity or being, the class of all classes, having its although elusive but fundamental reality. This means that both the classes that ordered and the function that relates, matches, and pairs them represent something real in the world, certain classes of things, objects, properties, and processes. And it seems that the very thing by which other things are related and connected is missing in the mathematical models of relations.

Formally, a binary relation may be distinguished by the number of terms, the sort of set theoretical matching, by its connectivity and cardinality, and by its formal properties of transitivity, reflexivity, or symmetry. But the determining factor remains the nature of the things related and ordered, the character of constitutive terms, since it can make all the difference in assigning this or that formal property. For example, the temporal relation of *before* can be symmetrical, asymmetrical, or nonsymmetrical, depending on the ontological models of time, linear, circular, or something else. Another illustration is the reality of reflexive relations, the identical relations of entities to themselves. Many still doubt to view such relations as realities, assigning them only to the domain of formal objects. For the time being, due to discovery of the new forms of substantial relations between living beings, this concern can be cast away. For instance, today, genetic cloning allows replicating a multitude of clones sharing identical genes from one and the same organism. Much as it has been long practicing in horticulture, where each clone is the same as the original in substance, it is not a replica, copy, facsimile, reproduction, or imitation, but rather a duplicate or counterpart made precisely like the prototype. Because of formal equivalency of mathematical relationship sets and their de facto counterparts, very often real relations are erroneously reduced to functions and predicates of a finite arity, as it is done in the CYC ontology and SUMO ontology. Whereas, in fact, the formal relations, both nonfunctional and functional relations with non-propositional functions and propositional functions, just belong to a sort of mathematical and logical relations in the EIS global taxonomy of associative entities, or relationships.

The Formal Ontology of Relationships: N-Relational Model of Reality

The EIS UFO as the RRR ontology supports the theory of relations in terms of mathematical concepts, and it is purposed to engulf the basic mathematical relations as its logical extension to the domain of formal relations. Within its general entity framework, the formal analysis of relations can be performed by applying the Cartesian product to the world variables, as in:

$$\Re = \langle \mathbf{W} \times \ldots \times \mathbf{W} \rangle = \langle (\mathbf{O}, \mathbf{S}, \mathbf{C}, \mathbf{R}) \times \ldots \times (\mathbf{O}, \mathbf{S}, \mathbf{C}, \mathbf{R}) \rangle \quad (2)$$

We come to the complex notion of **the universal class of multi-relations** of finite cardinality covering all sorts and kinds of external and formal relations, thus not omitting any possible relations. In reality, this ontological structure is embodied as the totality of relations (interrelationships, interactions, bonds, links, couples, connections, and ties) on the class of all things, thus accounting for the structure of the world. Again, by its nature, relations are not only the ways of ordering things standing to each other, or classes of ordered classes, or simply classes of classes of things, they are distinct kinds of reality along with other types of realities. The general relation formula shows that for its complete analysis we need to take into account the bases of relations, what sorts of mathematical entity are related. Given an ontological base and arity, we can analyze the totalities of multi-relations with different bases and denumerable arity into separate relational systems whose elements are binary relations based either on substances or states or actions, or specific relations.

So, in the simple case of a binary world, the universal class relations, $\mathbf{R}_W = \mathbf{W} \times \mathbf{W}$, will be described by the Entity-Entity Relationship Matrix structure consisting of 16 classes of ordered classes of variables of objects, states, changes, and relations, and represented as a tensor quantity fully described by its four components, $\mathbf{R}_{II} = \{O_j, S_k, C_l, R_m\} \times \{O_j, S_k, C_l, R_m\}$. The components of the **universal relative tensor** can be arrayed in a block:

O : O	**O : S**	**O : C**	**O : R**
S : O	**S : S**	**S : C**	**S : R**
C : O	**C : S**	**C : C**	**C : R**
R : O	**R : S**	**R : C**	**R : R**

All ontologically possible sorts of relations are encompassed in the structure, both the heterogeneous and homogenous relations among classes of entities of different kinds. The specific sorts of relations may comprise as its domains and ranges all the basic world's variables ranging over classes of substances, states, as qualities and properties, changes (processes), as actions, doings, occurrences, activities, and acts, or relations themselves. It should be emphasized that the general relation formula also permits all sorts of cross-connections between the classes of substances, states, actions, and relations.

As it has been indicated above, the nature of relationship types is determined by the type of related entities kinds, such as substances, states, changes, and associative relations. When the relational terms are substances, we deal with the first-order relationships among objects, when the relata are states (properties), we have the second-order relationships among states of all kinds and sorts. In case the relata are processes, we are reaching the third-order relationships among changes, and if the associative members look like relations, we are coming to the fourth-order links among different relations. That entails that entity-entity relation classes compose a hierarchy of relation types, like entity classes form a hierarchy of entity types, and that a relation class can be viewed as a member of a relation class only at the next higher level in the hierarchy. Then the parts of a relation, a relator and a relatum, are replaceable but with the order-changing result: the relationship may become the term, and vice versa, by changing the whole order of relation, like it is represented in the recursive formula: $R_{i+1}(X_i, Y_i)$, where the subscript i stands for relational order. The relationship may become a term (or relatum) $R_{i+1}=X_{i+1}$ or Y_{i+1} but only in the relation of a higher order complexity $R_{i+2}(X_{i+1}, Y_{i+1})$, while the relatum becomes a relator not within the same relation where it was an associative entity but already relative to the relata of a lower order complexity, $R_i(X_{i-1}, Y_{i-1})$, where $R_i = X_i$ or Y_i (Gunther, 1970). To climb the hierarchical ladder step-by-step, we start with the first level, when i =1 and $R_1(X_0, Y_0)$. Such a first-order relationship pairing substances (objects) belongs to the large class of substantial relations:

$$\mathbf{R_o} = \langle \mathbf{O} \times \mathbf{O} \rangle \tag{3}$$

This class characterized by that both the domain and the range of a **substantial relation** are things in the sense of substances, objects, agents, living beings, and persons. Instances of substantial relations are ranging from the part-whole relationships of physical, chemical, and biological substances (agents) to all types of relations among human beings, social, political, business, and human relationships. The latter can be illustrated by all sorts of connections possible among human beings, such as personal relationship (bonding and obligation), friendly relationship, acquaintanceship, sexual relationship, love affair, and family kinship, as marital relationship, marriage affinity, and blood kinship, or consanguinity, such as parentage, fatherhood, motherhood, brotherhood, sisterhood, childhood, and lineage; of all of which the most general relationship *is* kind, mankind, or humankind.

Another subclass of the substantial relations is the spatial relations of physical objects (bodies, places) with reference to a place-position or place-direction in space; what is expressed in natural language by spatial prepositions, adverbials of place, and verbs. According to the sorts of spatial magnitudes (dimensions), there are four types of relationships in space:

a. In relation to a point (*at, to, away from, back to, for, toward, through, by, past, beyond, e.g., north, south, west, east* as the four cardinal points)
b. In relation to a line (*onto, off*)
c. In relation to a surface (*on, off, along, across*)
d. In relation to an area, or *space* (*in, inside, out of, outside, through*)

Additionally, there are spatial relations among bodies, such as the relation of contiguity or proximity ("standing-next-to"), or the positional relations of "being {in front of, behind, beside, opposite, with, over, round, between, among, above, below (beneath, or under), far away from, to the north of, to the west of, to the left of, to the right of, etc.}." Of directions, there are six species of spatial relations independent of human perception and having real meanings: up and down; left and right; before and behind (Aristotle, 1990, Physics, book IV, ch. I).

Formally, the spacing of objects (their relative position as the parts of the whole or as the whole objects) can be described by spatial (object separation) function f_S with relation to the class of ordered reference-objects (coordinate system, the 3D space manifold, or the set of ordered parts of space) O; f_E: $O_1 \times O_2 \times ... \times O_n \rightarrow < 2^O, \leq >$. Then the object space can be constructed as a composite of the class of basic things serving as the reference frame of bodies and the extension (position) function, measuring the intervals between bodies, or assigning to each physical object a position in space in relative order. For instance, for the case of three individual objects, we can deduce the betweenness relation as a partially ordered ternary relation where one term is interposed between the other two: $o_1 \leq o_2 \leq o_3$ or $o_3 \leq o_2 \leq o_1$. Thus, all objects are placed (put in a particular relation) in space as to order of its parts, portions, or points, associated with the things spacing, like place, position, post, locality, location, site, situation, spot, or station.

Apart from this, the substantial relations of things can be expressed by a set of predicates and propositions expressing such relationships between two or more things, as in: P_O: $O_1 \times O_2 \times ... \times O_n \rightarrow S_O$. P_O is the n-ary substantial predicate, that which is stated (predicated, affirmed or denied) of an object (or objects), where n is the number of substance variables, O_i is a set of objects, and S_O is a set of substantial statements. A domain of the predicate functions is a class of substantial things, while its range is a collection of statements, for example., about the relative position of objects in space, like as "objects are in space." The value of P at the point $<o_1, o_2 ... o_n>$ constitutes a particular substantial statement referred to a particular object (e.g., place), like as "Moscow is north of London." In the general statement, "are in" is the predicate, while in the particular statement the expression "is north of" is used as the predication. In all other cases, the set of predicates is determined by the sorts of substantial relations.

Next, those relations that are matching states fall under the stative relationships incorporating both qualitative and quantitative relations, and known as the flow or transitional relationships between states. The states related can be as different as states of affairs, conditions, substantial properties, qualities, and quantities or magnitudes. A second-order **stative relation** may be formulated as a relational difference of generic states:

$$\mathbf{R}_S = \mathbf{S} \times \mathbf{S} \quad (4)$$

The world's variable **S** stands for the entity state space constructed as the totality of state spaces of all entities, where each state space is the aggregate of lawfully possible states of a thing; while a state of a thing is a collection of its properties (qualities and quantities). In that the genus of state constitutes the basic part of a property's definition, the meanings of property and its species, quality and quantity, will be determined as follows: property is an essential state (condition) shared by all members (subclasses) of a class; quality is a distinguishing state of entity type ("the differentia of the essence"); quantity (quantum, amount, measure) is an essential state that is subject to measure. The sorts (projections) of the state entity space are as different as the types of spaces: physical space (one, real and objective), mathematical spaces (multiple and conceptual), and mental (cognitive) spaces (multiple, real but subjective). The class of stative relations constitutes the functional laws describing various sorts of property relations, magnitude relation of quantities, functional dependency of states, and qualitative relations. This means that the functional laws of reality are relations between states formally represented by state functions. For instance, all possible relations of physical magnitudes (spatial magnitudes, size, shape, mass, weight, temperature, pressure, etc.) are specific manifestations of the stative relations of quantities marked by equality or inequality.

The quantity relationships may be of three types: ratio, rate, and scale. The ratio as a sort of relation in respect of size between two magnitudes of the same kind is expressed as a quotient and includes proportion of a part to the whole. Two magnitudes of a continuous quantity have to one another the ratio that a number has to a number. It may be as different as *Mach number, mechanical advantage, relative humidity, index of refraction; stoichiometry and mass ratio; intelligence quotient; turnover, efficiency, productivity, price-to-earning ratio;* and so forth. The rate is a rate of change over a time and shows the relative changes as a function of time. The sorts of rate are as follows: *tempo, flux, frequency, speed, acceleration/deceleration, flow; date rate; pulse rate, rate of respiration, metabolic rate; growth rate, inflation rate, index numbers or indicants, stock index, business index, price index, leading indicator; mortality and fertility.*

In terms of functional logic, the state relations of things can be expressed by a set of predicates (functional) P_S and propositions expressing the relationships between two or more states, as in: P_S: $S_1 \times S_2 \times ... \times S_n \rightarrow S_S$. P_S is the n-ary functional predicate, assigning to the class of ordered states a set of lawful statements. If a domain of the predicate functions is the Cartesian product of state spaces, its range is a collection of general statements about their interrelationships. The value of P at the point $<s_1, s_2...s_n>$ constitutes a particular statement referred to a specific state.

If the stative relations account for all forms of functional laws of reality, the third-order relations among changes, **change relations**, are accountable for causal laws and temporal relations:

$$\mathbf{R_C = C \times C = R_S \times R_S} \qquad (5)$$

Being responsible for the workings of the world, the (causal) relations of changes appear to be the most fundamental relations of reality; therefore, the topic needs a careful examination, which will be given later. Now two remarks are necessary: first, it is important to discern a causal connection between change entities and its possible instantiation, singular or particular causation. Second, the reason why causal laws of reality usually are formulated as functional laws of quantities can be seen from the formula for the causal relations reducible to the terms of states. In the language of change relations, we can give an account of temporal relations, a significant species of the **ontological relation of causality**, as measured by the change duration function: f_D: $C_1 \times C_2 \times ... \times C_n \rightarrow < 2^T, \leq >$. The duration function is mapping from a sequence (succession) of changes to time intervals of the time span of a reference clock—process of change, a fundamental time scale T, which is a partially ordered change space $C_C = < T, \leq >$. Such a series of related changes is referred to an ordered set of times as being a time scale, that is, the total duration function is a finite sequence of functions, each of which represents the duration of a specific change $f_D(c_1)$, $f_D(c_2)$, $f_D(c_3)$ … $f_D(c_n)$ relative to a fundamental time scale (a potential clock-thing). Depending how distinct changes relate to each other, the temporal order of changes may be as simultaneity and precedence or succession, like priority (before relation) and posteriority (after relation), contiguity and non-contiguity; while the divisions of time are determined by the changes that have happened (past), the changes that will happen (future), and some distinguished change (present).

Thus, if there is no changing substances, no time as no substantial objects, no space. Or, changes in time are contained and measured by time as objects in space are contained and measured by space. Besides, time as a way of ordering of changes and space as a way of ordering of objects are unified as the four-dimensional space-time (continuum) described by the composition of the extension and duration functions: $f_E \circ f_D$: $C_1 \times C_2 \times \ldots \times C_n \rightarrow < 2^{O} \circ 2^{T}, \leq >$.

As to the logical representation, the relations of changes can be expressed by a set of change predicates P_C and action propositions expressing the relationships between two or more changes (processes, actions, or activities), as in: P_C: $C_1 \times C_2 \times \ldots \times C_n \rightarrow S_C$. P_C is the n-ary changes predicate, attributing to the class of causally ordered changes a set of action statements. If a domain of the predicate functions is the Cartesian product of change spaces, its range is a collection of general statements about their possible relationships. The set of causal statements is a set of change statement. The value of P_C at the point $<c_1, c_2 \ldots c_n>$ constitutes a particular statement expressing the connection of specific changes.

At the top of relations comes a composite, fourth-order relation of relations, meta-relation, exemplified by proportionality or analogy, matching substantial relations, stative relations, and causal relations between each other:

$$\mathbf{R_R = (R_O, R_S, R_C) \times (R_O, R_S, R_C)} \qquad (6)$$

In according to the formula, any analogous relation is of three kinds: "As one entity (substance, state, action) is related (or is in) to a second entity, so is a third related (is in) to a fourth." It is said that we can relate anything to anything on the basis on anything, but mostly interesting analogies are those showing likeness of things belonging to different classes, like in the four formulae: (R_O R_R R_S); (R_O R_R R_C); (R_S R_R R_O); (R_S R_R R_C), and vice versa, showing differences of things belonging to the same classes, as in (R_O R_R R_O); (R_S R_R R_S); (R_C R_R R_C).

It follows that the numbers of high-order relations are as different as the numbers of ways in which the elemental classes of things (substances or objects, states or properties, changes or actions, or relations of various sorts) can be related to each other, with regard to the order of their relations. With regard to the order in which things are related, the number of possible combinations (or permutations) of different classes of things is much greater then the number of possible combinations taken without regard of the order in which things are connected. When we have a finite sequence of kinds of entities of length n, which is an n-termed, n-adic, or n-tuple multirelation, the well-known counting formulas from probability theory for permutations ${}_nP_k$ and combinations ${}_nC_k$ can be applied, where n denotes the number of things connected, k is the number of things taken at a time as correlatives. For instance, using the formulas, the number of relational permutations of three fundamental classes of things taken two at a time is amount to 9 possible kinds of high-order binary relations, of which the number of different relations between substances, states and changes is 6, plus another three identity relations between the same classes of different species of things not covered by the counting formula. As a result, the number of possible high-order analogies will amount to the sum 72 + 3, of which 3 is the number of identical proportions (R_O R_R R_O); (R_S R_R R_S); (R_C R_R R_C) within the same classes. Increasing the number of really possible relations at least two times, altogether, we numbered 86 (9+75) higher-order relations between classes of things that comprise a numberless multitude of lower-order relations as their parts and copies presupposing their existence.

On the whole, it appears that a formal ontology language should involve as its essential part a general theory of relations constructed as algebra of classes, since the class of all relations **R** forms a Boolean algebra under the unary operations and binary operations over relations, plus composition and transitive enclosure:

$$\mathbf{R}_W \subseteq \langle P(R), \subseteq, \cup, \cap, ', {}^{\#}, \circ, {}^{\wedge}, \mathbf{0}, \mathbf{R} \rangle \tag{7}$$

Here $\mathbf{R}_W$ as the totality of relations is the universal class of multirelations of variable arities, with the power set lattice P(R) is isomorphic with 2^R, that is, the totality of relations includes all sorts of links, bonds, and connections among entities of various natures and kinds. The universal class of relations has the semi-lattice structure with least element **0**, a unique relation that is contained in every relation, and last element **R**, another unique relation of which every other relation is the part. There are six algebraic operations used in the multirelational system of 2-tuples, 3-tuples, or n-tuples. Two of them are unary operations, three are binary operations, and one an ordering relation. The unary operations over relations include both inversion $^{\#}$ and negation $'$, while under the binary operations over relations fall intersection (meet) operation $\cap$ and union (join) operation $\cup$ plus the operation of composition $\circ$, together with the special operation of transitive enclosure $^{\wedge}$.

In the view of the fact that the universe of being is made up of four kinds of entities, $\{\mathbf{W}, W_I\} = <\{\mathbf{O}, O_j\}, \{\mathbf{S}, S_k\}, \{\mathbf{C}, C_l,\} \{\mathbf{R}, R_m,\}>$, a dyadic being will number 9 ordered 2-tuples constructed of the pairs: $O_i : O_i$; $S_j : S_j$; $C_k : C_k$; $O_i : S_j$; $O_i : C_k \ldots O_i : C_k$, excluding meta-relations. Next, the power set of 9 binary relations will amount to $2^9 = 512$; the binary world looks to be not so complicated. But in a triadic world $W_I \times W_I \times W_I$, the combination of triadic relations, like semantic relations of names, constructs, and objects, had better be processable by machines; for having 27 ordered 3-tuples of things, the power set of triadic relations will number 2^7= 134, 217,728. Now taking into account that the constituents of the higher-order relations are the world's variables ranging over the classes of entities, one can figure out the number of possible lower-order relations among species and individuals of the entity classes.

The meaning of the mathematical operations over relations can be shown for the simple case of the dyadic world, where all the entities are associated with the relations of the dimension arity of two. This means that the universal class of relations can be realized as a lattice-ordered structure defined by binary operations. Then the ontological relational structure can be described as the power set lattice <$2^{\mathbf{R}}$, ⊆, **0**, **R**>, ordered by the inclusion relation ≤, with two universal bounds, the elementary (unit) relation **0** = $\mathbf{R_0}$, the negation of which is the universal relation **R**, which are the extremities of any relation **0** ⊂ R ⊂ **R**. The inclusion relation also can be expressed by binary operations: R ≤ **R** is equivalent either to R∪**R** = **R** or to R∩**R** = R. Besides having the universal elements, in the relational lattice each specific relation R has a complement (inversion) $R^{\#}$, both constructing the least upper bound R ∪ $R^{\#}$ and the greatest lower bound R ∩ $R^{\#}$ (see Figure 8). As to the universal bounds, the null relation **0** is characterized by two equivalent laws **0** ∩ R = **0**; **0** ∪ R = R, meaning that there exists the minimum relation entity embedding in any sort of relations, while **R** is defined by R ∩ **R** = R; R ∪ **R** = **R**, meaning that there exists the maximum relation entity such that every other relation is part of it. In the diagram, the lattice consists of four essential parts: two union-irreducible elements, plus the bottom and the top elements that are presented by the nodes of the graph, whereas the slope lines reflect the ordering relationship between correspondent correlatives. The bottom element corresponds to the null relation **0**, and the top element to the universal relation **R**, and any relation in such structure has an inverse, such that two properties hold: R ∩ $R^{\#}$ = **0**; R ∪ $R^{\#}$ = **R**. This general relational structure can be equivalently presented as the relational system [**R**, ⊆] or as the algebraic system [**R**, ∩, ∪], where **R** stands for the universal class of all particular kinds of links, bonds, ties, or

Figure 8. The lattice of relations

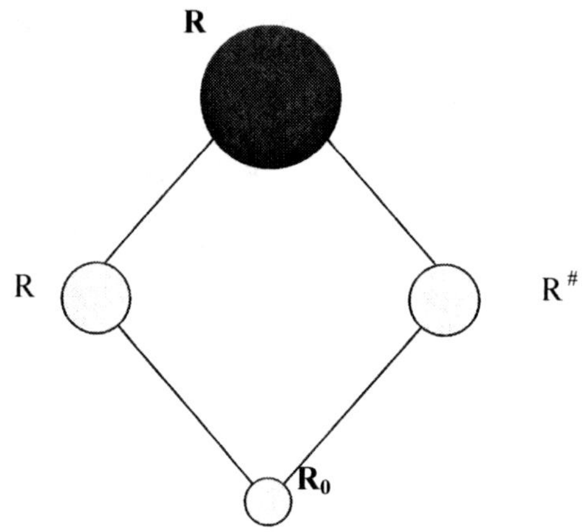

Table 2. The meaning of relative operations

∩	**0** R $R^{\#}$ **R**	∪	**0** R $R^{\#}$ **R**
0	**0 0 0 0**	**0**	**0** R $R^{\#}$ **R**
R	**0** R **0** R	R	R R **R** R
$R^{\#}$	**0 0** $R^{\#}$ $R^{\#}$	$R^{\#}$	$R^{\#}$ **R** $R^{\#}$ **R**
R	**0** R $R^{\#}$ **R**	**R**	**R R R R**

connections: spatial, temporal, causal, etc. Since for any set of the universe of relations it is true that $0 \le R \le \mathbf{R}$, the class of the forward relations R and the class of the inverse relations $R^{\#}$ are combined by the operations $\cup$ and $\cap$, as in Table 2.

The operations in the table give us some remarkable results: the universal class of relations is constructed as the union of R and its converse relation $R^{\#}$. The result of which yields a complete ordering in the form of a cyclic relation having the finite length 2n-1, if the finite sequence of the set of possible worlds is of length n, as in:

$$\mathbf{R} = R \cup R^{\#} = (W_1 \times W_2 \times \dots W_n) \cup (W_n \times W_{n-1} \times \dots W_2 \times W_1) \quad (8)$$

The formula means that universal class of relations is constructed as the union of all direct and inverse relations. For the simple case of binary relations, the full relation can be exemplified as the closed binary relation: $R_1 \cup R_2 = (X \times Y) \cup (Y \times X)$, which is defined as a partial ordering, whose members are mutually comparable. For the case of ternary relations $(X \times Y \times Z) \cup (Z \times Y \times X)$, we obtain a circular chain of three entities X, Y, and Z such that $X \le Y \le Z$ or $Y \le Z \le X$ or $Z \le X \le Y$.

Conclusion

The consistence of complex and dynamic reality comes from its nonlinear relationships, bringing order and organization to all the contingent things in the world, liable to the fundamental principles, invariants, constraints, and universal patterns of unvaried validity.

To have a complete description of existent relationships, any **relation** as **the interrelationship among things** must be distinguished as having (1) real properties as existence (whether a relation exists), nature (what the nature or meaning of it is), type (which natural kind it belongs), direction (what the direction of the connection is), and (2) formal attributes as degree, cardinality, connectivity, symmetricity, transitivity, reflexivity, and order type. Seemingly, none of current formal upper ontologies, as well as advanced knowledge and data modeling methodologies, is capable of using all these classifications.

Based on the systematic, comprehensive, and axiomatic account of relations, the EIS global ontology is able to flesh out its skeleton by embracing all the relevant relationship types proposed by extant lexical taxonomies and upper-level ontologies.

Constituting the core of the real world ontology and its representational language, a comprehensive formal account of complex relationships takes into account three basic dimensions: ontological, semantic, and mathematical, appearing as diverse aspects of the same thing, a polyadic real relation. As a result, the integral combination of a universal ontology and lattice theory may result in the standard relational model giving a consistent and complete theory of things endowed with an expressive representational power, deductive strength, and predictive capacity. The general structure of the N-relational ontology platform may serve as a representational and inferential frame for advanced knowledge-level software agents and cognitive machines able to sense the real world of causality, time, and space, where causality is the most significant kind of relationship, having as its components the interrelated causes and effects.

References

Aristotle, Topics. (1990). In M.J. Adler (Ed.), *Great books of the western world* (Vol. 1). Chicago: Encyclopaedia Britannica, Inc.

Aristotle, Physical Treatises: Physics (Physica). (1990). In M.J. Adler (Ed.), *Great books of the western world* (Vol. 1). Chicago: Encyclopaedia Britannica, Inc.

Chen, P. (1976). The entity-relationship model – toward a unified view of data. *ACM Transactions on Database Systems*, March 1(1) 9-36.

Degen, W., Heller B., Herre H., & Smith B. (2001). GOL: Towards an axiomatized upper-level ontology. In *Proceedings of the International C onference on Formal Ontology in Information Systems*. (pp.34-46). New York, NY, USA: ACM Press.

Fraisse, R. (1986). Theory of relations. In *Studies in Logic and the Foundations of Mathematics*, 118. Amsterdam: Elsevier.

Gunther, G. (1972). Cognition and volition: A contribution to the theory of subjectivity. In *Collected Works of the Biological Computer Laboratory.* Urbana, Illinois, USA: University of Illinois.

Heller, B., Herre H., Burek P., Loebe F., & Michalek H. (2004). *General ontology language (GOL): a formal framework for building and representing ontology.* Retrieved April 25, 2005, from http://www.onto-med.de/en/publications/scientific-reports/gol-no7-excerpt.pdf

Hayes, P., & Welty, C. (2006). *Defining N-ary relations of the semantic web.* W3C Working Group Note 12 April 2006. Retrieved April 22, 2006, from http://www.w3.org/TR/2006/NOTE-swbp-n-aryRelations-20060412/

IEEE P 1600.1 standard upper ontology working group. (2004). Retrieved April 25, 2005, from http://suo.ieee.org/

Open Cyc. (2003). Retrieved April 22, 2006, from http://www.opencyc.org

OWL web ontology language, overview. (2004). Retrieved April 22, 2006, from http://www.w3.org/TR/2004/REC-owl-features-20040210/

OWL web ontology language, use cases and requirements, W3C recommendation 10 February 2004. Retrieved April 22, 2006, from http://www.w3.org/TR/2004/REC-webont-req-20040210/

RDF Vocabulary Description Language 1.0: RDF Schema. (2004). Retrieved April 30, 2006, from http://www.w3.org/TR/2004/REC-rdf-schema-20040210/

Russell, B. (1990). The problems of philosophy. In M.J Adler (Ed.), *Great books of the western world* (Vol. 55). Chicago: Encyclopaedia Britannica, Inc.

Shreider, Y. (1971). *Ravenstvo, shodstvo, poryadok.* Moscow: Nauka.

SUMO ontology. (2004). Retrieved April 22, 2006, from http://ontology.teknowledge.com

SUO 4D ontology. (2005). Retrieved April 22, 2006, from http://www.tc184-sc4.org/wg3n-docs/wg3n1328/lifecycle_integration_schema.html

Chapter VII

What Determines the World: Causality as the Life-or-Death Relationship

Introduction

The world or reality is the totality of things diversified into a multitude of collections of sub-worlds with the characteristic constraints and boundaries; all determined by the fundamental causal laws. Accordingly, the logic of reality is ultimately one; for there cannot be many logics of the universe. Although many personal perspectives allowed, and particular truths, special things and meanings, the global logic of reality is a single universal consistent system relied on a set of ontological classes and relationships, fundamental rules and laws.

The universe is a deep, dark secret, mysterious and mystifying to human minds. And the fundamental challenge to the human mind is to provide a comprehensive account and model that explains in noncryprical terms how this unbounded environment changes and how its basic constituents causally related. And this is the legal responsibility of the UFO as a **causal dynamic ontology,** formulating the causal rules of world behavior as fundamental laws by establishing underlying regularities, uniformities, invariants, and correlations. So guiding science and technology, the task of dynamic ontology is to give us the overall structures, uniformities, patterns, laws, constraints, and invariants within which the many changes in the world take place, finding out what it is outside and inside, defining the natures and essences common to the denumerable multitude of individuals, by classifying the whole universe of things into the prime categories, classes, kinds and relationships.

The centerpiece of the universe is real causality and causation, the driving idea of the UFO; for comprehending the nature, meaning, kinds, varieties, and ordering of causes and effects amounts to knowing the beginnings and endings of things, to uncovering the implicit mechanisms of world dynamics, and to having the fundamental scientific knowledge. As to the real life reasoning applications, understanding of causality and its mechanisms amounts to possessing by the intelligent systems a real life world model enabling all significant inferences, implications, entailments, and predictions thereof.

Of the whole range of real-world relationships, the relation of cause and effect, or causality, happens to be not only the most fundamental relation of reality, but also the most cryptic, mysterious, and challenging issue to the mighty human minds for many ages. It appears, as its long history shows, the arcanum of this vitally important, life-or-death relation dwells in that a comprehensive and consistent analysis of its nature, structure, and meaning can be given only within the general framework of reality. The extreme importance of causality is explained by the eventual purpose of all human inquiry to achieve the causal knowledge about the behavior of the real world, thus providing for versatile intelligent computing systems the common basis for causal representation and reasoning about natural phenomena and mechanisms, social processes, and engineering artifacts.

Fundamentally, causality is a life-and-death matter because of its significance for forming the truthful world view and discovering a set of fundamental laws governing the world. From the social perspective, causality and causally aware machines have a life-or-death significance for the vital issue of humanity survival through existing and newly emerging global, worldwide risks (societal, geopolitical, technological, environmental, and economic) via understanding their causal roots, mechanisms, dynamics, and interrelatedness (Global Risk, 2007).

It should be mentioned that of all the relations in the world the most attention is commonly paid to the part-whole relations, which, however important, make up only one kind of possible ontological relations. It is necessary to escape a widely popular illusion in ontological engineering to identify the taxonomic hierarchies of entity classes with the real ontology of things. Without causal relationships, "the cement of the universe," any general system of things might look a mere empty extension to the logical theory of types.

Humans are beings of various desires and wishes, but only one of them makes us so unique and particular and intelligent, the never satiated thirst for knowledge, driving the mighty minds and intellects from the ancient times. "All men by nature desire to know" the underlying causes and fundamental principles, so that to achieve the cardinal truths and universal learning of the universe, and if not to master but to understand the invariant principles and rules of behavior of this dynamic, emerging, evolutionary, nonlinear, ever-changing world within a unified causal framework.

A Unified Causal Theory: Causality, Inverse Causality, and Causation

By the generally accepted definition, *causality* is the relation between causes and effects. By its primary sense, it is a special sort of relationship, a special category, form, class, or family of associative entities sharing an inherent sameness. The main part of the definition, "causality is a relation that. . .," clearly implies that whatever essentially inheres in the relationship-in-general also must be affirmed of causality as its distinguished species. That is, all the ontological axioms revealing the inherent properties of associative entities apply to causal relations as well:

- Holding between realities
- Being endued with order
- Being able to reverse the order
- Being embodied (instantiated, exemplified) in individual things
- Being comparable as identical, similar or different

Completed with the universal theory of relations, the set of axioms allows us to define *causality*, to disclose what it is, to reveal its essential nature and content, what the name really signifies, or what in reality is the thing signified by the general term of causality, many individual connections or one relational abstraction existing separately of them.

For instance, the **existence postulate** suggests the answer to the ever-first question, whether such a thing, like a causal connection, exists at all, and if yes, how it exists, actually or only as causal beliefs; if it connects, links up, associates, and ties up the real things in the real world, or only human thoughts and ideas. In a sense, the reality of causality enables the being of everything; for the structure of the universe (or any specific class of being) of things is determined by the totality of causal relations in the world (or in the domain of things).

The **order axiom** asserts that the causality relation connects entities in a set and fixed order, where there is the originating entity (named the cause) as well as the terminating entity (the effect).

Crucially, the **reverse postulate** states that every causal relation must have an inverse complement. Since, as with all other relationships, it is like the double-faced Janus, a bidirectional entity where the both directions must be identified. Shortly, *no reverse causality is established, no genuine causality*.

Next, the **instantiation postulate** holds that a particular, empirical causality is nothing else but a general, theoretical causality as applied to individual things, considered in concrete circumstances. It marks as different the **causality**, an abstract relationship between causes and effects, and the **causation**, a concrete relationship of actual causes to actual effects, which is a process, a set of operating changes put in a certain order, succession, or sequence (not all processes are causal, but any causation is a process). Following the last two assumptions, causality can be generally couched in the strict terms of necessary and sufficient conditions: an entity instance is the cause of another entity instance if and only if, there is a universal causal relationship (law or generalization) of the form: the first entity type is necessary and sufficient for the second entity type. The general causal rule subsumes the **reverse causality** formulation: *an event would cause another event just in case events of second type were to cause events of first type*. That is, the fact of reversing the order of entities, when either of the entity instances can be alternatively acted as cause or effect, is the key attribute for establishing whether a given connection is causal or not.

So, unlike the traditional approaches, the main point of the following argument is that, like all existing relationships, causality reverses the ordering, the order of cause and effect, making that which happens precede that upon which it followed. Another point is that if the knowledge of relationships, of their nature, properties, and rules of behavior, constitutes the basis for the whole reasoning about the world, a causal theory of the world will constitute the logical foundation for causal reasoning about all things, thus procuring the design criteria for building a powerful intelligent reasoning system.

The foregoing represents an axiomatic approach to the vitally important problem of our whole living asking for its decision before starting to build any general computing ontology language. Only the axiomatic description of causal relations, involving a general account of relations, can provide a solid theoretical foundation of causality and thus cover a wide range of views from different perspectives: instrumental or agentive, counterfactual, probabilistic, singular, functional, and so forth, putting too much emphasis on one favorite aspect of the law of causes. This wide variety of positions and views even makes some researchers recognize that the eternal problem of causality admits no solution, for it seems to defy a consistent explanation (Kim, 1995; Hulswit, 2004). In fact, the reason why the time out-of-mind issue allows a variety of interpretations can be excused by absence of a single general system of things, a common standard ontology, incorporating the most adequate ideas on causality and causation into the overall conceptual whole, thereby reconciling the particular differences and views.

Currently, major debates about causality revolve around the issue: whether it is a relationship among generic things (kinds, classes, properties, universals, types, and variables) or it is a connection among particular things (instances, individuals, particulars, tropes, or values of variables). Mostly such confusion comes from the misinterpretation of the definitive account of causes, their nature, kinds and numbers, given by Aristotle. Who first claimed that "All causes of things are beginnings" (Aristotle, 1990, Metaphysics); that we have scientific knowledge when we know the cause; that to know a thing's existence is to know the reason why it is (Aristotle, Posterior Analytics). Setting the guidelines for all the subsequent causal theories, by specifying its number, nature, principles, elements, varieties, order, and modes of causation, Aristotle's account of the causes of things is the most comprehensive theory up to now.

According to Aristotle's theory, all the causes fall into several senses, the total number of which amounts to the ways the question "why" may be answered by reference to:

- The matter or the *substratum*
- The *essence*, the pattern, the form, or the structure
- The primary moving *change* or the agent and its action
- The goal and the plan; the intention and the purpose; the *end* or the good or the final reason.

Consequently, the major kinds of causes come under the following divisions:

- **The material cause** is that from which a thing comes into existence as from its parts, constituents, substratum, or materials. This reduces the explanation of causes to the parts (factors, elements, constituents, ingredients) forming the whole (system, structure, compound, complex, composite, or combination) (the part-whole causation).

- **The formal cause** tells us what a thing is; that any thing is determined by the definition, form, pattern, essence, whole, synthesis, or archetype. It embraces the account of causes in terms of fundamental principles or general laws, as the whole (macrostructure) is the cause of its parts (the whole-part causation).
- **The efficient cause** is that from which the change or the ending of the change first starts. It identifies "what makes of what is made and what causes change of what is changed", and so suggests all sorts of agents, nonliving or living, acting as the sources of change or movement or rest. Representing the current understanding of causality as the relation of cause and effect, this covers the modern definitions of "cause" as either the agent or agency or particular events or states of affairs.
- **The final cause** is that for the sake of which a thing exists or is done, including both purposeful and instrumental actions and activities. The final cause or telos is the purpose or end that something is supposed to serve, or it is that from which and that to which the change is. This also covers modern ideas of mental causation involving such psychological causes as volition, need, motivation, or motives, rational, irrational, ethical, all that gives purpose to behavior.

Additionally, things can be causes of one another, causing each other reciprocally, as hard work causes fitness and vice versa, although not in the same way or function—the one is as the beginning of change, the other as the goal. Thus, Aristotle first suggested a reciprocal or circular causality as a relation of mutual dependence or action or influence of cause and effect. Also, Aristotle indicated that the same thing can be the cause of contrary effects, its presence and absent may result in different outcomes.

Besides, Aristotle marked two modes of causation: proper (prior) causation and accidental (chance) causation. All causes, proper and incidental, can be spoken as potential or as actual, particular or generic. The same language refers to the effects of causes, so that generic effects assigned to generic causes, particular effects to particular causes, operating causes to actual effects. It also is essential that ontological causality does not suggest the temporal relation of before and after between the cause and the effect, that spontaneity (in nature) and chance (in the sphere of moral actions) are among the causes of effects belonging to the efficient causation, and that no incidental, spontaneous, or chance cause can be prior to a proper, real, or underlying cause *per se* (Aristotle, 1990, Metaphysics).

All further investigations of causality will be consisting in imposing a favorite hierarchy on the order (priority) of causes, like as final > efficient > material > formal, or in restricting all causality to the material and efficient causes or to the efficient causality (deterministic or chance) or just to regular sequences and correlations of natural phenomena (the natural sciences describing *how* things happen instead of explaining the *whys and wherefores*).

If the assumption that causality consists in some kind of relationship among entities is accepted as an established fact, the issue of its distinction (what distinguishes causality from other relationships, which property makes it so special connection, and what sorts of things it relates) makes all the difference provoking many contrary views. As to the components (relata) of causal relations, they are thought as various as the kinds of entities: substances, object or agents; properties or states or facts; actions, events, or changes; alternatively viewed either as types, kinds, and universals or tokens, instances, and particulars; as variables or values of the variables. Some investigators hold that only agents, substances, or living things, may be efficient causes capable of bringing about changes of different sorts, or effects, thus projecting the agentive account of causal relations. Since each cognitive being instinctively perceives itself as the cause of his doings and actions, very often the agency theory reduced to intelligent agent considered as that which causes something to happen to somebody or something else, so that the happening of something is a cause of action of (somebody) something else. To put simply, somebody's changing (manipulating, control or preventing) a cause will result in the changing (manipulating, controlling or preventing) of an effect.

The larger group gravitates toward the so-called regularity theories of causality, introducing the more general idea of cause as the things that happen, events, and allowing it a law-like generalization. It is asserted that "causal laws are universal conditionals, and singular causal statements ought to be instances of them" (Davidson, 1967a). Another variation of this is that "causal dependence is counterfactual dependence between distinct events"; where 'event e depends causally on the distinct event c IFF, if c had not occurred, e would not have occurred' (Lewis, 1986), what was presented as the counterfactual account of causality holding for all possible worlds. More formal version of the regularity theory suggests using the basic notions of probability theory: sets of events (occurrences), relative frequencies, conditional probability function, interpreted as the likelihood of occurrence of some effect, given its cause. Yet the law of inverse probability usually is neglected in the probabilistic version of causality, just asserting that causes increase the probabilities of effects, all other things being equal, *ceteris paribus*. There are researchers who think that causes are facts, or state of affairs, and that the causal connection is a relation between state-of-affairs types, or universals, such as properties and relations, where a certain type of states of affairs gives rise to another type of states of affairs, and singular causation is thought to be the instantiation of a causal connection of universals (Armstrong, 1997). In addition to this, there is an interpretation of causality in terms of mathematical functions, where the independent, experimental variables are identified with causes, while the dependent variables are associated with effects, and where (structural) regression equations, described as an asymmetrical relation among variables, thus again overlooking the inverse functions, represent a causal ordering (Simon, 1952). In terms of variable quantities, there are attempts to give a more general account of causality (Hoover, 2000). For example, the counterfactual view is proposed to be generalized like as: a variable is a cause of another variable in certain circumstances if and only if each instantiation of first variable, which is possible, would cause some instantiation of the second one having the proper spatiotemporal relationships to the first one (Hausman, 2000).

At best, the received notions of causality are revolving around the classic (Kant's) definition involving laws (universality and necessity) and instantiated changes (individuality and temporal precedence). Following this interpretation, it is widely accepted that causality should postulate three necessary conditions: temporal antecedence, spatial contiguity, and general laws, whereby the happening of an entity (a physical object, phenomenon, situation or event) of a certain class is contingent upon the occurrence of an entity of another class (Born, 1949). Specifically, an event is the cause of another event if and only if, a universal law exists so that events of first type are necessarily followed by events of second type. Additionally to this, the absolute majority of researchers definitely incline to think that the notions of cause and effect involve change-like entities: actions, activities, events, processes, doings, and acts. So there is a general understanding that the terms of the causal relations are changes, rather then states or substances, and it is such a type of relations of processes that is both generic and genetic (productive of changes). Unlike other sorts of relative orders, in the causal relations, the things related (connected, bonded, linked, or coupled) are distinguished not only as action entities, but also as changing (causing, determining, effecting, producing, influencing, controlling, or preventing) each other, or at least as the one of them acting upon the other. Thus, generally, causality is thought as a relation of changes involving causation or causing, the production of an effect by a cause, a productive connection of change-like entities given in a certain fixed order of direction. This means that unlike other higher-order relations, the substantial relations of substance, the relations of state (of property, quantity and quality), the relations of change (actions, activities, operations, and processes), the meta-relations of relations, causality is a generic relation among kinds of changes (processes, actions, activities) ordered as causes and effects.

An essential point where there is a strong disagreement as to which of the following types of causality is fundamental:

- **Empirical** (contingent, particular, quantitative, temporal, domain-dependent, special-case, concrete, token-level, asymmetrical) **causality**
- **Universal** (necessary, generic, qualitative, timeless, spaceless, domain-independent, abstract, theoretical, nomological, type-level, symmetrical) **causality**

In the higher domain of entity types, causality relates change variables of different types, whereas in the world of particulars and concrete things, it has the character of a strictly ordered relation registered via the values of change variables (i.e., events, happenings, occurrences, or outcomes). Here, causality takes the form of an asymmetrical and genetic relationship between concrete events occurring in a localized time and space. Shortly, a singular or specific, or particular or token causality is all about how *this* causes *that*, how something concrete makes something particular happen, or how anything individual causes to do something specific. Under various guises, such a conceptually naïve interpretation appears to rein the extant AI applications of causal models, such as *causal ordering* (Iwasaki & Simon, 1986), *mythical causality* (Kleer de & Brown, 1986), *flow graphs* (Heise, 1975), *rational actions* (Ortiz, 1999), or *Bayesian networks* (Pearl, 2000, 2002), and *functional representation* (Chandrasekaran, 2004).

The popularity of local causality commonly justified by the influence of Humean position that the world is a series of contingent changes, events, connected by three types of relation: *Resemblance*, *Contiguity* in time and place, and *Cause or Effect,* where the cause is viewed as *"an object, followed by another, and where all the objects similar to the first, are followed by objects similar to the second."* (Hume, 1748, section VII). Of all the seven types of relations governing such a world, resemblance, spatiotemporal relations, causation, identity, contrast (contrariety), number, and quantity, the local causality was taken to be the most fundamental species of relation. Such proposition is seemingly grounded on the ostensive properties of causal entities—being spatially and temporally ordered and stuck to some particular thing, place, and time. Whereas "the understanding of causal relations requires the extension of our ontology beyond the spatiotemporal domain of objects and events. . ." (Vendler, 1967). This clearly suggests that any particular relations, causal or spatiotemporal, are just instantiations of more ontologically fundamental relations denuded of its singular correlatives. The discovery of which involved abstracting the essential natures from individual, concrete, and specific realizations by striping away the spatial-temporal limitations and non-relevant particular circumstances.

Ignoring a highly abstract (timeless and spaceless) perspective of causality ended up by neglecting the phenomenon of **reverse** (converse, reciprocal, transposed, reversed, contrary, or opposite) **causality**, where the entities of a given causality are reversing the being of their order, turning backward both order and effect, changing to the contrary its direction, as it happens with other real-world relations, and operating as **inverse causation** at the actual level of things. What can be illustrated with the mechanical *power/drive train* that transmits motion (power) from the engine of a car to the driving wheels, and conversely from the driving wheels to the engine.

There are many challenges to causality: universality and particularity, necessity and contingency, spuriousness and correlativity, prior variable, intervening variable, and control variable. But the most essential attribute is the fundamental fact of changing its direction, reversing, which is conspicuously missing from the modern causal theories. Although the idea of reversibility and circularity of causation has old metaphysical roots, like with many other great ideas, as nothing is new under the sun. The scholar who first intuited that all things "return upon themselves" either in number or in species or in genus, that whatever occurs also happens in a reverse order, was Aristotle. Concluding the work "On Generation and Corruption," he signified that in any natural sequence of occurrences, their link, tie, bond, or nexus must be necessarily reciprocal: "it follows that the coming-to-be of anything, if it is absolutely necessary, must be cyclical—i.e., must return upon itself" (Aristotle, 1990, p.440). In other words, the terms of any succession of phenomena so related that if one as the antecedent determines another as the consequent, then the other determines the first one; for "the nexus must be reciprocal" regardless that the sequence includes two or more members. Whence, it follows that "the coming-to-be of any real things is cyclical," due to the property of mutuality and reciprocality of their action or relationship.

Again, this fundamental fact is constantly encountered in many knowledge domains of contemporary science and technology. Browsing the Amazon.com inside book service, one can surface thousands and thousands of documents offering evidence of reverse causality in such diverse subject fields as Psychology, Education, Politics, Multimedia Technology, Communication, Medicine, Culture Studies, Business, Engineering, Archeology, and Sociology. Here are only some relevant phrases and passages: "the happening of a reverse order," "reciprocal causation," "causality from. . . to is found to be stronger than causality in the reverse direction," "reverse the direction of causality," "the causality reverses," "causality might be in reverse," "causality operates in the reverse direction," "the presence of reverse causality," "reverse the causality flow," "reverse the arrow of causality," "to reverse the normal order of causality," "a bidirectional perspective of a cause-and-effect relationship reverses the primary direction of causality," "reverse the normal order of causality," "reverse the cause and effect and see if you can generate a new relation making sense," and so forth. The examples and cases of such relations, where causality should be analyzed in both directions, are as different as follows: *global risks, warfare and natural disaster; exercise and fitness; parent's relationships and parental relations; population's IQ and nation's wealth (or poverty); R&D expenditures and patents applications; poor health and productivity; trade and exchange rate volatility; innovation and productivity; investment and labor; public policy and competition; (managerial) ownership and leverage; majority electoral system and two-party system; taxes and migration; poverty and knowledge; inflation and money supply; project and plan; care work and employment; governance and economic growth; military expenditures and growth of output; corruption and economic performance; investment and corruption; economic fundamentals (growth, interest, and inflation rates) and asset prices; market prices and investor expectations; human perception and the market reality; nationalism and modernism,* and so forth.

Which way does the causation run, either nationalism (patriotism and loyalty) causes modernity or nationalism is determined by modernity? Not immediately evident, but what is at once evident is that the scope of action of the phenomenon may run from the fund management techniques, as "reflexivity hypothesis" (Soros, 1998), to the global risk network (Global Risk, 2007) to the black hole information paradox (Hawking, 2004).

Causal Physics: Natural Processes, Forward, and Reverse

Again, there are remarkable laboratory experiments directly suggesting or testifying to the causal reversal principle for natural processes. And they are diverse as the test of Bell's inequalities with correlated photons, the asexual genetic cloning of a mammal using adult somatic cells, and the rescue of somatic cells from senescent death. At first sight, the experimental tests were designed to illustrate different truths of nature. "An experimental polarization of Einstein-Podolsky-Rosen type correlation" is thought to bear witness of spatial-temporal and causal non-localities of physical quantum reality. The nuclear transplantation test discovered that the genetic material of adult somatic cells does not undergo irreversible changes. The cell-aging experimental research established "a causal relationship between telomere shortening and in vitro cellular senescence"; what became possible by the discovery of the reverse transcriptase subunit of telomerase from eukaryotes. In fact, the laboratory experiments once more tested that the inherent quality of any natural processes to run both forward and backward also applies to both biochemical reactions and quantum physical processes. Bell's inequalities, the asexual genetic cloning of mammals, and the extension of somatic human cells' lifetime, all this does testify to the universal fact of Nature— *the causation reversal mechanism.*

The rule that *any natural process (phenomena) has a converse (relation)*, that any natural phenomenon (physical, chemical, and biological) caused by some previous phenomenon is to be reversed in its order of action, has a profound extent and depth. So the rule states that the opposite of physicochemical phenomena, where physical change produces chemical change, is the chemo-physical phenomena, where the chemical changing inversely produce changes in the states of physical systems. The same happens with biophysical and physico-organic processes standing in a converse relation.

The case of leading science, physics, studying matter and energy and their interactions, can be fundamental to the argument of **causal reversibility** or **convertibility**, though with some reservations. The causality reversal should not be confused with the physical notion of reversibility. For the latter implies reverse changes of a system from a final state to its initial position, like the reversible processes of a mathematical (frictionless) pendulum, where we only observe the inversion of the order of states involving one thing or the inversion of the order of changes again involving one and the same thing, simple or complex. In fact, a real process is to be identified with a causal relation between changes of distinct systems, or distinct processes. Things receiving a change, as passing-away, coming-to-be, alteration, or any another sort of change whatsoever, must be of different sorts; one system may undergo the processes of passing-away, another coming into existence, growth or degeneration, useful or harmful alterations, increase or decrease, forward motion or reverse motion. For the case of the real pendulum, the physical device consists of two distinct objects (systems) undergoing distinct changes, a compressed spring and a swinging pendulum

In physical sciences, theoretical causality, defined as a lawful relationship of general changes to matter and energy occurring to different objects in different places at different times, involves fundamental processes and physical effects, where the input variables function as causes and the output variables as effects. On the other side, physical causation, the working causality, will be the actual productive connections among distinct change occurrences happening to a specific object (or a system of objects) at a unique spatiotemporal location. Note that only this level applies to singular causation, deterministic or probabilistic, as a temporally ordered asymmetrical and irrefexive relation of individual events, acting in forward or backward directions.

The description of physical phenomena is rested on several basic concepts: matter or substance, energy, states, changes, interactions, and systems, the distinct collections of matter. To know the states of systems is to know all the possible properties, measured or observed, while to know the changes of states of systems is to know all the possible actions, alterations and modifications, manifested as different sorts of forces and fluctuations of force fields represented as changes of physical quantities. To know all the possible relationships among changes is equivalent to knowing the results and outcomes of all possible interactions among systems, and thus being able to explain or predict the behavior of complex physical entities (objects and force fields, gravitational, electromagnetic, nuclear) under the action of given forces.

All the basic physical phenomena, as well as the forms of energy, are divided as mechanical phenomena (mechanical energy), quantum-mechanical phenomena (e.g., atomic energy, activation energy, binding energy, and chemical energy), thermal phenomena (heat energy), magnetic phenomena (magnetic energy), electrical phenomena (electrical energy), and electromagnetic fields (radiant energy), each of which can act as a cause or an effect of a causal process. Then, excluding for now chemical changes of substances and taking the simplest case of binary causal relations, we obtain an exhaustive collection (network) of physical processes:

- Mechanical processes relating mechanical changes, motions and forces
- Quantum mechanical processes connecting quantum events
- Thermal processes relating thermal changes
- Electric processes converting electrical changes
- Magnetic processes interrelating magnetic changes
- Electromagnetic processes (radiation or waves) interconnecting electrical and magnetic fields

- Mechanothermal processes generating the differences in temparature by mechanical stresses and the inverse, thermo-mechanical phenomena producing mechanical forces with temperature changes
- Thermoelectric processes causing electricity with heat and electrothermal processes generating temperature change by electricity
- Electro-magnetic processes generating magnetism by the action of electricity and magnetoelectric processes producing electricity by the operation of magnetic forces
- Magneto-electromagnetic and electromagnetic-magnetic processes (including magneto-optical and optomagnetic processes) interrelating electromagnetic radiation and magnetism
- Mechanoelectromagnetic processes (including quantum processes and electromagnetic-mechanical processes (with reversed quantum processes) determining the interrelationships of mechanical energy and electromagnetic radiation
- Mechanoelectric processes (classical and quantum) and electromechanical processes where mechanical stresses generate electricity and vice versa when electricity effects mechanical phenomena
- Mechanomagnetic processes and magnetomechanical processes (quantum and classical) interrelating mechanical forces and magnetc forces
- Thermomagnetic and magnetothermal processes interrelating heat and magnetism
- Thermoelectromagnetic and electromagnetothermal processes interconnecting radiaton and temperature change
- Electro-electromagnetic and electromagneto-electrical interractions pertaining to conversion of radiation into electricity and converse phenomena, transforming electricity into electromagnetic waves

Each process as a relationship of cause (an input change or energy) to effect (an output change or energy) in order to occur requires a specific physical entity, object, system, material, or force field. Now, depending on the physical properties of a given material entity (size, mass, composition, weight, temperature, magnetization, etc.), a given process can be materialized as various physical effects subject to measurements and experimental observations. Being a specification of a physical process liable to the general principle that *any physical phenomena has a reversed process*, any physical effect has a converse, the reciprocal, reverse, inverse, opposite effect, where the change-cause and the change-effect are turned about in order or relation.

It is the universality of the convertibility of natural forces and thence their inherent unity inspired Faraday, Maxwell and Einstein to seek unification of all the physical processes by a single set of physical laws. It is the universality and completeness of scientific laws that enabled all the variety of modern technological applications. It is the universality and completeness of the conservation principles, like the energy conservation and transformation laws, that enable all kinds and manner of productive applications like energy-conversion systems, from chemical batteries to thermonuclear fusion reactors. For it is the conservation law of energy that determines all the energy changes from one form to another: nuclear, radiant, mass, gravitational, kinetic, thermal, elastic, electrical, and chemical.

Currently, there may be about 10,000 physical effects specified by a multitude of physical materials, systems, or force fields, of which the most part happens to represent only one-side effects. On the intuitive level, the idea of convertibility/reversibility in nature complete with the concept of unity of natural forces were guiding principles in Faraday's discovery of magnetic and electric effects. If such is the case, many inverse effects are to be discovered under particular experimental conditions, thus giving new physical laws for new physical devices, technology systems, and machines performing the transformation of physical changes and energy (mechanical, thermal, electrical, magnetic, electromagnetic, nuclear) into each other. Moreover, regardless of their multitude and variety, all the existent and not yet uncovered physical effects are falling into one or another of a few physical processes, as it is briefly indicated below.

The mechanical processes cover the mechanical effects ranging from the transformation of forces and motions to Hooke's law (including nonlinear law) to nonlinear interaction of acoustic waves (e.g., effects of self-action, generation of nonlinear waves, and so on).

The QM fundamental processes contain all the quantum-mechanical effects of the microparticles interactions (e.g., fundamental interactions, reactions between elementary particles, atomic collisions, scattering of photons, quasi-particles, quantum transitions, as well as nuclear reactions of type A (a, bcd) B and reversed processes B (bcd, a) A, here A - a target-nucleus, a - bombarding particles; b, c, d - emitted particles and B- remained nuclei in the forward process). For electromagnetic and strong (nuclear) interactions, the probabilities of forward and reversed processes are equal because of the reversibility principle; namely, the reversal symmetry of motion and the principle of detailed equilibrium for microprocesses clearly are the consequences of the causal reversal law.

The thermal processes embrace all the temperature changes to physical bodies (convection,cooling, heating), flow of heat, thermal conductivity, as well as the effects of thermal equilibrium radiation.

The electrical processes cover all the effects of the dielectric polarization of the material media and current electricity.

The magnetic processes include the effects of magnetization of solids, gases, atoms, molecules by an external magnetic field and also the phenomena binding changings of physical quantities characterizing magnetic phenomena in different physical media.

The electromagnetic processes include all the effects of electromagnetic waves, such as interactions and scatterings (including visible optical, X-rays and γ- rays bands) or the quantum field effects. Such effects are described by the nonlinear theory of fields.

The mechanothermal and thermomechanical processes comprise all thermodynamic, thermokinetic effects and also effects of thermal expansion in solids, gases and liquids; thermomechanical effects in superflowing liquids (He II), and so on.

The thermoelectric and electrothermal processes have to do with electricity generated by differences in temperature and vice versa, involving such physical effects as thermoelectric and electrocaloric effects in solid conductors (Seebeck effect, Peltier effect, Thomson effect, etc.).

The electro-magnetic and magneto-electric processes cover the law of Bio-Savare-Laplace, magneto-electric effects in antiferromagnetics, the law of electro-magnetic induction (in conducting circuits), and so on.

The magneto-electromagnetic and electromagnetic-magnetic processes include Faraday effect, Cotton-Mouton effect in liquids, solids, gases and plasma, Seeman effect in atoms (the main effect in magnetooptics that study an influence of constant magnetic fields on the optical properties of physical systems or objects). Inverse Faraday effect, inverse effect of Cotton-Mouton, and inverse Seeman's effect in the same objects (liquids, solids, gases and plasma) fall under these processes as well.

The mechano-electromagnetic processes include all the quantum electrodynamic effects, like as Compton effect and the inverse Compton effect, Mossbauer effect–nuclear γ-resonance. In the framework of Quantum Electrodynamics, the process reversal law expressed by the equality between the probabilities of forward and reverse transitions. For the classical mechano-electromagnetic processes, it is necessary to note the ponderomotive influence of light on a substance, acousto-optical effects (e.g., soundluminescence of liquids), dynamooptical effects in anisotropic media, effects connected with charged particles emission transporting through matter (Cherenkov effect, friction and transition radiation), external photoeffect, the effect of accelerating of charged particles (electrons) by intense electromagnetic fields—inverse Cherenkov effect , and so forth.

The mechano-electric processes (classical and quantum) and electro-mechanical processes processes include all the electrostriction's effects in solids (as piezoelectric effects in dielectrics, acoustoelectric effect in metals and semiconductors). In quantum physical limit, it is the quantum mechanical electrostatic Aaronov-Bohm effect, including inverse effect and so forth.

The mechano-magnetic processes and magneto-mechanical processes contain all the magnetostriction effects (piezomagnetism) in ferromagnetics and antiferromagnetics, gyromagnetic effects in ferromagnetics (Einstein - de Haas effect, Barnet effect). In quantum-mechanical limits, it is the magnetic effect of Aaronov-Bohm and the inverse to the magnetic effect of Aaronov-Bohm. In the classical limit, there are magnetohydrodynamic effects in conducting liquids (turbulent dynamo-effects), gyromagnetic instabilities in plasma (Weibel effects), and so forth.

The thermomagnetic and magnetothermal processes embrace the effects of changing of an object's temperature, or its internal energy induced by a change of its magnetic state (demagnetization or magnetization), especially well manifested in ferro-, ferri and antiferromagnetics (e.g., magnetocaloric effect). Here, it also is included as the less known thermomagnetic effects in solids and plasma: when the change of a thermal state of matter brings about the change (generation) of a magnetic field.

The thermo-electromagnetic and electromagneto-thermal processes include the effects of light scattering on the thermal fluctuations in gases, liquids, crystals, amorphous solids, and high temperature plasma; thermal (electromagnetic) radiation, Joul heating of plasma by intense electromagnetic field, effects of thermal self-focusing of an electromagnetic beam in different physical media, and so forth.

The electro-electromagnetic and electromagneto-electrical processes involve electrooptics studying the influence of a constant electric field on the optical properties of a substance and also the influence of light on media. Here must be included the Stark effect, Kerr effect, Pokels effect, nonlinear optical effects such as electrostatic self-focusing of an electromagnetic beam, and so forth; electroluminescence in solids and gases, the ventile photoeffect at the metal-semiconductor contact layer, Damber effect in a semiconductor, and recently found effects of electrostatic fields generation in plasma interacting with intense electromagnetic fields, and so on.

Now adding chemical change to the above physical phenomena, we obtain a distributed network of heterogenous physico-chemical processes:

- Mechanochemical processes
- QM- chemical processes
- Magnetochemical processes
- Electrochemical processes
- Thermochemical changes
- Radiation-chemical changes

Each part of the disparate processes is capable to act as the cause and the effect of a forward process, as well as the cause and the effect of the converse process. As an example, electrochemical phenomena consists of chemical changes of substances produced by electricity, while chemoelectrical processes represent the other way around. Then, here additionally to current electricity, static electricity, thermoelectricity, piezoelectric effect, photoelectricity and hydroelectricity we get galvanism, electricity produced by chemical action, as well as the converse chemo-electrical phenomena, when chemical reactions produce electrical fields. For instance, such significant processes for all living systems as photochemical reactions, where chemical changes produced by radiation, infrared, visible, ultravialet, fall into the causative couple of electromagnetic waves-chemical phenomena; both photosynthesis and vision in living things come from the photochemical changes of substances.

The same reasoning applies to biological, or organic phenomena occurring in living organisms causally interacting with physical processes and chemical changes in both directions, like bioelectricity and electro-organic phenomena, connecting electrical changes and biological processes, or thermogenesis and thermobiological phenomena, interrelating heat and organic actions.

Thus, all the forces of nature, mechanical motions and forces, gravity, heat, magnetism, electricity, radiation, chemical force, and biological actions, are causally interrelated, tied together, or mutually dependent, reversing the order of causal action in opposite directions, from the direct to the inverse. Real causality works both forward and backward, without breaching a temporal consistency between the past, the present, and the future. Postulated by the symmetric laws of nature, reverse causality can profoundly deepen all the established notions about the nature of causes and causation mechanisms. Moreover, in the understanding of the physical universe, a decisive role is to be played by reverse causality: a physical Theory of Everything that aimed to unify all the fundamental forces will involve the idea of inverse causality taking the physical processes to run in operation backward as forward.

Today all the news is coming from the so-called reverse science and engineering. As it is said, a dog biting a man is not a news event, the real news is quite reverse and unexpected, a man biting a dog. And it looks hardly to find a knowledge domain or practical sphere where the most innovational and ground-breaking ideas and strategies do not involve the reversing of conventional, normal, or primary order of causality. After all, the intuitive learning of the inverse relationship between asset prices and investor expectations allowed the hedge fund speculator, Soros to make a great fortune for himself and for his Quantum Fund's investors. Traditional economics, with its pricing models and modern portfolio theories, are stuck to the time-honored causal relationship from economic fundamentals to asset prices, while real markets more honor the reverse causality from the asset prices to the fundamentals (interest rates, growth rates, and inflation rates). First who learned that affecting the asset prices (market quotations of stock or foreign currencies) one can change the fundamentals or investor expectations (via buying and selling orders) gets a fortune without earning it, "the temptation to all dishonesty," by spiraling a destabilizing speculation instigating bandwagon effects, crazes, and bank-runs, ending usually in the securities market bubbles doomed to burst. So, both from theoretical and commercial perspectives, there is an urgent need in building a unified formal causal theory explaining the universality and necessity of reverse causality, the reversing principle of causes and effects; for forward and backward causation together produce everything that comes into existence, determining the consistence and order of the dynamic world.

Causal Mathematics: Formal Representation of Complex Reality

Mathematics as the study of quantitative relationships is well fit for a formal description both of ontological relationship in general and of causal relations defined as the productive relationships of (change) variables, $\mathbf{R}_C$. For the fundamental causal relations are the inter-relationships among specific kinds of entity variables, neither object variables nor state variables, but rather change variables, presented in the material world as physical forces involving changes in quantity (state) variables. Because of its nature, the most full formal description of causality can be given in terms of sets, functions, and categories (Mac Lane, 1971). Bearing in the mind that the algebra of causality is determined by the mathematics of ontological relations, a formal definition of a category of causality $\mathbf{R}_C$ is to be obtained by the following postulates:

- There is a class of disjoint change sets C_X, C_Y, C_X ...of a universal change set $\mathbf{C}$
- There is a class of transformations R_1, R_2, R_3, ... of a universal causal relationship set $\mathbf{R}_C$, named functions describing the causal relationships between change sets, namely from the domain change sets to the change range sets
- There exists an identity causal relationship $I_C = \Delta = R_C(C, C)$ which may be associated with each change set $\Delta_X = I_X$: $C_X \rightarrow C_X$
- To each ordered couple of change sets C_X, C_Y in R_C, it is assigned the set of one-to-one transformations R_1 (correspondences, morphisms, or maps) from C_X to C_Y, as well as the set of inverse morphisms R_1^{-1} from C_Y to C_X
- To each ordered triple of change sets C_X, C_Y, C_Z in R_C, it is ascribed a composition function (a law of composition) $R_1 (C_X \times C_Y) \rightarrow R_2 (C_Y \times C_Z) = R_3 (C_X \times C_Z)$

Then it follows that the category of causation is to be figured as a pre-ordered category (with involution $^{\#}$) of two disjoint change sets, C_X and C_Y, together with four morphisms: Δ_X, Δ_Y, $R(C_Y / C_X)$, and $R(C_X/C_Y)$. The structure corresponds to a full graph depicted as a symmetrical graph with all the possible interrelations (arcs) between nodes (loops), as in Figure 9.

Within the universal ontology formalism, the total of causal links can be constructed as the totality of ordered classes of changes. Seeing that causality belongs to the class of causally ordered processes, $\mathbf{R_C} \subseteq \mathbf{C} \times \mathbf{C} \subseteq \Pi \subseteq \mathbf{R}$, the entirety of its connections will constitute the causal structure of the universe:

$$\mathbf{R_C} \subseteq \langle \mathbf{C}, \mathbf{C} \times \mathbf{C}, \circ, \leq, ^{\#} \rangle \subseteq \mathbf{W} \tag{1}$$

The algebra of causal relations on a set of change variables involves the binary operations of set union and intersection, the unary operations of negation and inversion, the order relation, and the operation of composition, all subject to the laws and rules that are valid for relation in general.

In the language of class variables, **the causal relation of changes** (natural, physical, chemical, biological, psychological, and cultural) can be formulated as a system of interconnected change variables:

$$\mathbf{R_C} = C_i \hat{R} C_j \tag{2}$$

Here indexes i and j are subject to conditions: $i \cap j = \varnothing$; $0 \leq i, j \leq n$; R^2 is a composition of relation R such that $C_i R^2 C_j = C_i R C_k \circ C_k R C_j$. The causal formula indicates that the interplaying cause and effect always must be of distinct types of changes and processes.

The relation $\hat{R} = C_iRC_j \cup C_iRRC_j \cup C_iRRRC_j \cup$...being a complete enclosure of a finite class of change sets is represented by a full graph where all the nodes symmetrically linked up with each other by directed arrows. That suggests that the bondage or linkage of things

Figure 9. Causal relationship as a pre-ordered category

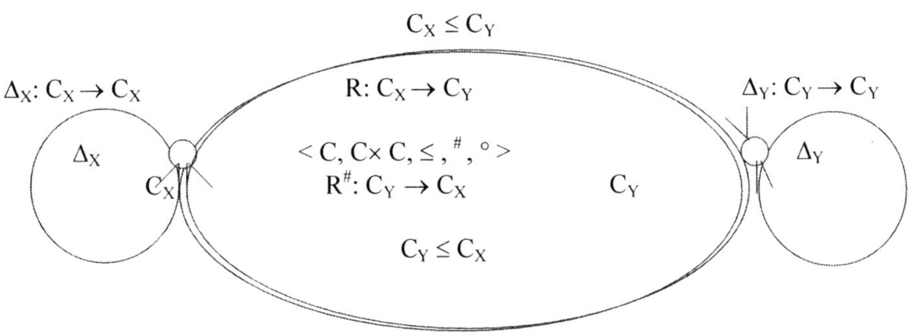

first can be measured as the set of all possible causal links (ties, bonds, and connections) among them, plus spatial and temporal relations. Also, it suggests that there is no such a strictly fixed priority of input causes before output effects like as given by asymmetrical expression C_X x C_Y, where the order of their being is assumed not to be reversed. In fact, things interact, every entity can change (act on) other entities or be changed (is acted upon by) by other entities.

Essentially, the class of causal processes as a species of the class of relations preserves a lattice-ordered structure, too, thereby constituting a partially ordered set of changes subject to the standard Boolean operations, and all other algebraic relations, along with the operation of composition and reversion.

The above rule allows a comprehensive and systematic analysis of causal relations in any knowledge domain due to the generality of mathematical account of causality. For instance, the causal formula for the totality of (causal) physical processes $\Pi_p \subseteq \mathbf{R}_C$ will be represented as $\Pi_p = C_i RC_m$; here indexes i, m and n are subject to conditions: i m=∅; 0≤ i, m ≤ 5 and n=5. The transitive enclosure satisfied due to *the fact of convertibility of natural phenomena.*

For example, in the important case of physics, a full network of physical processes will be expressed as in:

$$C_i \hat{R} C_m = C^M{}_i \, R \, C^T{}_j \circ C^T{}_j \, R \, C^E{}_k \circ C^E{}_k \, R \, C^{Ma}{}_l \circ C^{Ma}{}_l \, R \, C^{EM}{}_m \circ C^{EM} \, R \, C^M{}_i \qquad (3)$$

Where a symbol C^M represents mechanical changes, as motions and forces, C^T refers to all thermal changes and temporal differences, C^E stands for electrical changes, C^{Ma} symbolizes magnetic fluctuations, while C^{EM} is used to represent electromagnetic changes.

Mathematically, change of quantities (or of the state functions f(x), F (x)) is represented either as a differential d, increment Δ, derivative dy/dx with respect to x, or as a variation δ. The causal relationships among physical changes will be then expressed by symmetrical differential equations containing (change) differentials of a state function, such as:

- Maxwell's equations describing the changes of electromagnetic field
- Schrodinger's wave equation, the basic equation of wave mechanics
- Einstein's equations describing the changes of gravitation field

This means that, underlying all the formal interrelations of physical changes in the state variables, reciprocal causal relationships determine the structure of the physical world, its phenomena and mechanisms, as well as the physical artifacts, their structures and functions, designed with a view of realizing the required effects or processes. For instance, the operation and the components of such important devices as electric generators, transforming mechanical motions from primary sources to electricity, and electric motors, converting electric energy to mechanical phenomena using electromagnetic induction, is regulated by two principles—Ampere's law and Faraday's law. Because the same reversing causal process underlies both the laws and the action of the physical artifacts, both electric machines are complementary—motors may work as generators—and inversely, generators may function as motors.

The above examples additionally show that change is always either a cause or effect. And that only changes or processes, not substances or agents, states, qualities, or quantities, but the action of changing of integrity, state, form, shape, magnitude, direction, and so forth, make primary and precise causes of effects, whether resulting from the agency of intelligence and nature or by chance and spontaneity. In a general sense, causes are the active originators of changes brought about for the sake of another change (good, end, reason) by transmitting a form (structure, pattern, energy, motion, communication, information, content, ideas, etc.) from the acting cause to the effect. Such definition embraces all the cardinal types of causality and causation: physical, chemical, biological, mental, and cultural (economic, historical, etc.). And since every change must have as a cause another change, the number of the types of causes, *the matter, the form, the primary mover or agent, the end or the good*, is characteristically close to the structure of change, which principal elements are: *that which immediately causes change; that which is in change or under action; that in which change takes place; that from which change begins; that to which change proceeds* (Aristotle, 1990, Physica, p. 275, 304). That is, *the number and nature of the kinds of cause are determined by the number and nature of the kinds of change.* Generally speaking, there is nothing that can be out of the change influence, that cannot become different or vary and cannot be made different or altered, wholly or in some respect. We speak of a change in integrity, form, shape, magnitude, time, position, character, action, direction, number, possession, title, order, custom, policy, and so forth. The number of particular changes, variations, modifications, or events appear denumerable because of their relations to the class of things the elements of which are individuals or particular objects. Therefore, anything that occurs with individual things, simple or complex, is a contingent change, occurrence, outcome, possibility, happening, or event regardless of its scale, whether it is great historical events or trifling episodes and incidents of daily life. It looks that the numbers of changes are as much as the total of particulars.

Nevertheless, all the multitude of specific changes comes under few heads. For, in compliance with the general scheme of things, at the highest level of reality, change involves only four entities: *substance*, material or immaterial objects; *state*, its species, quality and quantity; as well as *change*, all sorts of actions and events; and *relation* of different types and kinds. Considering the fact that every change is always from something to something else (from entity to entity, entity to non-entity, and from non-entity to entity), the number of the kinds of change is correlated with the number of the kinds of entity:

- Change in substances, or substantial and material changes C_O
- Change in states, of quantity and quality, or qualitative and quantitative changes C_S
- Change of changes, or variable changes C_C
- Change of relations, or relational changes C_R

To the key type of the substantial form of change will refer the radical transformations, such as generation or creation, destruction or annihilation (e.g., deep transformation of matter, birth and death, transmutation of elements, chemical reactions, and making of novel artifacts). To the state change will refer, transition and change of form; various changes of physical states and mental states; development, evolution, aging; action or agency; energy change, increase and decrease in quantity; alteration of quality; change of position or place, locomotion or movement and rest; transportation, transfer, communication; change in time, variation. Major types of change of change, or the manner of action are all sorts of change in motion, force, action, or event, a rate of change. In turn, relational changes include the changes of possession, title, ownership, bonding, family ties, and so forth.

Again, as powers relate to possible effects, generic causes and effects should be assigned to generic changes, thus constituting two modes of generic causation, when the like cause the like: C_O causes C_O; C_S causes C_S; C_C causes C_C; C_R causes C_R. Or, when the unlike cause the unlike: C_O causes C_S and vice versa; C_S causes C_C and conversely; C_C causes C_R and vice versa; C_R causes C_O and vice versa, and so forth. The *vice versa* meaning here "with the causal order reversed" is determined by the **causal reversibility rule** formulized as the equation:

$$< C_X , C_Y > =\# < C_Y , C_X > \qquad (4)$$

Where the function $\# \subset \mathbf{R}^\#$ is *causality reversal operator* $\#$: $R_C \rightarrow R_C^\#$ that describes the operation of reversing the order of two disjoint change classes. The reversion rule can be formulated like as $\mathbf{C_X}$ is a necessary and sufficient for $\mathbf{C_Y}$, meaning:

A change of type C_X would bring about a change of type C_Y if, and only if, C_Y were to produce C_X.

This fundamental principle of change confirms that any fundamental process of reality stands in a relation of convertibility (reversibility), thus accounting for inverted, reversed, reciprocal and interactive phenomena in nature, mind, and society. Alternatively, the principle of inverse causes may be read as: whatever comes into being as a process just as well will come into being as the inverse process. Thus, the relation of causality is transitive, symmetric, and reflexive; for, additionally to the property of *transitivity* of changes (or actions) when *cause of cause is cause of effect*, there also is the other essential characteristic of causality reversibility: the inherent quality of all real processes to be causally reverted.

In the real world, due to the phenomenon of convertibility, processes are arranged in a circle of changes; thus making up the mutual, reciprocal causal relationships, as in: $R_C \cup R_C^\# \subseteq C_X \times C_Y \times C_X$, where the symbol R_C stands for a forward causal process and $R_C^\#$ for the backward causation. As a result, we obtain a full causal relation or a circular causal process made up of distinct changes, C_X and C_Y; of any kind, substantial, stative (quantitative, qualitative), variable, or relational:

$$(C_X \leq C_Y) \circ (C_Y \geq C_X) = C_X \leq\geq C_X \qquad (5)$$

The formula indicates that when the forward causation has come-to-be the backward causation must always come-to-be also. In other words, real causation is nothing but a mutual relation of cause-change and effect-change, that it is a reciprocal process between two or more kinds of changes. For such a tendency is retained for more complex cases of ternary, quadruple, quintuple, or n-tuple circles of causes. So, for the recurring series of actions of the length 3, which is a case of ternary relations $(C_X \times C_Y \times C_Z) \cup (C_Z \times C_Y \times C_X)$, we obtain a circular chain of three changes C_X, C_Y, C_Z such that× $C_X \leq C_Y \leq C_Z$ or $C_Y \leq C_Z \leq C_X$ or $C_Z \leq C_X \leq C_Y$ (i.e., forming a ternary loop or cycle of causes standing to each other in the relation of mutual intermediacy).

Thus, the **reverse causality** enables creation of complex relational entity, the **circular causality**, a reciprocal causal relationship, operating as a mutual causal process, nonlinear causal system, or reflexive, self-referent entity, which is defined as a set (group, Web, network, collection, or body) of changes (actions, activities, processes) that act on each other to form a single dynamic whole. Such a reflexive, self-referent causal entity may include a great number of causal loops, circles of causes, and networks of causal processes. Unlike linear causation marked by linearity and equilibrium, the dual, reciprocal, and mutual causality involves nonlinearity and disequilibrium, becoming the machinery driving the de facto world as nonlinear, ever-changing, and unstable.

Currently, the physical world is a world of nonlinear effects, self-organizing structures, and self-acting, synergetic phenomena, the world of economy is a reality of disequilibrium and nonlinear world, the information world of modern technology is increasingly a reality of self-organized and self-stabilizing systems, self-managed technologies and services. Again, the reversal phenomenon accounts for all vitally important natural cycles, such as geo-chemical, bio-geo-chemical, hydrological cycles, or carbon-carbon cycles, as well as for all sorts of nonlinear technological processes.

Altogether, the principle guides, controls, and regulates the intricate operations of complex dynamic systems, networks, and processes of different kinds and natures: the genetic networks, molecular networks, social systems, information networks, as well as the industrial and technical processes with feedback processes.

Noteworthy that in the differential calculus such causal nonlinear systems are described by simultaneous differential equations of two or more variables, like $dX_i/dt = f_i(X_1, X_2, \ldots, X_n)$, a field of activity of a mathematical theory of dynamic systems involving the notions of state space, phase trajectory, system, singularity, instability, attractor, stochasticity, and chaos. Besides, mutual causal systems as systems of mutual feedbacks are the main subject matter of the so-called second-order cybernetics, generalizing morphogenetic phenomena and morphostatic processes as "deviation-amplifying mutual causal processes" and "deviation-counteracting reciprocal causal processes" (Maruyama, 1963), although with limited understanding of causal reciprocality. Actually, the causal relation of mutual action or influence or dependence occurs not due to simultaneous feedback processes, negative or positive. This happens because of more fundamental process of causal reversibility independently operating in the whole universe and including feedbacks as a special case.

Underlying the mathematical, cybernetic, and AI models, the integrative theory of causality secures a real anatomy of the behavior of nonlinear causal processes as a circular ordering of distinct changes with n-tuples, eventually ending at the point of beginning and starting again at a new level of existence. Such totality of causes appears as an extensive system of circles of changes consisting of denumerable loops of actions, binary, ternary, quaternary, or n-ary cycles of causes, sustaining each other, directly or indirectly. And the character of causal influence in the circularly ordered sequence (series, succession, concatenation) of changes makes all the difference.

As it has been indicated, any entity-entity relationship (as well as causality, a change-change relation) can be described as a sign matrix with n rows (the number of cause-changes C_X) and n columns (the number of effect-changes C_Y). Here the elements c_i r c_j = r_{ij} of the n × n matrix are signs (+1, -1, 0), where the positive sign in square brackets [+] signifies the increase (growth) in the intensity (force, power, strength, degree) of a causal action, whereas the negative sign in square brackets [-] stands for the decrease of action. The symbol → will indicate the direction of a causal agency, while its cancellation will show that no causal relations among (or between) changes were established. At the same time, the positive sign in parentheses (+) points out that both changes occur in the same direction: e_i [+] $\xrightarrow{(+)}$ e_j [+], i.e., *growth causes growth.* Accordingly, e_i [-] $\xrightarrow{(+)}$ e_j [-] means that *fall produces decrease.* In a sense, the symbol (+) can be interpreted as a positive causal influence (effect, control, agency). On the other side, the symbol (-) indicates that *decrease results in growth,* e_i [-] $\xrightarrow{(-)}$ e_j [+], or that *increase brings to fall* e_i [+] $\xrightarrow{(-)}$ e_j [-]. Hence, the negative sign in parentheses (-) can be interpreted as a negative causal effect (influence, impact, action, change). Now, by computing whether increase in cause will produce decrease in effect or vice versa, the intelligent agent will be able to qualitatively establish the negative or positive character of complex changes.

Altogether, for an n-tuple causal relation, there can be two basic varieties of cyclic (causal) processes making the full circle:

Stable equilibrium processes, when the cyclical processes are characterized with an odd number N^S of negative (opposing, contradictory, or destructive) influences (actions, effects):

$$\mathbf{N}^{S}_{\otimes} = (2n\text{-}1)(\text{-}) \tag{6}$$

Unstable disequilibrium processes, when the causal processes are marked with an even quantity of negative influences 2n (-) or any number of positive influence n (+):

$$\mathbf{N}^{U}_{\otimes} = \{2n(\text{-});\ n(+)\} \tag{7}$$

In should be specially noted that the unstable cyclic processes fall within two essentially different subcategories of causal circles:

- "Vicious" circle of changes, which is based on a self-sustaining harmful cycle of cause and effect.
- "Virtuous" circle of actions, which rests on a self-perpetuating constructive loop of cause and effect.

Thus, additionally to the conceptions of **universality**, **necessity,** and **production** (the power to produce changes), the relationship of cause to effect necessarily involves:

- **Transferability, transitivity,** or **compositionality:** Cause of cause is cause of effect
- **Reversibility,** or **convertibility:** The property of real processes to be reversible in either direction

Not surprisingly that James regarded the axiom of transferred relations as the most fundamental law of human thought: "skipping intermediary terms leaves the relations the same." Again, the reversibility of causation, or that all real processes stand in a relation of convertibility, accounts for prevalent circular processes vital for all sorts of existence (the close examples of causal cycles in nature may be geochemical or hydrological cycles, carbon-carbon cycles, etc.) including the processes with feedback in technical systems.

But the decisive thing is here what type of influence, negative or positive, will dominate in a causally ordered sequence (series, succession, concatenation) of different changes; whether an increase in the first cause of a chain process produces a decrease or an increase in the final effect. Directed by the sign rules, the deductive engine of machine intelligence will be able to qualitatively establish the negative or positive character of causal effects. The examples when a resulting change has a destructive character extend from a massive genomic instability of chromosomes leading to cell aging as far as to the explosion of a star. At the same time, a variety of illustrations when circular processes have a constructive character can range from the growth of healthy cells to the formation of a new star.

Since the cycles of changes, as causal circles (loops), constitute the core of the real changing world, all kinds of being (nature, human society, human body, and the mind) do operate because of action of causal processes, stable or unstable. And, it is the great task of ontological inquiry to analyze the both types of processes, their nature and forms, inside structure, all possible chains and cycles, mutual interrelationships of the loops, and, especially, the cause transmission mechanism which as the drive (power) train is transmitting changes along the causal lines of influence in widely distributed networks of causes (see Figure 10).

Thus, unlike linear irreversible processes suggesting a temporally unidirectional cause-effect relation, in the real world of cyclic causal processes, the operation of causation comes back to itself through other causes (changes). For now each change (event, process) has effect on all other changes (events) either directly or indirectly, simultaneously or alternately. In other words, causal lines bending back on themselves constitute the closed causal lines, like Godel's time-like world lines (1949). It appears that the causal circles as the cycles of changes make up the engine of the world's change, where the causation mechanism of any complex dynamic system is tantamount to the sum of functional relationships of changing variables in the given system of things. As far as the world parts and domains are operated either with vicious, destructive cycles of changes or with virtuous, constructive loops of changes, the search of the circular ordering of causes is to be the end of all our inquiries.

Causal Sociology: Formal Representation of Social Reality

An interesting domain to prove the all-embracing totality of causality and the universality of its theory can be social reality shaped by social change with its complex causal workings, and described by social sciences, or sociology in the most general sense. As far as the causal relationships of changes make up complex processes, the connections of social changes form a global network of social processes $\mathbf{\Pi}_S \subseteq \mathbf{R}_C$, as in:

$$\mathbf{\Pi}_S = C_i \hat{R} C_j \tag{8}$$

The change variable C_i includes all the major causal factors producing social change:

$$C_i = \{C_N; C_D; C_T; C_E; C_C; C_P\}$$

C_N represents *changes in the natural environment* {e.g., soil erosion, air pollution, natural resources exhaustion, climate worsening, deforestation, epidemics, as well as their positive counterparts}.

C_D marks *demographic changes* {e.g., the birth rate, the mortality rate, population growth, migration, colonization, and urbanization}.

C_T signifies *technological changes* {e.g., the use of fire, the cultivation of soil and plants, the domestication of animals, the inventions of novel substances, techniques, processes, machines, such as minerals, alphabetic systems, number systems, the plow, gunpowder, the steam engine, the internal-combustion engine, the atomic bomb, the space rocket, the computer, the intelligent artifact}.

C_E suggests *economic changes* {e.g., commercialization, industrialization, collectivization, labor division, market formation, the alteration of property relations, the change in labor organization, or in economic fundamentals}.

C_C refers to *conceptual changes* {e.g., novel scientific, philosophical, religious, ethical, political, and social ideas about the world, nature, life, human society, and man}.

C_P indicates *political changes* {e.g., polity change, structural change, change of leaders, policy change, or war policy, political terrorism}.

In the case of uni-directional social evolution, the sequence of the major determinants of social change proceeds in series: the causation transmission mechanism operates along one-directional line of social changes of increase:

$$C_N[+] \rightarrow C_D[+] \rightarrow C_T[+] \rightarrow C_E[+] \rightarrow C_C[+] \rightarrow C_P[+] \tag{9}$$

In turn, the linear mechanism of translating the social changes of decline, resulting usually in political revolutions, follows the opposite way of decrease:

$$C_N[-] \rightarrow C_D[-] \rightarrow C_T[-] \rightarrow C_E[-] \rightarrow C_C[-] \rightarrow C_P[-] \quad (10)$$

As it stands with all real processes, in cyclic social processes as well, each social change has an effect on all other social changes either directly or indirectly and either simultaneously or alternately. Accordingly there are two basic varieties of social processes, regimes, or systems—stable and unstable. The stable social regime is when the cyclical process is characterized with an odd number of negative social changes. The unstable social order is when it is characterized with an even quantity of negative influences or any number of positive influences. The unstable dynamic social system can consist of vicious causal circles, self-sustaining sequence of causes and effects, when a resulting change has a destructive character (e.g., ecological catastrophe), as in the circle of changes:

$$C_N[-] \rightarrow C_D[+] \rightarrow C_T[+] \rightarrow C_E[+] \rightarrow C_C[+] \rightarrow C_P[+] \rightarrow C_N[-] \quad (11)$$

Or, it is composed of virtuous causal circles, when a system of processes has a constructive character (e.g., the preservation of the natural environment), as in the complex heterogeneous cycle:

$$C_N[+] \rightarrow C_D[+] \rightarrow C_T[+] \rightarrow C_E[+] \rightarrow C_C[+] \rightarrow C_P[+] \rightarrow C_N[+] \quad (12)$$

To demonstrate the efficiency of the causal theory of social processes, we shall see how the intelligent system can qualitatively reason about the causes of economic growth.

There is now general agreement on the primary variables of economic growth. Under the wealth-influencing factors come:

- Natural resources
- Raw materials
- Labor force
- Machinery
- Capital investments
- The fertility of land
- The money supply
- The rate of interest on money
- Inflation rate
- Government expenditure/ borrowing
- Rules or breach of law
- Rate of population growth
- Trade
- Rate of technological advance
- Political stability (the conditions of war or peace)

But there are differences as to how these factors act on each other. The answer, it seems, lies in searching for how *changes communicating causal mechanism* works in economic systems. Given the general theory of processes supported with economic statistical data for the growth-affecting causes, a possibility arises for building a large-scale ontological engine on economic processes.

If to make a summary of the main points of classical, neoclassical, and contemporary models of growth, *the fundamental economic factors (categories, ingredients, variables, causes, determinants, or forces)* can be classified as the following:

- **Labor (L)**, or human capital, measured by its amount and level of knowledge (education, skills).
- **Capital (K)**, measured by the amount of investments, savings , natural resources (mineral resources, sources of fuel, the fertility of land, etc.) or physical assets of all kinds.
- **Government policy (G)**, involving spending on education, health care, public administration, the armed and police forces expenses, investment by tax administration, export credits, saving rates, natural resource exports/imports, taxes on exports, domestic market liberalization, corruption, peace state, warfare, privatization of infrastructure, nationalization, civil-service reform, corporate tax rates, interest rate on money, the money supply, government borrowing, the rules of law or breach of law.
- **Technology (T)**, determined by intellectual resources - science and engineering.
- **Financial market** operating by interest rates **R**, on equities bonds and other debt instruments issued in return for the investment funds.
- **Unemployment of population (P)**, measured by its amount and durability.
- **Consumption, use of goods and services (C)**, determined by basic needs and the expectation of future incomes.

The causal relationship between basic categories of economic change and total production output **Q** can be expressed by the structural equation:

$$\mathbf{f: \{L, K, G, T, R, P, C\} \to Q} \quad (13)$$

Then the change of a national income, finite or infinitesimal, may be represented by the full differential of the total national outcome:

$$\mathbf{DQ} = (\partial \mathbf{Q}/\partial \mathbf{L})\mathbf{dL} + (\partial \mathbf{Q}/\partial \mathbf{K})\mathbf{dK} + (\partial \mathbf{Q}/\partial \mathbf{G})\mathbf{dG} + (\partial \mathbf{Q}/\partial \mathbf{T})\mathbf{dT} + (\partial \mathbf{Q}/\partial \mathbf{R})\mathbf{dR} + (\partial \mathbf{Q}/\partial \mathbf{P})\mathbf{dP} + (\partial \mathbf{Q}/\partial \mathbf{C})\,\mathbf{dC} \quad (14)$$

The first partial coefficient denotes the change rate of **Q** per unit change the amount of labor force **L** while the other variables are held constant. So the alteration of **L** contributing to a national output growth amounts $\delta\mathbf{L} = (\partial\mathbf{Q}/\partial\mathbf{L})\,\mathbf{dL}$. The second coefficient denotes the rate of change of **Q** per unit change of capital **K** while other factors remain unchangeable **K** =

$(\partial \mathbf{Q}/\partial \mathbf{K})\, d\mathbf{K}$. The same procedure applies to the rest variables contributing to the economic output **Q**. Then the basic differential equation for the total outcome may be rewritten as the aggregate of economic variations, as in:

$$DQ = \delta \mathbf{L} + \delta \mathbf{K} + \delta \mathbf{G} + \delta \mathbf{T} + \delta \mathbf{R} + \delta \mathbf{P} + \delta \mathbf{C} \quad (15)$$

The ingredients interactions are to be represented by communicating the appropriate, qualitative differentials of variables $\delta \mathbf{X}$ via causal circular lines. What is here decisive, not the numerical values, but whether these structural coefficients are zero, positive or negative:

$$\delta \mathbf{X} = \{+, 0, -\} \quad (16)$$

The method of qualitative variations allows us to compute changes in quantity and quality. To obtain a full system of structural equations describing the interaction of economic categories and subcategories, it is necessary to add the sign transitivity rules and the equilibrium conditions, such as equation for investment $\mathbf{K}_I$ and saving $\mathbf{K}_S$ and the equality for money demand $\mathbf{M}_D$ and supply $\mathbf{M}_S$:

$$\Delta\,(\mathbf{K}_I - \mathbf{K}_I) = 0 \quad (17)$$
$$\Delta\,(\mathbf{M}_D - \mathbf{M}_S) = 0$$

The rules of change transmission mechanisms also will allow the intelligent program to identify the types of loops (stable or unstable; virtuous or vicious circle) and to advise how to avoid an undesirable taking most preferable scenarios.

Again, the interaction of the main ingredients of economic changes are diagrammatically represented as a full, directed, and symmetrical graph whose vertex set is equal to a number of variables {1 ... m} and edge-set is a number of negative or positive influences $\{(i, j) \mid i \neq j \;\&\; a_{ij} = \pm 1\}$, where a_{ij} stands for the sign matrix of a structural coefficient discussed above. From a total flow graph of causal relations of quantities, the inference engine can select various types of processes—chain-like, reversed, or interactive. For example, among the chain-like series of events, when each change is strictly caused by the previous one, there are the money communicating mechanisms and the income and money models of growth.

a. Money communicating mechanisms make a complex causal process:

$$\delta \mathbf{R}\,[+] \xrightarrow{(+)} \delta \mathbf{K}_S\,[+] \xrightarrow{(-)} \delta \mathrm{C}\,[-] \xrightarrow{(+)} \delta\, \mathbf{I}\,[-] \xrightarrow{(+)} \delta \mathbf{Q}\,[-] \rightarrow \ldots$$

Here the first event denotes the rise of *interest rates*; the second, *savings return*; the third, *consumption fall*; the fourth signifies *investment downfall*; the fifth denotes *output shrinkage*. On multiplying the signs of the structural coefficients, the chain will be reduced to one link:

$$\delta\mathbf{R}\,[+] \xrightarrow{(-)} \delta\mathbf{Q}\,[-]$$

An alternative developmental pathway for the monetary transmission mechanism proceeds through the exchange rate:

$$\delta\mathbf{R}\,[+] \xrightarrow{(-)} \delta\mathbf{M}\,[-] \xrightarrow{(-)} \delta\mathbf{P}_E\,[+] \xrightarrow{(-)} \delta\mathbf{P}_I\,[-] \xrightarrow{(+)} \delta\mathbf{Q}\,[-] \rightarrow \ldots$$

In terms of natural language, *interest rate rise* causes *money appreciation* that effects *exports price growth* that, in turn, induces *imports price fall* bringing about *national output dampening*. The short-cut version is again boiled down to:

$$\delta\mathbf{R}\,[+] \xrightarrow{(-)} \delta\mathbf{Q}\,[-]$$

b. The income and money models of growth, being combined together, form a reversible process:

$$\delta\mathbf{K}_I\,[+] \leftrightarrow \delta\mathbf{Y}\,[+] \leftrightarrow \delta\mathbf{M}_S\,[+]$$

Here the first change is the *investment shifts*, the second is the *income changes*, the third denotes the *money stock* growth. A mechanical reasoning thence consists in a meaningful manipulating of symbols representing such qualitative changes as *unemployment fall* $\delta\mathbf{U}[-]$, *money supply growth* $\delta\mathbf{M}_S[+]$, *capital stock increase* $\delta\mathbf{K}[+]$, *share prices upsurge* $\delta\mathbf{R}_E[+]$, *consumption jump* $\delta\mathbf{C}[+]$, and so forth.

The fundamental difference of *the economic growth-influencing variables* from *natural causes* is that a rational agent is able to control or manipulate these factors. But we cannot avoid the fundamental fact of reality that *any cause is an effect at the same time.* It implies that all economic factors may be both the causes of growth and the effects of the growth. Therefore to give the full list of basic economic causes is hardly possible.

Now, in accordance with the ontological principle of reverse causality, any developed economy is a system of interrelated economic categories where causation runs in opposite direction as much as forward. Being an element of a complex nonlinear causal system, the wealth growth becomes the cause of changes in technology, labor and financial markets, government policies, labor force, the supply of capital, inflows of investment, general domestic product growth, and export growth.

Therefore all economic influences are both the causes and the effects of economic development (or backwardness). In a nonlinear dynamic system, as a modern economic system is, therefore, each factor becomes crucial, as an official corruption, mal-administration and government incompetence may result in a dismal decay of a whole social system.

Causal Reversibility as a Mechanism of the Real World, or the All-Embracing Totality of Reverse Causality

Representing reality as an unbounded dynamic nonlinear system of processes, forward and reversed, with one input and many outputs (results, consequences, upshots, or events), closely reflects the real state of affairs of the world. The cases taken from natural and social knowledge domains clearly demonstrate that we need to review the long-cherished beliefs that "like causes set up like effects, and unlike effects are effectuated by dissimilar causes." For, in fact, "similar causes can bring about completely dissimilar results" due to the principle of **reverse causality**. As it happens with the butterfly effect, when seemingly a nonessential change in a complex system can result in large, global effects affecting the whole system, like as a regional warfare might end up in unforeseen worldwide repercussions. Real world as an effective nonlinear reality is a dynamically chaotic world, or chaotically dynamic universe of entity-causes, originating from Nothing (Chaos) to develop to full Being (Cosmos), by the agency of Real Causality.

It strongly appears the behavior of complex systems and structures in the world, as well as the general patterns of change, is regulated by reversing backward causality making up with the forward counterpart the unique enigmatic phenomenon of **circular causation**. All flow of energy and matter in nature, all kinds of energy transformation and transfer are subjected to the all-comprehensive rule: any fundamental real process is liable to reversing its primary ordering of changes; or the order of operation of real processes turns backward in effect, whether it is the physical processes of water evaporation or a biological process of photosynthesis, or some other natural process. By causal necessity the opposite processes as water condensation or air respiration will complement these processes so as to form cyclic processes. Essentially, the natural mechanisms as the cycling of energy (light and chemical bond energy) and matter (carbon, oxygen, nitrogen, sulfur, phosphorus, and water) determining all the life on Earth are possible due to causal reversion.

So far we have been dealing with the sorts of entities, which viewed as unaffected and independent of time (duration) and space (extension), as existing out of the space-time continuum. But in fact, all particular things are to be placed in a spatial or temporal order, beside of being connected causally. They are always either in place or in time, which, when taken together, make up a special class of relations, spatiotemporal relationships, although with some specific distinctions. Space as a continuous quantity consists of the parts (places), which have position, or rather bear a relative position each to each, having common boundary between contiguous parts. Time is viewed as made up of what is gone by (past), what is at present (now), and what is to come (future); it is a number (measure) of change in respect of the change that has happened (the before relation) and that will happen (the after relation). That is, time is not a change but rather a kind of entity measuring out the whole process of change by means of some scale (fundamental) motion.

Those things that are in place are particular objects, that are constrained, measured, and contained by the place; and those things that are in time are events, happenings, occurrences, or outcomes, that are restricted, affected, measured, and contained by the time. In time, all particular things occur, act, do, and change; in space they exist, dwell, abide, endure, and continue. So, when instantiated or embodied into actual existence, the causal space will be transformed into the space-time (particular) causality R_P, which is the totality of generic causes $\mathbf{R}_C$ endowed with the natural classes of quantities—time-relations R_T and space-relations R_S, as in:

$$R_P = <\mathbf{R}_C; R_T; R_S> \quad (18)$$

As a matter of fact, the singular causation embodying the generic causality becomes a specific relation between concrete events in different individual things marked by a temporal precedence, spatial contiguity, and mass/energy transfer, as a typical individual causation. In the singular causation, causes and effects are particular events or individual changes. Being distinct entities, they involve concrete substances, times, spaces, but generic changes. For, in its essence, any event, or particular change, is an occurrence, actualization, or realization of some generic change. More generally, a set of particular changes, or events, E_X (e_1, e_2... e_n), is a complexity including an n-tuple of particular substances, a kind of change, a time, and a place, like in the equality:

$$E_X (e_1, e_2 ... e_n) = <o_1, o_2 ... o_n, C_X, t, s> \quad (19)$$

This corresponds to **the existence condition** of any event:

an event takes place (occurs, happens, goes on, passes, passes off, falls out, comes about, comes up, comes to pass, or comes into being) when the particular substance (or a system of objects) o_1 *undergoes (or effects) the process of change* C_X *at time t (about, during, within, over a period of time) in (at, by, on, near) place s.*

An event, happening, or occurrence, is then a change in some respect that occurs at a given place and time to some particular object. Or, it is a temporally and spatially expanded change going through states (stages), thus having beginnings, middles, and endings, and transforming itself from one stage to another stage within the same generic change. That is, events not only hold some temporal properties (duration, or length and temporal order, or arrangement in time, pace, or rate of change) standing in temporal relationships to other events, like *before, after, or simultaneity* relations. They also are liable to some changes, moving, spreading, growing, and so forth. Also, events are distinguished by their magnitude consisting of intensity (strength, degree) of phenomena and extent (reach or influence) of actions. If insignificant events are superficial and trivial incidents, the class of significant events includes those that have important effects, consequences or impacts, designated as influential, large, remarkable, signal, noteworthy, meaningful, catastrophic, prodigious, monumental, momentous, epoch-making, world-shaking, or profound changes.

Events then identified as the same when they are defined not only by the same causes and the same effects or by the coincidence of a time or place, but rather by a kind of generic change to which the events considered belong as individual members of the class. Involving substances, states, times and places, events always come from the few basic generic changes, coming as their embodied exemplifications, instantiations, or realizations in certain material conditions. Its particularity and numerosity arise from infinite numbers of individual objects, specific times, and places concerned. Therefor, in order to see through the nature of events as well as the causal relations in which they enter, we have to lift up beyond the domain of particular substances and spatiotemporal relations.

Particular causality is, therefore, commonly defined as a set, series, or succession of events, acts, or modifications ordered temporally and tied up with individuals things (simple or complex; individual, aggregate, or system). That is, in the case of actually operating causes, *the action of transforming, translating, transporting, communicating, transferring, conveying, transmitting changes* refer to specific material substances and immaterial objects, whether it *involves transfer of energy; motion, disease, heat, feelings; information, relation, rights; data, title, ownership, possession, intelligence, or ideas.* The strict formal definition of such a principle of local causality is typically given like this: a change in a thing $x \in X$ at time t, c(x, t), is a cause of another change c′(x′, t′) in a thing $x' \neq x$ at time $t' \geq t$ providing there is energy transfer from x to x′ during the interval [t, t′] (Bunge, 1982). Or, a specific causation is a relation between changes in different things, of which some act as the causes and the rest fall under the effect, where the cause both precedes and produces effect. In more general terms, particular causation takes place whenever a particular change (effect, event, outcome, alteration, mutation, or variation), $c_2 \in C$, is caused by the agency (influence, action, operation), $c_1 \in C$, of a particular substance, an intelligent agent, or by any natural force, $x \in X$.

The actual world is sometimes modeled as a universe of stochastic processes and chance events and probable changes regulated and structured by linear probabilistic causation. In fact, the idea of reverse causality well extends on the case of chance events restricted to particular things, places, and times. To describe a universe of chance changes, the probability metrics is required to be introduced. Usually, particular effects and causes are assigned to contingent changes (events) that happen or come to pass by chance. A typical definition of such a probabilistic causation is " 'event $B_{t'}$ is *prima facia* cause of event A_t if and only if (i) t'<t, (ii) $p(B_{t'})>0$, (iii) $p(A_t/B_{t'})>p(A_t)$, here $p(A_t)$ - the probability of event A in the moment t; $p(A_t/B_{t'})$ - the probability of event A in the time t under the existence of event B at the moment t' " (Suppes, 1984).

Evidently, it is hard to discover the effect of causal reversibility using the concept of time/place restricted singular causation in any its interpretation, deterministic or probabilistic. Although, there are some attempts to look into the reversing of causal ordering in Baycsian networks (Druzdzel & Leijjen, 2001).

As far as the reverse causation admits extension to the domain of chancy changes, probabilistic or contingent events, it is possible to construe a generalized algebra of events in terms of an axiomatic theory of probability (Kolmogorov, 1968). There central concept is a probabilistic space (E, S, P), where E is an event space, S designates a collection of subsets of the set E, and P is the distribution function of elementary events $e \in E$ on S.

Then *the process space* for non-deterministic events can be represented as *the probabilistic process space* where the ontological projection (mapping) function **F** is replaced by the distribution function P. In the mathematical theory of probability, the fundamental law for the mutually connected events $E_X \cap E_Y$ is written as:

$$P(E_X \cap E_Y)=P(E_X/ E_Y)\cdot P(E_Y)=P(E_Y /E_X)\cdot P(E_X) \quad (20)$$

The sense of the functional relation P consists in the quantitative measure of sets of possible events satisfying the conditions: $0< P_X, P_Y < P_{XY} \leq P_{YX} \leq 1$; P_X and P_Y denote the probability of a kind of event (a generic change) E_X and a kind of event E_Y, respectively. Now the probability formula can be transformed as the Probabilistic Law of Reversible Processes:

$$W(\Pi^{\#}) = P_{yx} = P_{xy}\, K_R \tag{21}$$

$W(\Pi^{\#})$ is the transition probability of the process Π (X/Y), the reverse of the given process Π(Y/X).

P_{YX} is the conditional probability of event E_X caused by event E_Y:

$$P_{YX} = P(E_X / E_Y) \tag{22}$$

P_{XY} is the conditional probability of event E_Y produced by E_X.

The ratio K_R measures the degree of process reversibility (convertibility) being a correlation coefficient:

$$K_R = \frac{P_X}{P_Y} = \frac{P_{YX}}{P_{XY}} = \frac{W(\Pi^{\#})}{W(\Pi)} \tag{23}$$

Here two extremes can be selected: i. when the ratio $K_R << 1$, the process Π (X, Y) is *quasi-irreversible*; ii. if the correlation coefficient $0<< K_R <1$, then the process is *reversible*.

For $E(X) = \{e_i\}_{i\in N_n}$ and $E(Y) = \{e'_i\}_{i\in N_n}$, the transition probability of the process Π(X/Y) may be represented as the matrices of elementary transition probabilities W:

$$W(\Pi) = \begin{array}{c|c|}
 & p(e'_1) \;\; \dots \;\; p(e'_n) \\ \hline
p(e_1) & w(e_1, e'_1) \dots w(e_1, e'_n) \\
 & \dots \\
 & \dots \\
p(e_n) & w(e_n, e'_1) \dots w(e_n, e'_n)
\end{array}$$

Where $w(e_n, e'_n) = p(e_n/e'_n) \cdot \dfrac{p(e'_n)}{p(e_n)}$

and

$$W(\Pi^{\#}) = \begin{array}{c|ccc|} & p(e_1) & \ldots & p(e_n) \\ \hline p(e'_1) & w(e'_1, e_1) & \ldots & w(e'_1, e_n) \\ & & \ldots & \\ p(e'_n) & w(e'_n, e_1) & \ldots & w(e'_n, e_n) \end{array}$$

Where $w(e'_n, e_n) = p(e'_n / e_n) \cdot \dfrac{p(e_n)}{p(e'_n)}$

It is noteworthy that the fundamental quantity used for description of quantum mechanical processes—scattering matrix (S_{fi}), or an amplitude of transition of the quantum mechanical systems from one state to another due to their interaction (scattering), correspond to the transition probabilities of forward ($W(\Pi)$) and inversed processes ($W(\Pi^{\#})$).

It emerges that the concept of singular causation, deterministic or probabilistic, as a temporally ordered asymmetrical and irreflexive relation of individual events (in different things) is fundamentally defective, although very popular. For, the particular causation, first of all, is lacking the very essence of causality: to be reversed in the order of action relating the wholly distinct change existences, and thereby opening the possibility of **circular causality**, *the engine of the world.*

Then the principle of local causality or, rather, causation should be formally reformulated as follows:

a change in (of) a substance $o_1 \in O_X$ at a time t and place s, $c_1(o_1, t, s) \in C_X$, is a cause of another change $c_1'(o_1', t', s') \in C_Y$ in a substance $o_1' \in O_X$ at a time t′ and place s′ if, and only if, changes of type C_Y causes changes of type C_X. What formally can be written as in:

$$E_X(e_1, e_2 \ldots e_n) \rightarrow E_Y(e_1', e_2' \ldots e_n') \mapsto (C_Y \rightarrow C_X) \tag{24}$$

This means *a causal connection [between any changes in the entity variables (as object, state, action, and relationship variables)] were to be established (had been established), just in case the inverse relationship would be determined (would have been determined).* Otherwise, the connection between the would-be cause and effect is spurious, contingent, regular succession, or non-causal correlation. In other words, causes bring about effects by necessity, or the antecedent change has come-to-be the consequent must always come-to-be, only if the causal relation is convertible, reversible, reciprocal, and eventually cyclical.

The above suggests that we have to decline a linear account of the world involving an empirical singular causation. Since it seems there is no greater illusion to human thought than thinking in terms of linear causality even embellished with all sorts of probability or graph theories addition. Since the very definition for the conditional probability together with Bayesian law of inverse probability positively indicates the universality of the phenomenon: **the reversibility of cause-effect relationships**. Again, there is strong scientific evidence that all fundamental laws of nature are temporally symmetric, like as the natural phenomena and processes described by such laws are reversible in the order of action; from physical effects to biological changes to socio-economic phenomena, all eventually are liable to order reversal. For car-minded readers it, again, may be well to remember how the transmission power train can work.

It is time to discard our false beliefs about causality, for any theory of the world based on this, mathematical, physical, or philosophical, falling short the reversible causal mechanisms, is substantially fallacious, and thereby deeply distorts our understanding of the world.

So, we need to accept that *causal processes run backward*, and that *the direction of causality turns the other way round*, and that the linear causal ordering was only a deep-rooted, time-honored but a confusion of the human mind, another myth of philosophy, sometimes wrongly presented as the consequence of the Aristotelian and Cartesian logical systems.

Conclusion

A causal theory of the world is evolved where the understanding of cause and causality presupposes three key assumptions:

- Necessity of the connection with an effect
- Universality of causality as the fundamental law of reality
- Totality and ubiquity of reversibility or convertibility

The fundamental principles and rules to be employed as the basic premises and axioms of causal ontology are as follows:

- The world is the totality of things ordered by causal relations
- Causation exists really, actually, and mentally
- All changes are subject to causal ordering
- All causal relations are in(re)vertable, reversible, or convertible
- Causality is a reversible relationship of changes

References

Abdoullaev, A. (2000). The ultimate of reality: reversible causality, In *Proceedings of the 20th World Congress of Philosophy*. Boston: Philosophy Documentation Centre. Retrieved April 20, 2006, from Paideia Project On-Line, http://www.bu.edu/wcp/MainMeta.htm

Aristotle, Logic (Organon): Categories. (1990). In M.J. Adler (Ed.), *Great books of the western world* (Vol. 1). Chicago: Encyclopaedia Britannica, Inc.

Aristotle, Logic (Organon): Posterior Analytics. In M.J. Adler (Ed.), *Great books of the western world* (Vol. 1). Chicago: Encyclopaedia Britannica, Inc.

Aristotle, Physical Treatises: Physics (Physica). (1990). In M.J. Adler (Ed.), *Great books of the western world* (Vol. 1). Chicago: Encyclopaedia Britannica, Inc.

Aristotle, Physical Treatises: On generation and corruption. (1990). In M.J. Adler (Ed.), *Great books of the western world* (Vol. 1) Chicago: Encyclopaedia Britannica, Inc.

Aristotle, Metaphysics. (1990). In M.J. Adler (Ed.), *Great books of the western world* (Vol. 1). Chicago: Encyclopaedia Britannica, Inc.

Armstrong, D. (1997). *A world of states of affairs,* Cambridge, UK: Cambridge University Press.

Born, M. (1949). *Natural philosophy of cause and chance.* NY: Dover Publications.

Bunge, M. (1982). The revival of causality. In G. Flostad (Ed.), *Contemporary Philosophy, a New Survey*, Vol. 2, Philosophy of Science. The Hague/Boston/London.

Chandrasekaran, B. (1999) Functional representation and causal processes. Retrieved April 24, 2006, from www.cse.ohio-state.edu/~chandra/FR-Paper-Advances-in-Computers.pdf

Davidson, D. (1967). Causal relations. *Journal of Philosophy*, 64, 691-703.

Druzdzel, M., & Leijen, H. (2001). Causal reversibility in Bayesian networks. *Journal of Experimental and Theoretical Artificial Intelligence, 13*(1), 45-52.

Global Risk 2007: A global risk network report. (2007). World Economic Forum. Retrieved January 10, 2007, from http://www.weforum.org/pdf/CSI/Global_Risks_2007.pdf

Godel, K. (1949). An example of a new type of cosmological solutions of Einstein's field equations of gravitation. In S. Feferman (Ed.), Kurt Godel's Collected Works. (1990). Vol. II. (pp. 190-198). New York: Oxford University Press.

Hausman, D.(1998). *Causal asymmetries.* Cambridge: Cambridge University Press.

Hausman, D. (2005). Causal relata: Tokens, types, or variables? *Erkenntnis, 64*(1), 33-54.

Hawking, S. (2004). The information paradox for black holes. In *Proceedings of the 17th International Conference on General Relativity and Gravitation*, 18-23 July, Dublin, Ireland.

Hoover, K. (2000). *Causality in microeconomics.* Cambridge: Cambridge University Press.

Hulswit M. (2004). *Causality and causation: The inadequacy of the received view.* Retrieved May 1, 2006, from http://www.library.utoronto.ca/see/SEED/Vol4-2/Hulswit.htm

Hume, D. (1990). An enquiry concerning human understanding. In M.J. Adler (Ed.), *Great books of the western world* (Vol. 36). Chicago: Encyclopaedia Britannica, Inc.

Kim, J. (1995). Causation. In R. Audi (Ed.), *The Cambridge Dictionary of Philosophy* (pp. 110-12). Cambridge: Cambridge University Press.

Kleer, J. de, & Brown, J. (1986). Theories of causal ordering, *Artificial Intelligence,* 29, 33-61.

Kolmogorov, A. (1968). Logical basis for information and probability. *IEEE Trans. Inform. Theory*, 14, 662-664.

Mac Lane, S. (1971). *Categories for the working mathematician.* Graduate Texts in Mathematics. Springer-Verlag.

Maruyama, M. (1963). The second cybernetics: deviation-amplifying mutual causal processes. *American Scientist*, 51, 164-79.

Ortriz, C. (1999). Common sense language about causes and rational action, *Artificial Intelligence*, 111, 73-130.

Pearl, J. (2000). *Causality: Models, reasoning, and inference.* New York: Cambridge University Press.

Pearl, J. (2002). Reasoning with cause and effect, *AI Magazine*, 23, no.1, 95-111.

Simon, H. (1952). On the definition of the causal relation, *J. Philosophy,* 49, 517-528.

Soros, G. (1998). *The crisis of global capitalism [open society endangered].* Public Affairs: New York.

Suppes, P. (1984). P*robabilistic metaphysics*. Oxford: Basil Blackwell.

Vendler, Z. (1967). *Linguistics in philosophy*. Ithaca, New York: the Cornell University Press.

Chapter VIII

How to Reason about the World: The Common Reasoning Platform

Introduction

Along with substances of all kinds, states of all manner, and changes of all types and exemplifications, relationships of all sorts and instances appear to be among the prime constituents of the universe as a whole and its realms, regions, and domains as the world of nature, the universe of society, and the domain of minds. Hence, knowledge of relations, as the cause-effect relationship, constitutes the basic core of real knowledge and, consequently, the logical fundament for all basic kinds of reasoning about the world. All true and effective reasoning upon reality, its particular classes, parts and features, is eventually to be founded on the underlying relations of substances, states, changes, and analogies, as well as on the meta-relations of whole-part, comparison, contrast, identity, resemblance, and difference.

To adequately represent and consistently reason about reality is vitally important not only for human beings but also for prospective intelligent machines driven by the ontological models of the world comprehending the logical models of possible worlds. A widely practiced logical tradition to represent the world in terms of abstract classes, properties, and relations or purely mathematical objects, functions, and relations looks to be a main conceptual obstacle to creating effective reasoning systems. Since the likewise artificial conceptualizations of the world are missing out the core of things, their nature and reality, providing the ontological ground and making true the truth. These would-be reasoning systems will not work effectively because of their built-in incapacity to work out any complex real problems or situations or challenges. Above all other things, such intelligent systems will be unable to see the difference of physical, mental, or social objects so that to recognize their attributes, qualities, properties, states, changes, and relations.

Then the challenging questions arising is how to design a versatile reasoning application both able to represent generic entities as well as capable to compute the meanings of their species and instances, such as places, times, words, wars, geological periods, justice, irrational numbers, artifacts, and so forth.

First and foremost, for a machine to become a really intelligent transducer, it should have the capacity to classify world things or tell apart such fundamental modes of reality as substances, quantities, qualities, places, times, positions, states, events, actions, processes, and relations. That is, it has to be endowed with the capacity of ontological reasoning about the world and its parts by "representing real-world objects, actions, and relationships internally as interconnected structures of symbols, and applying symbol manipulation procedures to those structures" (Newell & Simon, 1976). In other words, the key for designing reasoning machines is an adequate representation of the basic kinds of entities and relationships, their subordinate kinds, species, and individuals with the basic features, properties, characters, attributes, and traits.

So, what is laid down in AI as *the physical symbol system hypothesis* (the digital computing machine as a physical symbol system has the necessary and sufficient means for general intelligent behavior (action) through such an entity-centered representation and reasoning) is just another appeal to the world knowledge representation and reality-centric, ontological reasoning. Since real ontological dynamic modeling involves both encoding world knowledge and organizing reasoning rules by using the fundamental axioms and truths about relationships.

The whole point is here that a true and actual and valid reasoning about things in the world implies dynamic ontological reasoning. For, as a real science, Ontology deals with the substantial parts of Discourse about Anything or Everything, while Logic as a formal science is concerned only with the formal parts of it. And Logic considers the elements (the terms, propositions, inferences, or syllogisms) of the whole discourse aside from their reference to the world (or their real semantics). Then logical reasoning covers the formal patterns of discourse common to any reasoning, having little to do with reality, real significance, or real meanings. By contrast, ontological reasoning addresses the matter and content, the real components of discourse about anything, so forming the fundamental basis for any formal logical reasoning and rules, thus also underpinning the Web rules, a major part of the Semantic Web (A Proposal for an OWL Rules Language, 2003; the Rule Markup Initiative, 2007; Semantic Web Topic Hierarchy, 2007).

The Real Logic of Things: The Kinds of Human and Machine Thinking

To find out what ontological reasoning (argument, deduction, or inference) is, it is of necessity to address first to the content of the conception of reasoning as abstract thought standing as the general class to its numerous species and kinds, like as abstraction, analysis, argument, argumentation, demonstration, deduction, synthesis, computation, conjecture, analogy, regress, inference, cognition, and so forth. Such diversity is connected with the complex nature of reasoning. By its essence, as all cognitive species of mind, reasoning is a hybrid, heterogeneous process of cognition and knowledge, which is partly natural (or mental) and partly logical.

As a mental process (cognitive operation), a kind of thought, it has causal relationships with other mental acts such as sense, perception, memory, conception, judgment, imagination, emotion, and intention. Again, as a mental process, it is born and performed by the intelligent agent, the reasoner, the thinker: a human being, an animal, a machine, or an extraterrestrial being. Accordingly, we speak of human reasoning, animal intelligence, mechanical reasoning, or alien reasoning. As a logical entity, reasoning differs from other forms of thinking or kinds of thought by its conformity to some standard of truth, cogency and validity, by its formally valid deduction, without making a point of the truth of its premises.

Premises, reasons, causes, or antecedents, as well as the conclusion, effect, implication, or consequence, the result of reasoning consequent upon the premises, all enter into the logical line, chain, sequence, succession, series, consecution, or ordination of reasoning as its extreme elements. Another distinction of reasoning then comes from the fact that the kinds of its premises can be as different as the forms of knowledge. They may be as follows:

a. Axioms, principles, maxims, and truths
b. Hypotheses, assumptions, and postulates
c. Percepts and concepts, images or abstractions, perceptual, conceptual or symbolic representations
d. Concrete facts or abstract ideas

Therefore, the types of reasoning can be induction proceeding from a collection of particular cases (instances or examples), deduction proceeding from general principles (assumptions, definitions, or axioms), or hypothetico-deductive argument starting from the *ad hoc* hypotheses.

Apart from the extreme parts, premises and conclusions, here enters also the idea of a logical relationship, as proof or inference, in the forms of a process (motion, movement, transition, transformation, succession, association, connection, recursive calculation, mathematical demonstration, algorithm, etc.). When discussing the nature of reasoning, we speak of it as the process of thought, a motion from one reason to another, a step-by-step movement from one proposition to another, logically ordered relations of ideas, or thoughts. In such a formal, logical interpretation of reasoning, its truth or rather validity is defined only by the cogency and consistency of patterns of demonstration, by formal correctness of the structure of argument.

Another complexity comes from the referent class of reasoning, its universe of discourse. Like the objects of knowledge, the universe of reasoning may consist of any kind of entity and relation:

1. Substance, object, matter, and mind
2. State, quantity, quality, space, time, and situation

3. Change, event, cause and effect
4. Relations of all kinds and sorts

Consequently, there may be substantial, qualitative, quantitative, spatial-temporal reasoning, reasoning about change, actions, causal reasoning, reasoning by analogy, and so forth. On the ground of its close connection with knowledge, reasoning is often classified in line with the types of knowledge: (1) according to the types of things knowable or represented; (2) by the mechanisms, methods, techniques, and modes of reasoning process. Therefore, we speak of philosophical, scientific, formal, dialectical, rhetorical, and practical or experimental reasoning.

Thus, as its cognate species, knowledge, the process of reasoning possesses a manifold of dimensions: natural, mental, cultural, semiotic, and logical. For the time being, therefore, there is the bulk volume of definitions, explanations, interpretations, descriptions, and expositions aimed to specify and determine the boundaries of reasoning, both human and machine. For instance, the definitions of human reasoning are as diverse as:

- Abstract thought that is logical and coherent
- Modifying perceptual representations of the world
- Stringing linguistic categories according to the syntactic rules and the laws of grammar
- Causal processes of sentence-like entities symbolically represented, syntactically structured, and physically realized
- Manipulating causally the cognitive images of sentences
- Stimulating consecutively physico-chemical, neural-like events ending in a motor response
- Collective quality arising from the interactions of individual minds
- Purposeful behavior as a heuristic search of a problem space
- An interpretation of signs of some kind, likeness, indications, or symbols

Together with the human reasoning, the nature of machine reasoning, the mechanism of mechanical and automated processing of its meaningful symbolic structures of the world's elements is now among the hottest topics in AI and Semantic Web. Since finding the answer to these challenging questions is widely thought to be almost tantamount to revealing the nature of mechanical intelligence. As it stands, there also is no generally accepted definition of machine reasoning (Davies, 1998). Therefore, among regular topics of AI conferences and Semantic Web meetings, one meets the plenitude of all sorts of reasoning and rules: abductive, automated, case-based, causal, common-sense, geometrical, inductive, model-based, non-monotonic, probabilistic, or qualitative reasoning, reasoning about actions, reasoning under uncertainty, resource-bounded reasoning, situated, spatial, temporal reasoning; derivation rules, reaction rules, inference rules, transformation rules, business rules. As a consequence, the types of definition of machine inference vary as widely as of human reasoning. For example, machine thinking may be described as:

- Connecting machine perceptions by the relationships of similarity, space/time contiguity, and local causality
- Formal, logical calculation of abstract symbols in terms of the predicate calculus
- Following the axioms of probability theory
- Processing the qualitative differentials as to the principle of local causality
- Transforming quality spaces as logical spaces endowed with a metrical structure, and so forth

The main reason of such variety lies in the controversial conceptions of human reasoning partitioned into two wide groups: the class of deductive argumentation, passing from universal premises to particular things, and vice versa, the class of inductive inference, proceeding from particular facts to empirical generalities. Or, it is as reasoning of the forms and formal relationships between statements as distinguished from a way of reasoning involving the substance (content) and connections of the statements of facts. The former is exemplified by the formal logical approaches: the proposition calculus, the predicate calculus, theory of syllogism, modal logic, set theory, and metalogic. The latter suggests empirical inference, such as statistical, cognitive, societal, economical, and psychological approaches to reasoning.

What looks evident from our brief overview, machine reasoning needs such a conception extensible to all the possible types, underlying all sorts of inference, argument, and demonstration: perceptual or experiential, inductive or logical, rational and deductive, intuitive or discursive, scientific or practical, dialectical or rhetorical.

As a matter of fact, despite the widely varied descriptions, explanations, and interpretations of human and artificial reasoning, all the above-mentioned positions and standpoints tend to implicitly assume or imply a common fact. The fact is that that **reasoning** consists in *identifying and mapping the consistent order and relationships of entities, in representing the ordering relations among the elements of reality in terms of truths (facts, reasons, arguments, or causes)*, where the truth of the antecedent premises causes the truth of the consequent conclusion. The understanding of reasoning as been determined by the kinds of relationships allows us to give a broader definition of an intelligent, thinking, or reasoning machine, as a computing entity endowed with ontological rules and reasoning, knowing (understanding) of relationships of all sorts and manner. This enables mental-like processes (automatic intellection, mechanical thinking, logical information processing) such as construction, reasoning, explanation, planning, problem solving, converging and diverging intellection, and prediction.

There are thus two things conclusive for the nature of reasoning, natural or artificial: (1) to have absolute and fundamental premises pertaining to the nature of entities and relations; (2) to have an accurate description of the class of relations and its species, as a causal relationship. There is only one field of knowledge able to deal with it, the science of entity in general, or fundamental ontology, whose axioms and general facts are applicable to reasoning about any particular entity of the world. Furthermore, there is only one form of demonstration inferring a conclusion from the axioms of reality as the first principles, the basic truths, immediate statements, self-evident maxims, or absolute assumptions. It is a real **ontological reasoning**, *the fundamental reasoning about reality by means of veridical representation of states of the world and its domains*, underlying all sorts of real logics: situation logics,

logics of events, scientific logics, and domain specific logics, economic, political, and so forth. Incorporating all the best features of formal logic, such a **reality-centric reasoning** is marked by a combination of two virtues: focusing on the truth or falsity of premises together with the validity of proposition and inference forms, since valid reasoning alone is not enough to reality-minded cognition.

Common Reasoning Environment: World Reasoning Rules and the Web Rules Language

The ontological theory of relations allows us to analyze not only the nature and relations of real kinds and the ways the major kinds of entities stand to each other, but it also provides the ultimate, truth-making ground for the major kinds of reasoning. For the world relational structure can be mapped into the space of propositions (facts) using the isomorphism between the universal relation structure $\mathfrak{R}$ and its universal predicate $P^n_m (W_1 \ldots W_n)$ of degree n and order m, as in:

$$P^n_m = P: W^n = S \rightarrow \{1, \text{if } (w_1 \ldots w_n) \in W^n ; 0, \text{if } (w_1 \ldots w_n) \notin W^n \} \quad (1)$$

So any expression consisting of a predicate variable of order m and degree n followed by class (and individual) variables is a well-formed formula, from a logical point of view. By its structure the m-order and n-placed predicate $P^n_m (W_1 \ldots W_n)$ is a collection of n-ary predicates with the domain W_1 x...x W_n, the range S covering the totality of statements constructed from the predicate variables P, plus the world's variables of substances, states, changes, and relations. According to the definition, it comes up as the general formula for all kinds of meaningful propositions about the world. Then the ordered triple of the world as the universe of discourse W, the classes of ontological predicates $P^n_m (W_1 \ldots W_n)$, and the collection of a set of statements S can be defined as the all-embracing ontological context, the all-reasoning environment, or the world description framework:

$$K = \langle W, P^n_m (W_1 \ldots W_n), S \rangle \quad (2)$$

The elements of the ontological context are the tripartite onto-semantic entities having intension, extension (or reference) and truth-value, where both the meaning and the truth or falsity of theses elements depend on the fact of their correspondence to the entities and their relationships in the world. In other words, the meaning of any construct, its intension or definition, extension or reference class, as well as the proximate context consisting of the totality of its conceptual relatives, antecedents (purport) and consequents (import), can be revealed by using such a general context. Being a knowledge entity, the class of constructs is known to include as its species individual terms, predicates, propositions, and conceptual bodies (contexts and theories), where the predicates may embrace individuals, classes (or sets) and relations, thus coming up as the type of constructs predicated both of classes or individuals. Consequently, the nature and types of predicates are determined both by the relations referred to and the type and number of its terms, components, or arguments. As a result, within the universal framework, depending on the predicate reference class and the type of arguments predicated, we can select four broad categories of ontological predicates:

- **The first-order, n-placed predicate of substances (or objects):** $P^1 (O_1 \dots O_n)$, where the substantial variables occur in quantifiers and replaceable by names, or substantives.
- **The second-order, n-placed predicate of states (or properties):** $P^2 (S_1 \dots S_n)$, where the state variables occur in quantifiers and replaceable by abstract nouns of states, by adjective predicates, or state verbs.
- **The third-order, n-placed predicate of changes:** $P^3 (C_1 \dots C_n)$, where the change variables occur in quantifiers and replaceable by abstract nouns of changes, active adjectives, or action verbs.
- **The fourth-order, n-placed predicate of relations:** $P^4 (R_1 \dots R_n)$, where the relational variables occur in quantifiers and replaceable by relational names, relational adjectives, relational verbs, or function words.

In the EIS SUO, any meaningful proposition, which is possible to build up out of the predicates and their arguments (subject and object variables), will always be in one of theses broad categories of predicate and their appropriate species, like the classes of substantial, spatial, qualitative, quantitative, active, temporal, or causal predicates. That is, each of these kinds of predicate is either asserted of its class or outside of its category. Within the same category, it will state its essence, like "man is an animal," or "number is a quantity," where subclasses (child classes, derived classes, subdivisions, or incorporated classes) and classes (superclasses, parent classes, base classes, incorporating classes) exist within hierarchical inheritance relations. Again, if one class is predicated of another, not within the same ontological category, it is a property attribution. For what is predicated does not define the essence of a substance, but rather its state (attribute, property, quantity, or quality), change (action, activity, doing, or behavior) or relation.

To analyze the general predicate formula $P^n_m (W_1 \dots W_n)$, we had better to pass to the illustrative cases of unary and binary predicates. The unary predicates belong to the class concepts of four descriptions:

- **The class of substances:** $P^1 (O_1)$ attributing (non-relational) properties to substances
- **The class of states:** $P^2 (S_1)$ assigning (non-relational) properties to states
- **The class of changes:** $P^3 (C_1)$ ascribing (non-relational) properties to changes
- **The class of relations:** $P^4 (R_1)$ referring (non-relational) properties to relations

So the unary predicate refers to a class of things (objects, states, changes, relations) with a certain qualification, or to a totality of entities determined by certain properties, but not of the same class. For instance, the predicates of substances give a description of some kind of substance by ascribing to it some states, attributes, properties, qualities, quantities; changes, actions, activities, functions, behavior; or roles in relationships. That which is predicated of substances is variable contrary states in virtue of which the substances are changing, for example, heavy and light, white and black, sweet and bitter, hot and cold, dry and moist. The same logic of class description also holds true for a description of characteristic properties of states, changes, and relations.

The most essential of all possible unary predicate is the one-place ontological predicates to "be" or "exist" $\exists$ and not to "be" or not to "exist." Taken in the basic sense of having being, reality, actuality, existence, or truth, the former is applied to all sorts of beings, $\exists$ **W**, $\exists$ **O**, $\exists$ **S**, $\exists$ **C**, $\exists$ **R**, meaning: "the world exists," "substances have being," "states are," "actions have reality (take place)," and "relations exist." The latter, taken in the sense of nothing, nonentity, nonbeing or not having reality, actuality, existence, or truth, is also can be extended to any sort of entity as implying not any being or thing: substance, state, action, or relation. All other predicates are to be placed between these two universal elements: the existence (being, truth) and non-existence (nothing, falsity) predicates, like in a fuzzy mathematical logic in which the range of truth may assume a continuum of values from 0 to 1.

As a special sort of construct, any unary predicate of class has its extension (referents) together with its intension (definition), plus the collections of assumptions and consequents. As to the binary predicates associating or connecting things, they are also of four sorts:

- **Substantial predicates:** $P^1(O_1, O_2)$ comprising the ontological predicates of substances or objects, material or nonmaterial
- **Stative predicates:** $P^2(S_1, S_2)$ comprising the predicates of states, properties, qualities, quantities
- **Change predicates:** $P^3(C_1, C_2)$ comprising the predicates of changes, actions, and events
- **Relational properties:** $P^4(R_1, R_2)$ comprising the relational predicates

The binary and dyadic predicates stand for the relations between two classes of entities expressed in NL by collectives, class names, and mass nouns. They are what is asserted or negated of a number of things showing differences in kind, which may be objects, quantities, qualities, actions, or relations themselves. The scope of predications or predicate variables may range as wide as: the causal predicate, the inclusion predicate, the subsumption predicate or the predicate of implication, the predicate of inequality, the predicate of division, the pairing predicates, and so forth. But, the most general binary predicates are those referred to as a set of definitive predicates, as to "be identical with" **Pi** or "different from" **Pd** and as to "be {am (are, is), was (were)} a substance, a state (condition, quality or quantity), a change, and a relation." Such definition-related predicates concerned with the matter of sameness and difference, with the identity of things or the class membership, defining either a thing's essential nature or what its name means and signifies. No less extensive value are the property predication predicates stating about a being's property, non-relational and relational, like as to "be{the property of, in possession of, equal in value to, worthy of, of a certain length; in relationship to, the cause of, dependent on, a substitute for, a part of, true of, like, in place, in time}."

Another universal predicate is to "have" **Ph**, which ordinary meanings are spread as wide as: to have the ownership, contain as a part, possess a state, attribute, or quality, undergo a change or being affected with a change, cause to be, do something or engage in activity; possess a relation to or being in relationship to. The both sets, the definitive predicates and the property predicates, are to be used with reference to wholes and parts; substances and objects; states or properties, qualitative and quantitative; changes, actions and activities; relationships of all sorts, kinds, and manner.

Thus, the major species of the binary predicates are as much as the major kinds of binary relations: the meta-predicates, as the part-whole predicate **Pph** and its inverse the whole-part predicate **Php**, both expressing all the possible relationships between parts and wholes; the comparison predicate **Pc**, like identity **Pi** and resemblance **Pr** and difference predicates **Pd**; as well as the substantial, the stative, the causal, and the spatiotemporal predications. Naming relations among different sorts of arguments, the meta-predicates do not possess some fixed order varying according to the type of arguments (objects, property-states, actions, or relations).

In that the binary predicates as expressing relational properties stand for, name, or represent relations, all the relational rules and identities for relations are valid for predicates as well; for the lattice of relations and the lattice of predicates are subject to the rule of order preserving (i.e., isomorphic). In other words, the universe of predicates is nothing else but a realization (conceptual model) of the ontological relational structure:

$$\mathbf{P_W} = < 2^{\mathbf{P}}, \Rightarrow, \vee, \wedge, \neg, {}^{\#}, \mapsto, \mathbf{0}, \mathbf{P} > \quad (3)$$

The universe of predicates is the power set lattice of predicates $2^{\mathbf{P}}$ embracing all the possible combinations among them. In such a logical structure, the partial order, two binary and unary operations (complementation and inversion), composition, as well as top and bottom relations are respectively interpreted as the entailment (implication), disjunction, conjunction, negation, conversion, and transformation. The universal bounds of such **a lattice of predicates** are the universal predicate of all the arguments, while the null predicate representing the null relation is the empty predicate with no arguments.

The totality of the entity variables, ontological predicates as meaningful axioms (theses, atomic formulas, or well-formed formulas), formal rules of relationships, like the formation and transformation rules of relations, constitutes nothing else but the ontological predicate calculus (OPC). There the formation rules specify all the possible combinations among elementary formulas, while the transformation rules determine all the possible consequences, theorems (the specific statements of facts) of the axiomatic logic of reality. Typically, within the same order of variables, the atomic predicate formulas are combined with logical (truth-functional) operators. Or they are formed by the use of the quantifiers, which are nothing but one-place ontological predicates of existence, like as other primitive logical predicates having been ontological predicates at some initial point. Thus, if P (X, Y), Q (Y, Z), M (Z, U) express meaningful atomic predicates (atoms) of arbitrary arguments (or well-formed formulas), but within the same order, then the following expressions are also meaningful formulas: $\neg P$, $P^{\#}$, $P \vee Q$, $P \wedge Q$, $P \Rightarrow Q$, $P \Leftrightarrow Q$, $(\exists X)\, PX$, $(\forall X)\, QX$, $(\forall P)(\exists X)PX$, $(\forall Q)(QX \equiv QX)$. The formation rules for predicates are formulated as a realization of the corresponding relational rules such as:

The rule of entailment (or implication) based on the order relation rule:

$$(P \Rightarrow Q) \equiv P = P \vee Q \text{ or } Q = P \wedge Q$$

The existence of a converse predicate based on the reverse-the-order axiom:

$$P^{\#}(Y, X) \Leftrightarrow P(X, Y)$$

The existence of an identity predicate E =Pi based on the existence of the identity relation:

$$X\, E\, Y = X \Leftrightarrow Y$$

Additionally to the formation rules, the transformation rules of OPC, arranged as an ordered succession (consequence, succession, series, sequence, or consecution) of comparable predicate formulas, deal with the ontological atoms (elementary predicates and axioms) so that to derive from them more specialized facts or theorems. For example, among the transformation rules of OPC are:

The rule of composite deduction based on the transition of relations:

$$P \mapsto_{C} M = X(P \circ Q \circ M)U \equiv (XPY \wedge YQZ \wedge ZMU)$$

The non-monotonic rule of analogy:

$$P(X, Y) \mapsto_{A} M(Z, U) = P :: M$$

The symbol $\mapsto$ indicates that the predicate to its right is a valid deductive consequence to the predicate (s) on the left. Such a rule of composite deduction based on the transitivity of relations is of special value in the OPC. In the formal predicate logic, the deductive rule is a couple of ordered sets: an antecedent (a set of atoms, the body of a rule) A and a consequent (a set of atoms, the head of the rule) C, meaning that if the antecedent conditions are true, then the consequent conclusion must be true of necessity. The conditional rule is written in symbols as $A \mapsto C$. In the OPC the transformation rules are equivalent to the rules of ontological reasoning (demonstration or deduction), which actually is composed of four constituent elements:

- The antecedent premises, the definitions of things, the basic axioms *par excellence*, and the hypotheses of existence, the most general facts and basic truths about the world, the true, primary, and indemonstrable premises of reasoning acting as its primary cause, reason, or ground.
- The consecution (actual, causal, or logical) of formulas mapping a real order of things in the world.
- The consequent conclusion, effect, result, or consequence, or what is proved as novel truths, facts, or theorems.
- The classes (and kinds) of things (the subject-class) of which essential properties are revealed by the reasoning.

In the OPC, within the general ontological context, the species of reasoning consists in having conformable the order and connections of meaningful predicates with the appropriate species of relations, the relations of identity, resemblance, difference, part-whole, causality, space and time, and so forth:

$$A \mapsto_{K} C \qquad (4)$$

Of all kinds of axioms (truths) that can be the premises of reality-minded reasoning, logical, mathematical, empirical, semantic, moral, or political, the highest ones are the truths about reality itself. Namely, the ontological axioms on the nature and existence of things, like the following formulated as NL sentences:

1. Whatever exists is an entity, a part of being as a whole.
2. Whatever is, is; something cannot be and not to be.
3. Whatever is, is either in itself or in another.
4. Whatever has a property (or attribute) has all the properties (attributes) of that property (attribute).
5. Whatever changes, either turning into other substance or going into other state, action, or relationship.
6. All relatives have correlatives.
7. Whatever happens might as well happen backward.
8. All things exist in some relations to all other things, and so forth.

Unlike logical principles and formal axioms relating to the rules of formal argumentation, ontological axioms are the principles of knowledge; for, above all, they are about realities, their existence or nonexistence, properties, and relationships. Being the universal and necessary statements of the fundamental facts of being, of all that is and happens, they are the first principles of reality and of reasoning at the same time. Having universal premises about the basic kinds and classes of entity, added with domain-specific, particular facts, **the general reasoning entity** will be able to derive the most significant statements about the specific classes of things in the world. By its nature such the ontologically reasoning entity will underlie software agents, softbots, knowbots, intelligent agents, thinking objects, smart programs, personal assistants, or other would-be reasoning applications.

As to the actual consecution, the rule of composite deduction based on the transitivity of the predication relation is one of major rules of ontological reasoning. For instance, entailment reasoning can be formally figured as an ordered structure of reasons (causes, facts, or truths) arranged by the implication or predication relation, a sequence of successive predications, implying that whatever is predicated of a genus (class) is true of each of its species (subclass).

It should be noted that the universality of predication was first observed by Aristotle in his Categories: "When one thing is predicated of another, all which is predicable of the predicate will be predicable also of the subject" (Aristotle, 1990, p. 5). In the language of logic of classes, the principle of predication will sound like: if one class of entities is included in a second, and that second class is included in a third, the first class is included in the third. In mathematical logic, the principle is reformulated as the following: if predicate P implies Q, and Q implies M, then P implies M. At last, for causal relationships we have: whatever is the cause of a cause also is the cause of its effect, or "cause of cause is cause of effect." At the same time, the negative formulation of the principle will be stated as the following: whatever is universally denied of a class (a concept) also is denied of its subordinate species (subordinate notions), or what cannot be predicated of a predicate cannot be predicated of its subject.

No less significant species of reasoning is the inverse reasoning involving the *reversal axiom* stating that the original order of the terms of relations is always subjected to conversion (inversion, reversion, reversibility, reciprocity, or community):

$$(P \Rightarrow P^{\#}) \wedge (P^{\#} \Rightarrow P) \equiv P^{\#}(Y, X) \Leftrightarrow P(X, Y) \qquad (5)$$

Thus, it is an intrinsic property of relations to be capable of inversion; for *if some entity is a relation, then it will be capable of reverting, and if something be capable of inversion, it will be a relation.* Otherwise, such an entity is not a relation *per se*, but some other thing, an object or a state. Hence to predicate convertibility of correlative things is the permanent feature of all sorts of relational predications or assertions:

- The identity predicate **Pi**
- The difference predicate **Pd**
- The resemblance predicate **Pr**
- The part-whole predicate **Pph**
- The have-relation predicate **Ph**
- The substantial predicates P^1

- the stative (qualitative and the quantitative (equality and inequality) predicates P^2
- the predicates of changes P^3, like the predicates of causes
- the relative predicates P^4, like the predications of time

Of all the listed assertions, what is predicated of causes and effects appears as one of the most foundational predications; for to have knowledge means to know the cause at the first place, since "it is the cause that we seek in all our inquiries" (Aristotle, Posterior Analytics, p. 123) and since "all reasonings concerning matter of fact seem to be found on the relation of *Cause and Effect*" (Hume, 1990, p. 458). There are changes that are coincidently associated, without necessary or causal relations, like the serving of meal and the striking of a clock; but there are changes that are causally connected. The issue of issues taxing the great minds for ages how to define causality so that to discern it from contingent association of changes is hardly to be solved without having **the fundamental theory of relations**. For, as a species of relations, causality inherits all the inherent properties of relation:

- To be real
- To be ordered and directed
- To be inverted
- To be relative or comparative
- To be exemplified as singular causation

By this reason, if the causal relation holds between one kind of change and a second kind $P(C_1, C_2)$, it also holds between the second kind and the first kind:

$$[P(C_1, C_2) \Rightarrow P^{\#}(C_2, C_1)] \wedge [(P^{\#}(C_2, C_1) \Rightarrow P(C_1, C_2)] \equiv P(C_1, C_2) \Leftrightarrow P^{\#}(C_2, C_1) \quad (6)$$

It emerges that the convertibility of causal predicates is based on **the Law of Causal Reversibility**: if a kind of change B causally follows a kind of change A, then A follows B always and universally. Namely, *the reversal of causality happens always and universally, since a real cause is both a sufficient and necessary condition for its effect, as the true effect also is a necessary and sufficient condition for its cause.* Otherwise, the relationship is not in the elite category of causal relations. Other possible formulations of the binary causal predicate may be three types of contrary-to-fact statements, given in time present, past, and future: if A caused B, then B could (would) cause A; if A had caused B, then B could (would) have caused A; and if A were to cause B, then B would be caused by B.

The meaning of the composition rule of reasoning can be shortly demonstrated for the vital case of productively related changes, that is, causal relationships. Consider a couple of ordered changes (or events) undergone with two distinct things: $<S^0, S', F>$ and $< S', S'', G>$. Then the fact of connection of two changes will be functionally presented as $< S^0, S'', G \circ F>$. Or, in the form of the equation $S'' = (G \circ F)\, S^0$ asking for a description of causal relations four distinct states and two functions at least. In this functional equation, the related terms made of a set of initial states S^0 and a set of final states S'', while the composition of two functions (G∘F) stands for the coefficients of the state equation. Following this, the character of the composite X (G∘F) Z = X (G) Y and Y (F) Z means that there is always the middle term in reasoning that connects the major premises with their conclusion. Finding this middle term makes all the difference: "quick wit is a faculty of hitting upon the middle term instantaneously" (Aristotle, Posterior Analytics, p. 122). For, the "middle" is always concealed from us, as any cause and reason to be inquired. And, unlike the major and minor predicates, it is not sensible, it cannot be perceived, seen, felt, or directly measured, but rather conceived by the agent in the reasoning. From what is argued above, it is plain that the ultimate value of any reasoning system should consist in its representational power measured by the coverage of reality and deductive capacity or power or reasoning efficiency measured by the depth of causal explanations (explicatory potential) and the accuracy of predictions (predictive or oracular capacity).

Now that the sphere of interpretation of the world description model is more extensive then of any formal models, it can be used as a universal language both for formal systems of mathematics and logic and theoretical systems of sciences. For the fundamental formal ontology as a general theory of entities entails (and contains) as its models and applications abstract mathematical and logical structures, as according to the scheme: from the world to domain-specific knowledge through abstract mathematical structures and formal logical systems:

$$<\mathbf{W(orld), R(elation), N(othing), T(hing)}> \mapsto <S, \leq, 0, I> \mapsto <U, \subseteq, \Lambda, V> \mapsto <P, \Rightarrow, P_0, P_m^n> \mapsto \ldots \tag{7}$$

Practically, this suggests that the EIS SUO rendered as the universal ontology language endowed with the collection of transformation rules seems to be much more productive then the ontology languages based on formal logic, like the OWL ontology with its extension, an OWL Rules (Horrocks et al., 2003). First of all, the rule language is lacking any systematic classification of classes and relations (called properties), since, as in an axiomatic theory of sets, the primitive terms as class concepts and binary relations are assumed as undefined notions. As a consequence, it operates with few predicates: the class description unary predicate, binary predicates (as data type properties, the range value is RDF literals, and object properties), the sameAs predicate, and the differentFrom predicate, plus the definition of classes as sets of individuals. As the elements of the class description (axioms) are chosen equivalent classes, disjoint classes, subclasses and superclasses, unionOf, intersectionOf, complementOf, domain, range, allValuesFrom, someValuesFrom. Again, the correlatives of relations are considered to be only individuals, and all relations (or properties) are analyzed in terms of transitivity, symmetry, functional property, inversion, and inverse functional property. The most basic concept, Thing, is defined as the universal class of individuals, whereas Nothing as the null class. All the enumerated class axioms and predicates are applied as the atoms for an antecedent and a consequent of an OWL rules language. With this conceptual equipment, the language is presented as a universal Web ontology language capable to describe (formalize) the classes and relations (properties) of any domain, so that to secure machine reasoning about the classes and their elements, and thus providing interoperability (commensurability) and integration in the world of Web-based ontologies, data, documents, and applications.

Conclusion

The fundamental reasoning about the world is based on causal relationships, their inherent properties of composition, transitivity, or convertibility. The theory of relationships completed with the principle of inverse causality underlies the basic rules of thinking about things: the causal reasoning about reality, its parts and domains, such as the physical universe, the mental world, or the socio-cultural domains.

So machine understanding is equivalent to the knowing of (causal) relationships. Then for the real life intelligent systems, effective reasoning implies the knowing and understanding of the cause-effect relationship with all the important implications, entailments, and predictions thereof. It is thus developed an ontology-based reasoning environment to be used as a common intelligent platform for the semantic e-technologies, rule technologies, and rule-based agents, and versatile knowledge applications.

References

A proposal for an OWL rules language: Semantics and abstract syntax. (2003). Retrieved September 12, 2007, http://www.cs.man.ac.uk/~horrocks/DAML/Rules/

Aristotle, Logic (Organon): Categories. (1990). In M.J. Adler (Ed.), *Great books of the western world* (Vol. 1). Chicago: Encyclopaedia Britannica, Inc.

Aristotle, Logic (Organon): Posterior Analytics. (1990). In M.J. Adler (Ed.), *Great books of the western world* (Vol. 1). Chicago: Encyclopaedia Britannica, Inc.

Davis, R. (1998). What are intelligence? And why? *AI Magazine*, 19, 91-110.

Horrocks, I., Patel-Schneider, P., Boley, H., Tabet, S., Grosof, B., & Dean, M. (2005). *SWRL: A semantic web rule language combining OWL and RuleML*. Retrieved April 25, 2006, from http://www.daml.org/2003/11/swrl/

Hume, D. (1990). An enquiry concerning human understanding. In M.J. Adler (Ed.), *Great books of the western world* (Vol. 36). Chicago: Encyclopaedia Britannica, Inc.

Newell, A., & Simon, H. (1976). Computer science as empirical inquiry: Symbols and search. *Comm. ACM* 19, 113-126.

Semantic web topic hierarchy. (2007). Retrieved September 12, 2007, from http://wiki.ontoworld.org/index.php/Semantic_Web_Topic_Hierarchy

The rule markup initiative. (2007). Retrieved September 12, 2007, from http://www.ruleml.org/

Chapter IX

How the World is Signified: Real World Semantics, or What Meaning Relation is

Introduction

How reality or the world with its parts and levels might be truly symbolized and represented by emerging semantic technology and knowledge systems appears the most challenging topic in the field of top ontology and ontology engineering. Along with causality, knowing the relationship of meaning makes all the difference in true representation of the real world features, in understanding (sensing, reading, or resolving) the real meaning values of world knowledge representation and reasoning. A formal account of meaning (or significance) is becoming a decisive issue in the whole matter of the Semantic Web promising machine-based processing by means of advanced information technologies. For without understanding the nature of meaning, its critical dimensions, mechanisms, and algorithms of representation in computable forms, the whole enterprise of semantic technology is an otiose undertaking and expensive academic mystification. As far as computing ontology is viewed as a semantic model where the relationships among resources are to be identified, differentiated, or processed by automated tools [SICoP, 2005], the above meaningful topics presuppose creating the standard ontology framework. As far as the emerging Semantic Web is the universal medium for the exchange of information across users, systems, applications, and networks, the unified frame ontology is the universal semantic platform for a uniform organization of all human knowledge.

As for now, a great many of Web researchers and developers are trying to create the application programs capable to meaningfully process the Web data, services, and processes just by using syntactic markup languages, such as XML or RDF or OWL or RuleML tagging structures (Berners-Lee et al., 2001; Uschold, 2003; XML, 2004; RDF, 2004; OWL, 2004; RuleML, 2007). Though, it is plain that "meaningless data cannot acquire meaning by being tagged with meaningless data, and the ultimate source of meaning is the physical world" (Sowa, 2000b). Such illusive hopes come from a lack of understanding of meaning (or signification), its nature, types, modes, aspects, components, and hence the nature of semantics-based machine understanding and reasoning. It is an unjustified deprecation of the extensive field of semantics, when reducing it to a formal semantics resorting to the notion of abstract possible worlds (RFD Semantics, 2004; OWL Semantics, 2004).

As a matter of fact, the meaning (or significance) relationship involves as its foundation three substantial constituents:

- Real entities (things in the world)
- Constructs (representations and rules)
- Signs (symbol systems as static Web data and dynamic Web agents)

So to place the Semantic Web on a solid semantic base, instead of formal semantic languages, it is necessary to develop a real world, ontological semantics dealing with existent referents and relations as mapped to signs, their patterns and syntactic structures.

And it is most critical that ontology and semantics are inherently complementary. The former provides a set of basic constraints (truths) the world imposes on the languages, while the latter symbolically identifies the entities in the world, setting a set of semantic rules specifying how to assign a substantial significance, veridical meaning, or material signification to signs and their collections. It is significant that to understand meaning as such we need to realize that any sign is a "monster beast" of three different parts, like Chimera. Being a crossbreed of three sorts (icons or images, indices or indications, symbols or marks), "it is an (1) *entity* that represents (2) another *entity* to (3) an *agent* (human, animal, or robot)" (Sowa, 2000b).

Characteristically, from the semantic outlook, the universe of things can be marked by threefold division:

a. There are things that are merely things, all acting as the real world referents of meanings.
b. There are things that also are signs of other things (natural signs of the physical world and mental signs of the human mind).
c. There are things that are always signs, as conventional signs (words) and other cultural nonverbal symbols and social constructs.

In other words, there are things that MAY have meanings (the things of the external world, as all sorts of indications, evidences, symptoms, and physical signals); there are signs that ALWAYS have senses and meanings (the entities of the mind as ideas and images, thoughts and feelings); and there are signs that HAVE to get their meanings (as linguistic entities and cultural symbols). So, while natural signs serve as the source of signification, the cognitive agent is the agency through which signs signify naturally occurring things, such as objects, states, qualities, quantities, events, processes, or relationships. The types and modes of signification (meaning) vary with the types of signs (or symbols), natural and conventional, vocal and nonvocal, material and cultural, as well as with the kinds of things, for which the signs (symbols) signify or stand.

So the goal of real ontology is to analyze the significance relationship in order to yield a comprehensive semantic theory capable to merge the major definitions of meaning; namely: (1) thought, intention, or understanding; (2) intension or cognitive content or sense; (3) message or content or information or communication; (4) truth condition; (5) usage; (6) measurement or computation or operation; (7) extension or reference or denotation; (8) the composition of intension and extension.

Ontology and Semantics

As the science of meaning, semantics is traditionally divided into two broad parts: theoretical and empirical. The former is subdivided into general or basic semantics dealing with the Symbol-Construct-Thing triangle and special branches involving model theory (semantics of mathematics and logics), semantics of scientific knowledge, and semantics of commonsense knowledge. Together with the pragmatics of psycholinguistics, sociolinguistics, and anthropological linguistics, the latter encompasses the semantics of natural language (Bunge, 1974). But a more consistent view is when semantics is coming as the descending hierarchy of theoretical, special, and empirical levels. Then the theoretical semantics concerns with the entire semantic triangle of **sign domain** (sings, marks), **mental universe** (constructs), and **extra-linguistic and extra-mental external reality**.

The study of meaning relationship from the global worldview is the goal of **ontological semantics** relying on a set of ontological, epistemological, and formal semantic tools and methods. Among its subdivisions are special and empirical semantics researching the meaning triple by specifying its components as specified by particular knowledge domains of interest. Applied or special semantics encompasses the model theory (concerning a possible interpretation of mathematical symbols and structures); formal semantics (logical aspects of meaning of predicates, propositions, and theories); semantics of science (scientific knowledge); and semantics of ordinary knowledge (the sense of commonsense knowledge). The empirical semantics contains linguistic semantics studying natural language meaning; psycholinguistics dealing with the language user, its speech acts and contents and personal values; and cognitive semantics researching the cognitive aspects of meanings.

As the branch of linguistics, the language semantics consists in analyzing the linguistic items (elements, units, expressions, and constructions) and their meanings. In authoritative dictionaries, its definition varies as the study of the meaning of words, the theory of meanings of speech forms, the study of meaning in language forms, or the study of language meaning.

From a general perspective, semantics is about the relation between signs (symbols) and what they signify, whereas syntax analyzes only relations between (linguistic) terms, expressions, and constructions. The more completed definition of linguistic semantics will be as the study of language forms in connection with the extra-linguistic things they signify (denote, express, code, embody, mean, realize, etc.). The nature of the "what they signify" makes all the difference, for it is the content and substance of language, without which linguistics is completely meaningless and useless.

From the most general perspective, semantics appears to be the study both of meaning (relationships of signs and symbols to things) and truth relationship (conformity to realities of language constructions), so residing both in the things and in the intellect. Matching the associations of the symbols and signs with their constructs and entity classes, semantics organically completes ontology dealing with the fundamental kinds of things. The unique role of upper and domain ontologies, the universal reference ontology in particular, consists in giving this ultimate meaningful content to languages, natural or artificial, informal or formal, by imposing the entity type structures on lexicon as the underlying structure of a language, its meaningful parts, elements, and constructions.

It is essential that under the field of semantics comes not only the task of uncovering the semantic relations from the sets of symbols to the universe of constructs, from the sets of constructs to the kinds of realities; or directly from the symbol structures to the classes of things. The types of referents, or reference classes, the kinds of representations, the sorts of meanings, and all the forms of truth, logical, mathematical, factual, or philosophical, also belong to its objects of inquiry. But assigning a truth-value and significance to any body of symbols calls for fundamental axioms of entities, like as a discovery of the meaning and truth and sense, and reference of any well-formed body of knowledge. In other words, semantics inherently committed to world models, since any meaning always presupposes some sort of world views, explicitly or implicitly.

In addition to the theoretical issues, the state-of-art knowledge applications, like the Semantic Web and Digital Aristotle, ask for fruitful combination of the ontology paradigm and linguistic knowledge being equal to the emerging knowledge technologies. This combination may yield us the real world (ontological) semantics, onto-semantics, revealing a type of representation formalism applicable to different knowledge domains, sciences, natural and artificial languages, and formal systems. If to enumerate some of them, among the largest divisions will come:

- The onto-semantics of science (the natural sciences, the social sciences, the humanities), holding the positivistic and realistic semantics of science (Barwise & Perry, 1983).
- The onto-semantics of mathematics, embracing model theory or model-theoretic semantics (Barwise & Etchemendy, 1989).
- The onto-semantics of logic, encompassing logical semantics and metalogic (Bunge, 1984).
- The onto-semantics of common knowledge, covering all sorts of common sense knowledge models (Lenat, 1998).
- The onto-semantics of language (Stepanov, 1981; Nirenburg & Raskin, 2002), including nonsymbolic (noncognitive) language and symbolic (conceptual and pictorial) representations.

The fact that ontology and semantics are crucial for languages, natural or artificial, and for the building of language technologies and data machines, now looks to be a common place. There is hardly any way to find any deep inquiry devoted to the natural language study without declaring its commitment to some sort of cognitive semantics (Jakendoff, 1990; Talmy, 2000), linguistic semantics (Frawley, 1992), or linguistic ontology (Dahlgren, 1995; Bateman, 2001; Farrar & Langendoen, 2003).

The relationships between the words and the extra-linguistic world may be studied from two directions, as it is the case with any relationship, the part-whole, causality, resemblance, and so forth. One common way, widely practiced by modern semanticists, is to start from the grammatical categories and relations to the non-linguistics categories. Proceeding this way, the researchers use the syntactic classes (*part of speech, number, person, gender*) and/or grammatical relations (*anaphoric relation, aspect, tense, agreement, case, conjunction, complementation, coordination, inflection, modification, modality, subordination, semantic role, transitivity, intransitivity, voice*) for reconstruction a linguistically motivated top ontology.

As a result, the grammatical structure of the natural languages affords to distinguish the basic terms of the universe of discourse. They are the key concepts of thought and the fundamental categories of reality, such as *substances, objects, agents, properties, qualities, quantities, conditions, states, situations, causes, acts, events, changes, actions, processes, places, times, relationships*, and so forth. It is then stated that the entities represented by substantives are substances or objects; by adjectives, qualities or quantities; by verbs, states or actions; by prepositions, conjunctions, or cases, relationships. This grammatical approach might be had its urgency when the ancient ontologists had started the tall order of formulation a set of top-level entities. But this approach also much contributed to many confusions: traditionally such key word classes as verbs were neglected as being able to express relationships among things. Accordingly, many still think that a relation is not a real entity and so generally classifying verbs as indicating whatever, existence, state, action, occurrence, or change, except relational entities (Levin, 1993).

As a matter of fact, more fundamental and effective approach is quite the reverse, from the few higher entity classes to the grammatical entities and relations:

deep ontological structures → semantic representations and structures → surface syntactic representations and structures

Namely, we need to move not from the language to the world, but from the world to the language. As it happened with the **real mathematics,** where a top-bottom mapping from the substantial world to mathematical categories replaces a bottom-top mapping from set-theoretical elements to real world things. Instead of linguistic ontology, then, we have to speak of **ontological linguistics** as a general framework of languages, which key parts are **ontological grammar** (morphology and syntax) and **ontological semantics**. The first part of the onto-grammar, ontological morphology (inflectional, derivational, and compound) is concerned with the word formation rules in terms of top-level entities, while ontological syntax, with the phrase and sentence forming rules in terms of top-level relationships. At last, the ontological semantics is another name for the veridical semantics projecting the classes of things in the real world onto the objects in the language (data) world, instead of mapping linguistic entities (or objects of the computational world) onto the real world. If the meaning in the Semantic Web lies in the notion of real world semantics (Uschold, 2003), then it should be interpreted and defined not as a mapping of objects of the computational world onto the real world, but rather the opposite.

Ontological Linguistics: A Unified Theory of Language

In the grand semantic triangle of meaning composed of **language** (all the signifiers, I), **mind** (all the things and constructs of the mind, II) and the **real world** (all the entities and relations signified, nonlinguistic and non-mental, III), **ontological linguistics (grammar and semantics)** occupies one of the central positions (see Figure 11). In the scheme of meaning, **language** is the **WordSpace** made of linguistic entities (forms, units, elements or constituents) and their linguistic relationships. Mind is the knowledge (cognitive) space made of mental entities and their cognitive relationships. The external world is the totality of extra-mental and extra-linguistic entities and relationships, the universe of entity variables having material values, conceptual values, and semantic values. Here the world's parts and components with

Figure 11. The relationship of language, mind and reality

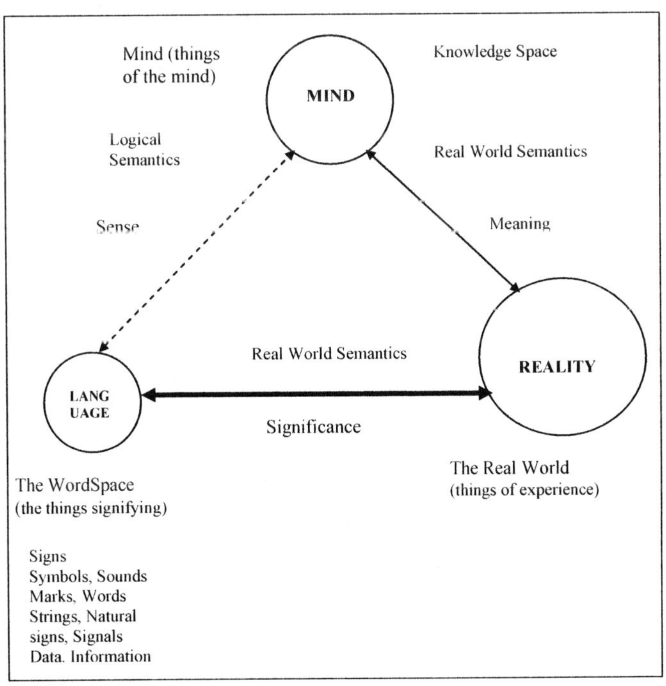

their functions and relationships serve as semantic values to the language elements, parts, and constituents. Altogether, there are six binary relationships of meaning:

- (I→II), from symbolic structures (signs, symbols, sounds, marks, words, words strings, signals, or data) to the domain of constructs (concepts, propositions, models, theories), the subject matter of formal, mathematical, and logical models of meaning
- (I→III), from the symbolic expressions to the extra-mental and nonlinguistic things of the real world
- (II→III), from the constructs, ideas, and thoughts to the entities and relationships of objective reality
- (II→I), from mental entities to symbolic descriptions (terms, predicates, and expressions)
- (III→I), from the world referents to the language elements, forms, and constituents
- (III→II), from the real things of the world to the mental representations

Other relationships, defined between a class and itself, are syntactic relations (I→I), a set of conceptual associations of entity sets (II→II), and a collection of real relationships (III→ III). The semantic relationships are marked by the characteristics inherited from a class of relations:

- Type of existence depending on the sort of arguments connected
- Bi-directionality, being reversible and reciprocal, each semantic association is complemented by its inverse related class
- Degree, the number of related components, as the binary and ternary with the number of arguments 2 and 3, respectively
- Cardinality, the number of associated entity instances one side of the relation refer to
- Connectivity, the mapping of associated entity occurrences in the semantic relationship (e.g., one thing-to-one concept or one concept-to-many words)
- Transitivity, a relation between three sorts of entity (sign, concept, thing) such that if it holds between the sign and the concept and it holds between the concept and the thing it must hold between the sign and the thing
- Existence dependency, a dependent, conditional relation as defined by having of origin (source, parent) on one side and issues (results, children) on the other

As far as the existence of the language is dependent on the existence of the world, the sort of semantic relations for further investigation will be the world-language associations (III→I), the subject matter of the ontological linguistics and its divisions, onto-grammar and onto-semantics. Practically, this choice is motivated with the possibility of enabling machine reasoning by semantic markup languages through using ontological descriptors, like semantic markups, metadata (XML) tags, or OWL classes. According to the real semantics, the ultimate meanings of linguistic expressions and constructions come from the universal class of entity, its top-level divisions, types, and kinds. As the inverse class to the sign-thing correspondence of linguistic ontology, the veridical semantic relationship is marked by the following attributes: its type of existence is the actual relations from actual things to actual linguistic objects; the connectivity type is one-to-many; the value of cardinality is many (multiplicity of representations). Since the determining direction of meaning is from the world items to the symbolic, linguistic items, the general class of (entity-entity) relationships is multiply expressed both by the syntactic classes and grammatical relations. Among the syntactic classes, such relation-bearing expressions are the word classes, or parts of speech, as the common names of relationships, relative adjectives, associative verbs as well as all the closed-class words of prepositions and conjunctions, or connectives. So, one category of relation is indicated by many verbs: "associate," "colligate," "connect," "link," "link up," "tie in," "relate," "interrelate"'; "concern," "come to," "bear on," "touch," "pertain," "refer to"; "have a relation to," "establish a relationship to," "attach," "bond," "hold," "involve," "affect," "regard," and so forth. To express a semantic relation, we may say "the signifier is linked to the signified," which is too general. More specific semantic verbs are "mean," "denote," "designate," "connote," "code," "embody," "express," "indicate," "name," "represent," "stand for," and so forth. Altogether the entity-entity relationships are represented by all sorts of lexical and grammatical relations: case relationships, qualifying relationships, clause-level connections of subordination or coordination, and so forth. Under the class of lexical relations fall a variety of relationship words (terms and expressions), such as the whole-to-part expressions, comparative, causal, spatial, or temporal constructions. The systematic ontological analysis of linguistic categories indicates that all the possible relations between linguistic forms and constituents consist of the linguistic relations determined by morphology and syntax, grammatical relationships, and lexical relationships, where the meaning of words depends on the nonlinguistic class of relations.

In all, *ontological linguistics can be defined as the study of language in terms of entity words and relationship constructions*. This asks for reviewing the traditional separation of words into two disjoint parts by the grammatical roles in sentences; namely, content, open-class meaningful words (nouns, verbs, adjectives and adverbs) and function, closed-class words (determiners, prepositions, pronouns, conjunctions, and particles) as lacking independent meaning. Instead, the words are split up into entity expressing word classes and relation-indicating word forms. Except particles and some pronouns, all function words are nothing else but relational terms getting their meaning from really existing associative entities. Again, all the basic parts of speech, or form classes, also can be carved up into meaningfully distinct categories referring either to the nonlinguistic entity and relationship types.

It emerges from the above that the most challenging issues, like what kinds of things are signified by universal words, what in reality makes sentences true, or what in the world do we mean by sentences, and how relations can be expressed in terms of language, all fall into the realm of ontological linguistics. To construct a comprehensive ontological grammar as the reality-committed theory of semantic values acting as *the universal theory of meaning relationships from things to sign patterns* (words and word strings, terms and constructions), we need to appeal to the key entity structures for assistance.

As Russell observed, with a few exceptions of proper names denoting particular things, almost all other words found in the encyclopedic dictionaries are general terms standing for universals shared by individuals. They are either the substitutes of the general classes of thing or stand for the subordinate classes. Since entity and relationship types act as the semantic values to all linguistic expressions, the classification of general words and terms is to be patterned after these underlying modes and forms of being as the basic sorts of realities. In line with the EIS scheme of things, these major entity universals as the semantic values of all the meaningful linguistic words and constructions include:

- **Entity or thing,** the whole being, existence, all, reality, the totality of entities or the universal class of all classes
- **Being-in-substance**, the primary class of substances, objects, and agents
- **Being-in-state**, the primary class of states, conditions, qualities, properties, and quantities
- **Being-in-change**, the primary class of changes, processes, actions, doings, events, occurrences, or happenings
- **Being-in-relation**, the primary class of relationships of various sorts, part-whole relations, similarities and differences, causation, space, time, and so forth

If natural language as a whole represents reality, then all its significant components and units (terms, phrases, clauses, and sentences) can be partitioned into the expressions of substances, states, changes or actions, and relationships. That is, *the number of elemental parts of speech is as many as the number of elemental forms of entity*. With that, any natural language has to comprise its linguistic forms and meaningful expressions, terms, phrases, and sentences as associated with the basic parts and elements of reality.

Unlike many formal ontology languages designed for machines, the starting-point for a universal ontology language $\mathbf{O}_L$ is the world itself, the largest class and aggregation of realities. Such world structure comes complete with the order-preserving mapping function **F** introduced in the causal mathematics of reality and specified now as a representation function assigning to every thing of extra-linguistic reality an expression from the universe of words **L**. Such specification allows us to obtain the world description (interpretation) function $\mathbf{\Phi} \subset \mathbf{F}$ assigning to the entities in the world their corresponding language expressions:

$$\mathbf{\Phi}: \mathbf{W} \rightarrow \mathbf{L} \qquad (1)$$

To make sense out of symbolic languages the cognitive agent needs to know not only about the knowledge domain and the language describing it, but also the nature of relationship between these two sides. As with all the associations of things, the relationship between the world and the languages is reversible and reciprocal, as it happens between expressions and things in natural language generation and understanding. To generate language, to speak or communicate, the agent has first to think of things in the world and only then to communicate, while to understand, the hearer or listener or reader is doing quite opposite, it perceives the spoken (written) words (strings) and only then starts thinking of the referents so that to understand their meanings and senses. Usually and traditionally, in studying the meaning of languages, grammarians and semanticists are bound to the inverse semantic relation, meaning or signification function (with two key parts of reference, denotation, extension and representation, or connotation). Instead, in our case we focus on the reality description function, such that the universe of possible worlds **W** represents a domain of entities (the independent variables), while the level of language **L** is a range of linguistic entities (the dependent variables). Constituting the main body of natural languages and speech, the universe of words **L** can be exhaustively divided into four linguistic categories. They are the substance (object) denoting terms $\mathbf{L_O}$, the state (condition) referring expressions $\mathbf{L_S}$, the change (action) indicating terms $\mathbf{L_C}$, all ordered, conjoined, or connected with the relationship expressing words $\mathbf{L_R}$:

$$L = \{L_O, L_S, L_C, L_R\} \tag{2}$$

As a consequence, the totality of the linguistic elements of speech with all the possible grammatical expressions and constructions, as phrases, clauses, and sentences, can be formally represented as the underlying formal structure of universal ontological language:

$$\mathbf{O_L} = < \mathbf{W, P(L); R_L; \Phi} > \tag{3}$$

Its main distinction is that the language frame unlike other formal language explicitly contains the world-language mapping function, whence comes its designation as the world model of language. Altogether, the onto-linguistic model comprises the following elements:

- The totality of things (entities and relationships) **W**, the range of values of the description function **Φ** corresponding to all possible interpretations and meanings of words, expressions, and composite constructions
- The universe of words presented as the universal class of words, **P(L)**
- The possible linguistic interrelationships $\mathbf{R_L}$, grammatical and semantic

In other words, the formal structure presents the language space (or WordSpace) as the totality of language constituents and linguistic relationships composed of grammatical relations and meaning relationships. By its general meaning and definition, the power set of the universe of terms describes all their possible combinations consisting in total of sixteen subsets of terms and their constructions:

$$\mathbf{P(L)} = 2^L = \{\{\mathbf{L_O}\}, \{\mathbf{L_S}\}, \{\mathbf{L_C}\}, \{\mathbf{L_R}\}, \{\mathbf{L_O}, \mathbf{L_S}\}, \{\mathbf{L_O}, \mathbf{L_C}\}, \{\mathbf{L_S}, \mathbf{L_R}\}, \ldots\} \tag{4}$$

As for the symbol $\mathbf{R_L}$ indicating associative constructions, it will mean linguistic relations defined on the finite set of general terms and expressions and thus accounting for their structuring by applying the syntactic formation and transformation rules bound up to the semantic function **Φ** responsible for imposing order and meaning on the classes of word.

When being extended to all sorts of linguistic objects (letters, morphemes, words, and expressions), the general language structure becomes the subject of mathematical linguistics, where relations are presented as a class of grammatical and syntactic relations, like as of tense, possession, comparison, number, gender, person, case, and so forth. But only when we take into account the description function **Φ** associating lexical units and their complex constructions with the corresponding elements, compounds, and complexities of reality, the language reveals its ontological and semantic structure.

A characteristic mark constituting the difference of the *ontological semantic grammar* from the traditional grammar is that it unifies grammar and semantics of natural language as a natural whole considering the resulting system as an extension of ontology, as a general language of things, a universal schema of any symbolic language. For it can admit of much wider extension by adding the alphabet of language, atomic symbols, signals, letters, or numbers, all the smallest lexical units, as morphemes, together with such complex expressions as sentences, composite or compound.

To pass to the more specific syntactical rules and principles of the general ontological grammar, we have first to outline the nature and character of the semantic function. As far as the classification of words and their modes of signification are determined by the fourfold taxonomy of entity, the signification function of meaning for terms can be specified as follows:

$$\mathbf{\Phi} = < \mathbf{\Phi}_{O}, \mathbf{\Phi}_{S}, \mathbf{\Phi}_{C}, \mathbf{\Phi}_{R} > \quad (5)$$

$\mathbf{\Phi}_O$ is **the object description function** mapping from the classes of substances (objects and agents) of all sorts and denominations into the class of substantives:

$$\mathbf{\Phi}_O: \mathbf{O} \rightarrow \mathbf{L}_O \quad (6)$$

$\mathbf{\Phi}_S$ is **the state representation function** (for mental states, the designating function) mapping from the real states of all manner and designations (conditions, qualities, properties, quantities) to the class of state expressions:

$$\mathbf{\Phi}_S: \mathbf{S} \rightarrow \mathbf{L}_S \quad (7)$$

$\mathbf{\Phi}_C$ is **the change mapping function** from all sorts of possible and actual changes, motions, movements, actions, acts, and occurrences into the class of change referring constructions:

$$\mathbf{\Phi}_C: \mathbf{C} \rightarrow \mathbf{L}_C \quad (8)$$

$\mathbf{\Phi}_R$ is **the relation mapping function** from the class of all possible relations in the world into the class of relational expressions and grammatical relations:

$$\mathbf{\Phi}_R: \mathbf{R} \rightarrow \mathbf{L}_R \quad (9)$$

The analysis of the description function demonstrates that it involves the signification functions (the denotation functions of the substantive expressions and the predication functions of other universal terms) as its inverse functions. This fact is crucial in finding the meanings of composite words, expressions, and constructions such as phrases and sentences typically constructed along the following scheme:

(Letters, Syllables, Morphemes)→ Word→ Phrase →Clause → Sentence (Simple, Complex) → Compound →Paragraph → Text → Context

In the hierarchy of linguistic units, a word is synthesized by the combination of lexical units, morphemes, while phrases and clauses generally are described by all the pairs of subsets of the universe of words:

$$\mathbf{L}_{PR} = \{\mathbf{L}_O, \mathbf{L}_S\}, \{\mathbf{L}_O, \mathbf{L}_C\}, \{\mathbf{L}_O, \mathbf{L}_R\}, \{\mathbf{L}_S, \mathbf{L}_C\}, \{\mathbf{L}_S, \mathbf{L}_R\}, \{\mathbf{L}_C, \mathbf{L}_R\} \quad (10)$$

If phrases are meaningful expressions lacking grammatical subjects and predicates, a full sentential formula makes a tripartite complexity of subject, predicate and object, expressing all the possible relationships among things, as presented in the matrix of entity-entity relationships:

$$\mathbf{\Phi}_{OS}: \mathbf{W} \times \mathbf{W} \rightarrow \Sigma \quad (11)$$

The meaning representation function $\mathbf{\Phi}_{OS}$ is assigning to every relationship a statement, an independently meaningful portion of language and speech. Accordingly, a sentence is not only a group of words having a complete grammatical structure, but also a string (collection, number, set, combination) of words expressing a relationship. In each sentence, as in any composite word, the meaning of the whole is determined by its parts. The meaningful ordering of words and the truth conditions of sentences come from their description functions telling us not only what the terms signify, but also in what relations they stand to each other: substance to attribute, the relation of consequence, the relation of class inclusion, or the relationships of time and space. In the hierarchy of sentences, the most significant statements are those that connected with the assertion or denial of the following:

1. Whether a thing exists and has reality, that is, the fact of a thing's being
2. Whether the connection of the thing with its property is a fact
3. Whether the thing has this or that state and modification
4. Whether it changes, acts or being acted upon
5. Whether it is related to something else, for example, as cause or effect

Additionally to possession of the meaning composed of constituent words and phrases, some sentences differentiated by having the truth-value. Such variety of composite words belong to the special sort of sentences, propositions or statements, the truth or falsity of which depends on the fact of the thing's being or not being. So, altogether, additionally to predicating of a thing's relationships, the sentence can record its quality (affirmative, negative, infinite), modality (necessity or contingency, possibility or impossibility), temporality (the past, present, or future), and quantity (universal, particular, singular).

Also, more composite sentences can be constructed out of the simple sentences, termed as clauses, by the agency of relational and logical terms. Usually such a composite construction is separated as a compound sentence with two or more co-ordinate clauses and a complex sentence with one main and one or more subordinate sentences. In the end of the train, the ordering of the composite expressions with a view to describe certain fragments and parts of the world generates an entire text as a subject of discourse, treatise, or speech, that is, a whole theme with its subordinate topics, divisions, points, heads, or questions describing content or meaning as a whole body of fact.

Any natural languages, either analytic using auxiliary words or synthetic (inflectional) employing inflections to show grammatical relationships, follow the structure of reality employing the five cardinal classes of word, the words of entity and existence, the terms of substances, the words of states, qualities, conditions, and quantities, the expressions of changes and actions, and the words of relations. Being an analytic language and using mostly particles and auxiliaries to modify the meanings of words and to express their relationships, Modern English may serve as a valuable paradigm and ready lexical resource for natural language-based knowledge machines and agents.

In the case of English, eight parts of speech are distinguished as distinct word (or form) classes, which are names (or nouns and substantives) **N** added with pronouns and determiners **D**, adjectives of various denominations **A**, verbs of every sorts and manner **V**, prepositions of several sorts **P**, adverbs of several types $\mathbf{A_D}$, conjunctions of all kinds **C**, as well as interjections. Additionally to its morphological properties and syntactic roles, each grammatical division also has a certain lexical significance determined by its reference to the things in the world. That is, by its ultimate meanings, all the eight parts of speech can be brought under five all-comprehensive rubrics: the words of entity, of substances, of states, of changes, and of relationships.

According to the meaning function comprising as its part the description and representation rules, all manner of substances will be the entities denoted by substantives, individual and general; while states are the sorts of reality denominated by adjectives, state verbs, and abstract nouns derived from the adjectives and state verbs. The class of changes is the sort of being represented by action verbals (verbal nouns), action verbs, intransitive and transitive, and participial adjectives; while the relations are expressed by relational words as prepositions, relative pronouns and conjunctions, as well as relational substantives, adjectives, verbs, and adverbs.

Names (individual and class, collective and mass), simple or complex, together with verbs, simple or complex, make up the elemental structure of any significant sentence: affirmative, negative, or interrogative. Along with its name, the substantive is that linguistic entity or subject of which is spoken, while the verb predicate as "a sign of something said of something else" (Aristotle, 1990, On Interpretation, p.25) is what is said about the entity. It is patent that without knowing the nature and property of namespace, it is hardly possible to erect the algorithmic model of WordSpace, its dimensions, attributes, and rules.

Conclusion

The real world semantics is developed as the world modeling ontology language. It is shown that all true representational languages must have ultimate ontological ground in the nature of real entities. In the framework of ontological linguistics, the status of natural language is lifted as a RRR system. This enables powerful intelligent computing systems capable to analyze, understand, and generate natural language without resorting to formal logical languages. Many common imperfections inflicting natural languages, like the ambiguity of meanings, the material and formal fallacies in discourse and reasoning, the abuse of words, also can be effectively reduced. Formulating basic meaning assumptions, which provide clear and accurate correspondence of linguistic categories to the classes of realities, makes possible to process the significance of words, sentences, and utterances. In the real world semantics, to find out the meaning and truth of a sense-bearing sentence is equivalent to discovering its corresponding states of affairs in the world.

In addition, the UFO ontological language becomes a most general automata-oriented language generating infinite grammatical constructions from a simple set of rules. As a consequence, the theories of generative and transformational grammars implying the subject-predicate logic and mostly dealing with phrase-structures rules are engrossed by the ontological grammar as special cases.

References

Aristotle, Logic (Organon): Categories. (1990). In M.J. Adler (Ed.), *Great books of the western world* (Vol. 1). Chicago: Encyclopaedia Britannica, Inc.

Aristotle, Logic (Organon): On Interpretation. (1990). In M.J. Adler (Ed.), *Great books of the western world* (Vol. 1). Chicago: Encyclopaedia Britannica, Inc.

Barwise, J., & Perry, J. (1983). *Situations and attitudes.* Cambridge, MA: MIT Press.

Barwise, J., & Etchemendy, J. (1989). Model-theoretic semantics. In M. Posner (Ed.), *Foundations of cognitive science.* Cambridge, MA: MIT Press.

Bateman, J. (1995). *The generalized upper model.* Retrieved April 24, 2006 from http://www.fb10.uni-bremen.de/anglistik/langpro/webspace/jb/gum/index.htm

Bunge, M. (1974). *Semantics I: Sense and reference; Semantics II: Interpretation and truth.* Dordrecht, Boston D.: Reidel Publishing Co.

Bunge, M. (1974). The relation of logic and semantics to ontology. *Journal of Philosophical Logic, 3*(3), 195-209.

Dahlgren, K. (1995). A linguistic ontology. *International Journal of Human-Computer Studies, 43*(5-6), 809-818.

Farrar, S., & Langendoen, T. (2003). A linguistic ontology for the semantic web, *Glot International, 7*(3), 97-100.

Frawley, W. (1992). *Linguistic semantics.* Hillsdale, NJ: Lawrence Erlbaum Associates, Publishers.

Jackendoff, R. (1990). *Semantic structures.* Cambridge, MA: MIT Press.

Levin, B. (1993). *English verb classes and alternations: A preliminary investigation.* Chicago, IL: The University of Chicago Press.

Lenat, D. (1998). *Leveraging CYC for HPKB intermediate-level knowledge and efficient reasoning.* Retrieved April 25, 2006, from http://www.cyc.com/hpkb/proposal-summary-hpkb.html

NL based knowledge representation: New perspective (2005). The 18th International FLAIRS Conference. Retrieved May 1, 2005, from http://ranger.uta.edu/flairs05/

Natural language and knowledge representation. (2006). The 19th International FLAIRS Conference. Retrieved May 1, 2006, from http://www.cs.miami.edu/~geoff/Conferences/FLAIRS-19/Schedule.shtml

Nirenburg, S., & Raskin, V. (2002). *Ontological semantics.* Cambridge, Mass.: MIT Press.

OWL web ontology language semantics and abstract syntax. (2004). Retrieved May 1, 2006, from http://www.w3.org/TR/2004/REC-owl-semantics-20040210/

Resource description framework. (2004). Retrieved April 22, 2006, from http://www.w3.org/RDF/

RDF semantics. (2004). Retrieved May 1, 2006, from http://www.w3.org/TR/2004/REC-rdf-mt-20040210/

RDF vocabulary description language 1.0: RDF schema. (2004). Retrieved April 30, 2006, from http://www.w3.org/TR/2004/REC-rdf-schema-20040210/

Russell, B. (1990). The problems of philosophy. In M.J Adler (Ed.), *Great books of the western world* (Vol. 55). Chicago: Encyclopaedia Britannica, Inc.

SICoP. (2005). Semantic interoperability community of practice (SICoP). White Paper series module 1. Introducing semantic technologies and the vision of the semantic web. KM. GOV. USA. Retrieved November 6, 2006, from http://colab.cim3.net/cgi-bin/wiki.pl?SICoP

Sowa, J. (2000). Ontology, metadata, and semiotics. In B. Ganter & G. Mineau (Eds.), *Conceptual Structures: Logical, Linguistic, and Computational Issues*, Lecture Notes in Artificial Intelligence # 1867 (pp. 55-81). Springer-Verlag.

Stepanov, Y. (1981). *Imena, predikaty, predlojeniya: Semiologicheskaya grammatika,* Moscow: Nauka.

Talmy, L. (2000). *Towards a cognitive semantics: Typology and process in concept structuring.* Cambridge, MA: MIT Press.

The rule markup initiative (2007). Retrieved September12, 2007, from http://www.ruleml.org/

Uschold, M. (2003). Where are the semantics in the semantic web? *AI Magazine, 24*(3).

XML, extensible markup language (XML) (2004). Retrieved April 25, 2006, from http://www.w3.org/XML/

Chapter X

How to Represent the World: Ontology-Controlled Natural Languages

Introduction

As far as human knowledge about the world is commonly given in NL expressions and as far as universal ontology is a general science of the world, the examination of its impact on natural language science and technology is among the central topics of many academic workshops and conferences. Ontologists, knowledge engineers, lexicographers, lexical semanticists, and computer scientists are attempting to integrate top-level entity classes with language knowledge presented in extensive corpora and electronic lexical resources. Such a deep quest is mostly motivated by high application potential of reality-driven models of language for knowledge communication and management, information retrieval and extraction, information exchange in software and dialogue systems, all with an ultimate view to transform the World Wide Web into a machine-readable global language resource of world knowledge, the Onto-Semantic Web.

One of the practical applications of integrative ontological framework is to discover the underlying mechanisms of representing and processing language content and meaning by cognitive agents, human and artificial. Specifically, to provide the formalized algorithms or rules, whereby machines could derive or attach significance (or signification) from coded signals, both natural signs obtained by sensors and linguistic symbols.

Of all the valuable outputs, the most significant application of ontological knowledge seems to be computer processing of natural languages, as man-made systems of linguistic forms and categories expressing thoughts and representing reality in all its complexity. The reason why language engineering has been failing to produce intelligent computer systems understanding and generating human language is because of the underestimation of the role of fundamental ontology in the study of natural language (theoretical linguistics), and its main parts: grammar, syntax, and semantics.

One of the misconceptions widely spread both in the KR communities and NLP interest groups is that a human's knowing of the world is basically performed without natural language, and that linguistic communication only marginally involves thinking, "language is froth on the surface of thought," as J. McCarthy states. This view resounds another bad confusion: natural language cannot be knowledge and reasoning representation language, but only an interface, for it is non-algorithmic, ambiguous, vague, and unsystematic. However, such a serious misunderstanding has been met with stiff opposition: "natural language is the ultimate knowledge representation language," and that "the richest source of ontological categories is the vocabulary of natural languages" (Sowa, 1995, 2000a).

Besides, interesting works have appeared promoting NL as a knowledge representation and reasoning system, attributing to it both representational and inferential features and so introducing a new class of large-scale, general-purpose knowledge systems based on natural language (Iwanska & Shapiro, 2000). This NL paradigm of intelligent systems has made the special tracks of international conferences devoted to the theme "NL based Knowledge Representation" (NL, 2005, 2006). Endorsing the language role in human and machine acquisition and processing of information about the world, we support this natural position by devising a formal unified theory of language as a single body of formal rules of grammar and semantics and ontology.

Within the general semantic framework provided by the ontological entities, the formal description of natural language, its elements and structure, is to be based on a consistent set of linguistic expressions and rules:

- Morphologic formulas for the formation of words as linguistic signs and symbol strings expressing content and real meanings
- Phrasal rules for the formation of word strings denoting concrete individuals
- Syntactic rules (or sentential functions) for forming linguistic strings mapping associations and relationships of things
- Linguistic rules of inference forming sentence sequences according to real logic or ordering of entities

Suggesting a general theory of signs and languages within the ontological paradigm, we establish the following: (1) human language in its nature is tractable algorithmically and systematically; (2) it is expressive and precise; (3) NL is a natural (material) edition and variation of thought; (4) NL is a variant representation of the world knowledge. As much as ontological primitives are knowledge standards, NL symbols and signs are primitives for various knowledge constructions. When considered as a general knowledge and representation language, an opportunity for various NL-driven meaningful machines opens up: NL reasoning applications, NL representation systems, NL answering systems, NL Web search engines, and many other linguistic intelligent programs.

Any powerful knowledge system or intelligent application will be employing a dynamic global model of reality, serving as a common code of real meanings and fundamental axioms, rules, principles, laws, and standards, represented in NL constructions. With the built-in universal semantic framework, the NL-driven ontological technology is to embody the essential mechanisms of knowledge, intelligence, reasoning and intelligent behavior.

Universal Namespace and Web Namespaces

In natural languages, each and every name stands for (or names, symbolizes, refers, identifies, denotes, expresses, applies to, points to, refers, signifies, or designates) some sort of thing, while not all things and constructs are denoted by names. There are names of individuals, of classes, of objects, of properties, of actions, or of relations. But the name of names is the word *entity (entitas, esse, be, what is)* first introduced as a general noun meaning *being* and *existence* in the most unqualified sense, and later modified as a concrete name meaning either a particular substance, the being of a thing in a given state or a distinct conscious existence. Complemented with the indefinite noun "not-being" or "nothing," the universal negation of all forms of being, the unqualified term "being" serves as the mother name for the nominal expressions signifying the major classes of entity: substance, state (property, quantity, quality), activity (passivity), and relation (place, time, position). With that all other names, nouns, and substantives are likely to be various derivations or modifications of the universal names: "thing," "entity" or "being."

To be consistent with the fourfold structure of things, the names of entities **N** expressing **being-as-entity** or **entity-as-being** are to be divided into the following categories:

- **The names of substances (objects) N_O**, as beings, objects, matters, units, parts, systems, and agents, inanimate or animate
- **The names of states (properties) N_S**, as conditions, qualities, quantities, or properties
- **The names of changes (processes) N_C**, as acts, actions, activities, agency, or doings
- **The names of relations N_R**, as causal, comparative, spatial, temporal, logical, and semantic relationships

The term "name" is used here to its fullest extent regardless of the fact that there generally are three types of entities:

- The whole classes of things (universals), each of which is taken as one mass or aggregate
- The classes of things as a whole
- Their parts, kinds, and examples, such as individual things (particulars), single members, instances, or occurrences

When taken in the broadest sense, "name" refers not only to an individual entity, but also to the whole class of things. As a result, it covers both individual and proper names denoting units or individuals, as well as general names (appellatives, or common nouns) denominating a class of entities. In such usage, the term "name" will embrace all other words of the same family: denomination, designation, appellation, and epithet. Also, it includes proper names differentiated by specifying one unit of a class of things.

Thus, names as class nouns and mass substantives refer to the substances (materials), states (conditions), changes (activities), or relations (time, place) in general, where the particular names as proper nouns refer merely to special examples, separate parts, or particular instances of those mass entities. Altogether, there are three types of class names: count class-names of entities, collective class names of things, and mass class names of beings, plus the compound names made of count, mass, or collective nouns.

A word as a mass name refers to the entity in general; as a count name it refers to special entities, kinds or individuals; and as a happy medium collective name it stands either for the entity in general (as a single entity) or for the individuals composing it.

It emerges that the collection of the entity names together with their grammatical (morphological and syntactic) connections and semantic relationships will constitute the universal namespace of natural language. It makes an essential part of word hyperspace, or WordSpace, the key structure of natural language applications. The universal namespace can be embodied as a physical, conceptual (mental or semantic) or virtual entity, like the XML namespace defined as "a collection of names, identified by a URI reference, which are used in XML documents as element type and attribute names" (eXtensible Markup Language, 2004).

To examine the WordSpace, we need to analyze its complex composition as elements and their possible combinations, that is, to divide the names inside of the namespace into distinct sets of name types, as in:

$$\mathbf{N} = \{\mathbf{N}_O, \mathbf{N}_S, \mathbf{N}_C, \mathbf{N}_R\} \quad (1)$$

In traditional linguistics, the names of entities are separated into two or three disjoint divisions of the same rank and status:

- Concrete names or terms of particular individuals (as contingent things)
- Collective names of collections, classes, or sets of concrete things (as concrete universals)
- Abstract names of such things as states, qualities, relations, numbers, classes (as universals or necessary things)

According to the standard theory of all entities, a deeper approach to names is achieved when they constitute a hierarchy of nouns (the abstract > the collective > the concrete) starting with most general name of a class of things and then successively enriching its content by descending to particulars.

At the concrete level, the modifiers introduced, specifying, qualifying, and determining the meaning of the general name by means of determiners and qualifiers, as in:

$$\mathbf{N_P} = < (\mathbf{D} + \mathbf{A} + \mathbf{N}\ )\ \mathbf{N}\ > \qquad (2)$$

By its definition, the phrasal name denotes a particular example of a class of thing, that is, such particular entities as physical objects, persons, particular properties, or events. Keeping in mind the ontological assumptions, we can pass to the formal systematic analysis of names of entities, their nature, taxonomy, phrasal structures, as well as morphological and syntactic rules of behavior.

As for the basic types of general names, the key element of universal namespace is object names, the names of substances, or substantives. The aggregate of substance names makes up the class of nominal substantives embracing, apart from proper names, general or common names, such as mass substantives, names of classes, and collective nouns. If the general name refers to a substance, mass, body, object, or material in general, as a whole part of reality, the particular term denotes a separate kind, part, example, instance, or unit of that substance, mass, body, object, or material. With that the whole class of the *substance denoting names* $\mathbf{N_O}$ constitutes the totality of common names of substances (objects) $\mathbf{N_O} \subseteq \mathbf{W_O}$, with three essential modifications:

i. The collective name referring to the collection (grouping or assemblage) of objects as one aggregate, whole, or unit (collective singular) or to the members of the collection as separate persons or units (collective plural) $\mathbf{n_c}$

ii. The class name denoting either an example (instance) of a class of substance (a person or an animal) or the class as a whole (person or animal) or one example (a part, portion, piece, unit, item) of a mass $\mathbf{n_i}$

iii. The mass name including abstractions and referring to the (physical) substances, bodies, materials, liquids and gases in general, of which special objects or separate units are composed of, $\mathbf{n_m}$

The range of values of the description function for the class names of substances identifies the scope of all actual referents of natural language:

$$\mathbf{\Phi_O}\{\mathbf{O}, \mathrm{O_j}\} = \mathbf{L_N} \qquad (3)$$

Here the variable $\mathrm{O_j}$ ranges over the classes of objects, physical and conceptual, reaching all sorts and manner of substances, beings, objects, bodies, and agents.

As it is well-known, the sum of the meanings of the component parts of the word (or sentence) contributes to the meaning of the whole word (or sentence). Therefore, the real content or reference of words often can be determined morphologically through morphemes, the smallest lexical units: combining forms, as from *acantho-* to *-zyme* and *zymo-*, and affixes: prefixes with negative prefixes, from *a*-to *ultra-*, infixes, and suffixes (or postfixes with inflectional endings and terminations) for nouns, adjectives, and verbs. The listing of such morphemes for English mainly includes Greek and Latin elements, and well-tabulated with all the corresponding meanings [Greek and Latin Elements in English, Combining Forms; Webster Dictionary].

Seeing that the constituent parts of the word make up its significance, the names of substances can be distinguished by their morphological markings. For the case of substantive names as the first illustration may well serve a special variety of the class names, *agentive nouns*, indicating three sorts of agents:

1. Biological agents as active causes, persons, animals
2. Physical and chemical substances as exerting some force or effect
3. Instruments and various artifacts, machines and intelligent agents

An agent is thus a sort of substance that does, exerts power, produces changes, or performs actions, or is capable of producing certain effects, events, or results, like a causal agent or agency. Its designation marked by a member of the set of typical agent morphemes acting as a classifier (signifier or descriptor) indicating the distinct semantic class in which the agentive noun belongs:

-AN, - ANT, -ARY, -ENT, - ER, - OR, - IST, -EE, - AR, -IAN, - ICIAN, - ITE, - STER, -TRIX, -IC, - IUM, - IDE, -LYTE, -GEN, -GOGUE, -ETTE, -EN, -ESS, -URE

The following typical agent names may serve as instances: *American, attendant, notary, president, computer, editor, scientist, refugee, liar, physician, mobster, aviatrix, critic, uranium, oxide, usherette, chicken, actress, ligature, hypnogogue, virucide, mitogen.*

On the other hand, typical endings for the general names meaning a collection of (or aggregate or class or kind) objects are as follows: -ADE, -AGE, -ERY, - HOOD, -SHIP, -DOM, -AD, -TY, as *cavalcade, leafage, machinery, jewelry, priesthood, readership, Christendom, triad, twenty*.

Special groups of words constitute the names of living things and the branches of knowledge and life; they can be easily differentiated by the name endings. To form the names of classes and orders of animals, the lexical descriptor *-acea* is used, while the names of superfamilies are marked by *-oidea*, of families by *-idae*, of subfamilies by-*inae*. The endings *-acean* and *-id* are applied to a member of a class or a family of animals, as in *crustacean, arachnid* or *leporid*, while *-ad* is used to name plants, *cycad*. The names of family of plants end in *-aceae*, like as in the family names of *pinaceae* or *vitaceae*. But still many class names of animate beings should be just kept in memory as having arbitrary endings.

Things look simpler with the names of sciences, arts, systems, and practices commonly marked by the ending *-ics,* as in *mathematics, linguistics, politics, robotics, athletics*, or by *-logy*, as in *biology, ontology, technology*, or by *-ism*, as in *nationalism*, or *capitalism*. Though, to relate this type of names fully to the class of the substance names would be a mistake.

Thus the denomination class of substances amounts to a wide-variety of things expressed by the substance names. Among such things are concrete persons, objects, and places named by proper names; an individual example of a class of thing or the class as a whole, both named by class substantives, which have in turn the subdivisions: a collective name referring to a collection or aggregate of individuals and a mass noun denominating the whole class of things, such as physical substances and materials (taken as a whole and in general) of which particular objects are composed.

To identify and distinguish among names as between mass, uncountable names and countable nouns, a special class of words is used as determiners or determinatives. Determining the referents and meanings of names, they comprise identifiers **I**, definite and indefinite, identifying a class of thing or an example of a class and quantifiers **Q,** definite and indefinite, expressing a quantity:

$$\mathbf{D} = \mathbf{I} + \mathbf{Q} \tag{4}$$

The indefinite identifiers include the indefinite articles and the zero article. So, the indefinite article *a* or *an* is assigned to a class of thing as a whole or to an instance of such a class; whereas the zero article, absence of any article, another indefinite article, can be applied to a class as a whole, to the whole class of thing, such as human beings in general. The definite identifiers cover the definite article, some pronouns (the demonstratives, possessives, anaphoric pronouns), and ordinal numbers. The definite article, *the,* identifies the particular examples of a class of thing, as well as a whole class of animate and inanimate substances. Of other determiners, cardinal numbers make definite quantifiers, whereas indefinite quantifiers are expressed by uncertain quantities, such as *a few* or *a little*, *many* or *much*, *a great number* or *a great amount of, several, all, plenty of, a lot of enough, some, any, every, no. All* refers to the whole number; to all the elements or units, *every*; to all taken separately, *each*; to one example of class of all, *any*; and to no one of all, *none*. Thus a group of quantifiers consists of the definite quantifiers, numerals, the names of numbers, ordinal and cardinal, and the indefinite quantifiers added with indefinite pronouns representing a thing generally or indefinitely. Unlike the logical quantifiers (*some*, *all*, or *no*), the grammatical quantifiers embrace *much, many, more, most, few (a few, fewer, fewest), little (a little, less, least), several; all, another, any, both, each, every, either, neither, none, one, other, some, such; everything, everyone, everybody, anything, something, somebody, someone else, somewhere, somehow, sometime, somewhat.*

Beside the determiners, the substance names can be modified (described, defined, qualified, specified, or restricted) in the meanings by other modifiers, the universal words describing the characteristic attributes, qualities, or properties of a substance, namely, by other nouns, adjectives, or by simple sentences, clauses.

Standing in the subordinate relation to a head noun, such modifiers occupy their position according to whether they describe essential or nonessential attributes of a substance; thus forming a noun phrase $\mathbf{N_{OP}}$ with the strict order of the modifiers, as in:

$$\mathbf{N_{OP}} = < (\mathbf{D} + \mathbf{A} + \mathbf{N}\,)\, \mathbf{N_{OP}} > \tag{5}$$

The symbol structure $\mathbf{N_{OP}}$ stands for *the object naming phrases* where the head noun is modified by other determiners, as well as by attributive adjectives and nouns signifying substance, state (property, quantity, quality), change (action, influence), and relation (place, time, cause, purpose). The strict order of modifiers in the object nominal group is started with determiners and continued with epithets, like participial adjectives (the present participle, and past or perfect participle), expressing some nonessential features of the entity referred to by the head name. This is followed by a major word class, adjectives, describing some particular properties of the substance denoted by the head word, like the quality adjectives (admitting of variation of degree and of contraries) describing mental states and attitudes (habits and dispositions, or inherent powers), sensible qualities (physical properties, color, temperature, weight, sound, taste, smell, etc.), or quantitative qualities of size, figure, and shape, and relational properties (spatial, temporal, etc.).

The established order of the words is so: after the determiners, come the adjectives of epithetical characteristics, then size; shape; time or age; color; place or situation; substance, stuff, or material. Most close is what forms with the head noun one lexical unit, a compound noun. So the end side of the sequence of adjectives is occupied by the adjectives or nouns signifying substances, the substantive adjectives. The closest position to the head noun is occupied by the nouns bringing a specific meaning for the head word, which are the relational adjectives, as classifying adjectives designating a subclass of a thing, like as *spatial* or *temporal entity*. The role of modifying names is either to indicate its purpose, possession, or subclass, or to show the material of which the substance is made. As a result of such nominal modification, we get a special kind of names, compound names, written as one word, with a hyphen, or as two or more words, not discussing here a group of compound names made up of verbs, adjectives, and prepositions, like *man-of-war*, *take-off*, *come-to-be*, *pass-away*, or *notary public*.

The rule of dividing substances into subordinate classes and kinds by the agency of modifying nouns applies also to all other types of names. The more nouns attached to the head word, the narrower its meaning; typically the number of modifying nouns exhaustively specifying any name amounts to five or six words, including the determiner. Formally, the substantive name compounds can be described as an ordered set of nouns, where the sequence and the stress pattern make all the difference, and the possible cardinal number of elements corresponds to the range mentioned. Some examples are as follows: *the ontology knowledge database management system, the first-order predicate calculus system, the Semantic Web reference ontology, the World Wide Web agent markup language*. While decoding the meaning of the last compound, the brainy machine should be able to infer, albeit mechanically and to limited extent, that "a markup language" is a subclass of "language," "agent" indicates the purpose of the language, and that "World Wide Web" tells it the place of origin or situation of the language. This capacity can be enabled by encoding the general rule of constructing the substance naming expressions, where adjectives as a special qualifier and modifier, much contribute to the meaning of substantive names.

By its meaning, adjectives express the values of properties (attributes) of things, such as substantive properties, stative properties, active properties, and relational properties. So the significant semantic role of adjectives is to express the values of properties of entities as it is shown in the descriptive formula:

State (or property, quality, quantity, and attribute) Name (Entity Name) = Adjectives

Symbolically, the general rule for constructing and analyzing adjectives is formulated as in:

$$N_S(N)= A\{A_O, A_S, A_C, A_R\} \quad (6)$$

It follows that the adjectives can be ontologically divided into four semantic groups:

- Substantive $\mathbf{A_O}$
- Stative (descriptive of conditions, qualities and quantities) $\mathbf{A_S}$
- Active (participial) $\mathbf{A_C}$
- Relational (classifying) $\mathbf{A_R}$

This grammatical formula reflects a general ontological rule that a property (attribute) is formally represented as a function mapping from an entity (relationship) set into a value set (or a cross-product of value sets). The classification of adjectives as substantive adjectives indicating permanent characteristics of objects, state adjectives, action adjectives, and relational adjectives appears more useful and systematic than dividing adjectives only as descriptive or qualifying, relational or classifying, comparative and superlative. Besides they are analyzed according to grammatical properties: positive or negative, gradable or ungradable (standard or absolute), upgraded or downgraded, attributive or predicative.

Grammatically, the meaning of adjectives can be obtained by typical morphological affixes, the prefix (*a-*) and the suffixes having the specific meanings:

- (-ABLE, -IL,-ILE; -ABLE, -IBLE, -IVE) meaning "able to," "able to be" or "tending to" (*applicable, sensible, ductile, irritable, productive, attentive*)
- (-INE, -IN, -ING, -IVE, -ORY) connoting "like," "being the nature of" (*bovine, boring, native, desultory*)
- (-AC, -IC, -ACEOUS, -AL, -AR, -ARY, -ILE, -IL, -INE, -ORY) indicating "relating to" (*scientific, curvaceous, historical, circular, temporary, promissory*)
- (-AL, -ATE, -IAL, -ID) meaning "characterized by" (*jovial, approximate, martial, flaccid*)
- (-FUL, -OSE, -OUS, -IOUS) suggesting "full of" (*plentiful, verbose, poisonous, spacious*)
- (-ANT, -ENT; -FIC, -FICENT) suggesting "doing, agency or action, causing, and making" (*errant, dependent, terrific, sufficient*)
- (-AN) indicating "coming from," "following," or "belonging to" (*suburban, Republican, crustacean*)
- (-LESS) meaning "without" (*useless, helpless, etc.*)
- (-ISH, -Y, -LY) suggesting "having the quality of," or "recurring" (*childish, stony, muddy, dusty, lonely*)
- (-AC; -ED) implying affected by (*maniac; amazed, frightened, pleased, surprised, etc.*)

The addition of the prefixes DIS-, IN-, UN- to participial adjectives (*ing*- and -*ed*-participles) and to their derivative adverbs suggests "not" or "reverse of."

It emerges that for automatic semantic processing of adjectives in order to determine a kind of thing, which this part of speech indicates, it must be originated by a stemming algorithm removing all its affixes so that to reduce to the root, base, theme or radical, while for meaningful processing of names the morphemes turn up more important than roots.

The next core element of the namespace is *the name of state*. It indicates **being-in-state** or **state-of-being** taken in its general reference and meaning as abstracted from all sorts of specific states, conditions, states-of-affairs, situations, circumstances, and cases. Semantically, such words are very close to the substance names interpreted as the totality of things having common state or condition, that is, to the names denominating the whole class of things. But, grammatically, in contrast to the substance names, the state names, $\mathbf{N_S} \subset \mathbf{L_S}$, belong to a subclass of state words mostly derived from adjectives (of attributes) or state verbs by adding the state markers (descriptors or classifiers):

-AGE, -ANCY, -ENCE, -ENCY, -ETY, - ITY, -IBILITY, -CION, -SION, -ION, -TION, -TUDE, -SHIP, -HOOD, - NESS, -TH, - ACY, - OSITY, - Y, -ISM, - CRACY, -DOM, -ERY, -TY.

The *state name variables* standing for the state (condition, quality, property, or quantity) of entity includes all nouns of state, condition, quality, quantity, and property, as the instances: *bondage, arrogance, existence, activity, reversibility, condition, magnitude, friendship, falsehood, happiness, truth, strength, fallacy, verbosity, orthodoxy, skepticism, theocracy, wisdom, slavery, cruelty*. It should be noted that in some cases the prefixes may be as important in defining nominal meanings as the suffixes. So are the prefixes designating the multiples and submultiples of physical units, such as length, weight, temperature, sound, power, frequency, or electrical current. For instance, the prefixes of multiples of physical units are: deca-, hecto-, kilo-, mega-, giga-, tera-, peta-, exa-, while the prefixes of submultiples are deci-, centi-, micro-, nano-, pico-, femto-, atto-.

As it happens with the substance names, each stative noun can be modified, defined, limited, or described by other nouns, adjectives, clauses, and determiners (identifiers or quantifiers), thus forming a state noun phrase: $\mathbf{N_{SP}} \rightarrow$ pre-modifier + $\mathbf{N_S}$ + post-modifier.

The nominal variable $\mathbf{N_{SP}}$ indicates a kind or an instance of a class of states, that is, the state or mode in which some entity (thing, object, person, or agent) exists (a condition, situation, position, circumstance, etc.). Namely, a state in general can be exemplified by the specific attributes of particular things, their conditions, qualities, properties, and quantities, as in the examples: *the darkness of sky, the softness of water, the absence of brains, the importance of decision, the activity of mass, the existence of life, the resemblance of quality, the truth of sentence, the health of body, the extension of body, the hostility of enemy, the misery of war, the ambiguity of language, the hardness of being.*

The formal account of the *state presenting nominal expressions* $\mathbf{N_{SP}}$ can be taken from the common noun phrase formula (2) for the substantive names, where the head noun signifies the state of being:

$$N_{SP} = < (\mathbf{D} + \mathbf{A} + \mathbf{N})\, N_{SP} > \quad (7)$$

Together with determiners and adjectives, the symbol structure $\mathbf{N}_{SR}$ includes all the possible kinds of compound nouns made with the state names as the head nouns subject to modifications, $< \mathbf{N}_O, \mathbf{N}_S >$; $< \mathbf{N}_S, \mathbf{N}_S >$; $< \mathbf{N}_C, \mathbf{N}_S >$; $< \mathbf{N}_R, \mathbf{N}_S >$, where $\mathbf{N}_S$ is a set of the class of state words $\mathbf{L}_S$. When the state name variable is replaced by the indications of various sorts of states, while the substance name variables are substituted by the names of objects, we obtain the name phrases for a countless number of particular states.

In the case of human beings, the examples of the state name compounds suggesting specific mental states can be constructed from the mental dispositions and attitudes of a person, $\mathbf{N}_O$ = (*person*), such as $\mathbf{N}_S$= {foolishness, braveness, cleverness, cruelty, generosity, gladness, goodness, kindness, silliness, sadness, stupidity, wickedness, wisdom, willingness, reluctance}. For physical objects, the state indicating phrases can be constructed from $\mathbf{N}_S$ = (*properties, quantities, quality*) and $\mathbf{N}_O$ = (*body, nucleus, electron, atom, molecule, gas, water, solid, plasma*). To construct particular phrases describing the specific attributes of bodies, more specific property words are to be used: *magnitude, extension, multitude, number, size, shape, figure, weight, mass, temperature, humidity, solidity, texture*, and so forth. As a result, we obtain the state-related expressions like *body size* or *body temperature* or *air humidity* widely adopted in modern science.

As for the names of change aggregated as a set of verbal nouns (verbals), they indicate **being-in-change** and **change-of-being**. This means that the collection of the change and action and activity names $\mathbf{N}_C \subset \mathbf{L}_C$ is a subset of change-related words, and mostly derived from verbs by adding the change suffixes, the change nouns designators, descriptors, or classifiers, so as to express the meaning of specific changes, as acts, actions, events, or processes:

-ANCE, -ANCY, -ENCE, -ENCY, - ING, - AGE, -ERY, - MENT, -CION, -ION, - SION, -TION, -XION, -IZATION, - URE, - FACTION, - AL, -ADE, -ISM, -OSIS, -ITIS.

Beside the *-ing* participle and the *-ing* as expressing gerund, some of these onto-morphological designators can be mistaken with the descriptors of the state names that express both action and condition: -ANCE, -ANCY, -ENCE, -ENCY, -MENT, -CION, -ION, - SION, -TION.

But, it should be observed, most of the verbal nouns are natural names, **non-derived verbal nouns**, such as *fall, fear, love, hate, influence, cause, change, effect, impact, aim, change, move, order, wonder, trouble, work, plan, dream, desire, fear, smell, taste, sense*. A number of verbal names are derived from their verbs while preserving the meaning of the stem verbs, such as the following **derived verbal nouns**: *performance, existence, aging, wishing, breakage, discovery, management, decision, cognition, pleasure, satisfaction, approval, metamorphosis, encephalitis.*

Altogether, a set of change names $\{\mathbf{N}_C\}$ consists of the following subsets:

- *Naturally given (non-derived) change substantives* $\{\mathbf{n}_n\}$
- *Derived verbals* $\{\mathbf{n}_{d,}\}$

- *Gerundial names* $\{\mathbf{n}_g\}$
- *Infinite verbals* $\{\mathbf{n}_i\}$ commonly marked my the infinitive marker *to.*

As a result, the totality of change names is equivalent to the sum total of all verbal nouns, as in:

$$\{\mathbf{N}_C\} = \{\{\mathbf{n}_n\}, \{\mathbf{n}_d\}, \{\mathbf{n}_g\}, \{\mathbf{n}_i\}\} \quad (8)$$

An inherent difficulty with the Change Names or Change Nouns consists in that a scope of their connotations is not as simple as that of other names, for they do not carry only the meaning of their root words. The ambiguity and multiplicity of their meanings come from the intricate composition of change. That is, such significance complexity can be explained by the number of the elements any change or action usually involves. Apart from *that which changes* (the matter of change), *that to which* (final state or property) and *that from which* (initial state or property) the change takes place, it includes *the action itself, the state of being in action* and *the process* by which such change is carried out. The whole meaning range of the change names, therefore, covers:

- *The change, as action or process, of what is expressed in the verb root* (reading, reflection)
- *The product or result of an action* (reading, connection, achievement)
- *The state (quality, condition) of being* (astonishment, completion)
- *The means (instrument) by which the process of change is accomplished* (a covering)
- *The material connected with a process* (washing, reading, piping, flooring)
- *The power, quality, or faculty of* (mental acts, perception, judgment)

To get a specific meaning for a verbal noun is much easier when it is combined or compounded together with other words. As with the state names, the *action name constructions* also can be obtained from the name modification construction by substituting the proper head name of actions, as in:

$$\mathbf{N}_{CP} = < (\mathbf{D} + \mathbf{A} + \mathbf{N}) \, \mathbf{N}_{CP} > \quad (9)$$

The change phrasal expressions (for various sorts of changes, doings, or actions) consist of determiners indicating whether the subjects of change are some concrete objects or substances in general, attributive adjectives, and the permutations with the head change name: $\mathbf{N}_{CP} = < \mathbf{N}_O, \mathbf{N}_C >; < \mathbf{N}_S, \mathbf{N}_C >; < \mathbf{N}_C, \mathbf{N}_C >; < \mathbf{N}_R, \mathbf{N}_C >$, where $\mathbf{N}_C$ is a subclass of the change words $\mathbf{L}_C$. The symbol structure $\mathbf{N}_{CP}$ embraces all the verbal nouns modified by other nouns, adjectives, determiners, as well as clauses: $\mathbf{N}_{CP} \rightarrow$ pre-modifier + $\mathbf{N}_C$ + post-modifier. The change phrase formula reflects the name description of kinds, cases, and examples of change, particular actions, activities, or specific events and occurrences.

As regard to the associative grammatical elements or components of the namespace structure, the relation names, the topic appears most complicated and entangled. This is mostly because of the long-lasting ambiguity of the nature and meaning of relationships, their true place in the world of things as associative entities. In traditional English grammar, this results in the so-called function, or closed-class words, covering various relative words, prepositions, conjunctions, some adverbs, auxiliary verbs, and relative pronouns, as a sort of signs having neither meaning nor content but only grammatical function. In line with the EIS standard ontology, we recognize the reality of connective entities clearly manifested by such things as causal relations or spatial and temporal orderings, and so forth. A relation has a reality in and of itself, which is not identical to the correlative terms related by it. With that the things indicated as relations are different to their relatives. This means that the relatives as *being of or related to something else* may be designated by the substance, state, or change names, while the relations *per se* are connoted by relative names, as well as by relational verbs, adjectives, pronouns, prepositions, conjunctions, or disjunctives. So there are relative words expressing relatives, as *whole/part, cause/effect, heat/cold, master/slave, double/half, knowledge/ignorance, good/evil peace/war, parent/child, premise/conclusion.* And there are relational terms expressing the relations *per se.* The relational terms here are understood as the indication of all possible ways, whereby the relatives can be related, compared, associated, united, or connected: in nature, in meaning, in thought, or in discourse. Unlike the relative terms correlated by the relationship as such, the types of relations can be as various as the following: part-whole relationships, causal relations, relationships in space and time, comparison, similarity or diversity, analogy, proportion, opposition, contextual relations, mathematical relations, logical connections, semantic, and grammatical relationships.

Then, to name relations, we should not refer to a couple of words naming things that have a certain relation to each other. Instead, the names of relations indicate **being-in-relation** or **relation-of-being**. This suggests that the collection of the relational nouns $\mathbf{N_R} \subset \mathbf{L_R}$ is the subset of relation-related words consisting of some non-derivative names, as *hierarchy* or *analogy*, and mostly derivatives from proper adjectives and verbs by adding the (abstract) noun suffixes, such as:

-ANCE, -ANCY, -ENCE, -ENCY, -TUDE, -SHIP, -ITY, - ING, - AGE, -ERY, - MENT, -CION, -ION, - SION, -TION, -XION.

As an illustration of relative names may serve a collection of general names of relations: *affinity, analogy, arrangement, association, bondage, causation, classification, comparison, contiguity, connection, contrast, correlation, correspondence, dependence, equality, identity, interaction, hierarchy, link, mutuality, opposition, position, possession, parallel, ratio, reciprocity, reference, resemblance, similarity, situation, structure, system, tie*; *truth, knowledge, attitude, parentage, friendship, kinship, and so forth.* These general names express classes or categories of relation under which fall more specific words naming the variety of species, the subordinates of relations. For instance, relative identity and diversity involve for their description such relational names as *same* and *other, likeness* and *unlikeness, sameness* or *difference, equality* and *inequality, similarity* and *dissimilarity*; all implying that something is the same or different only when being related to something else.

As for *time*, it is above all a relational entity and only then a quantity, measurable and measured duration of change, motion, and movement; hence, accounting for the temporal relation of a verb, a grammatical tense, present past, future as well as progressive and perfective with their present, past and future tenses. Temporal names expressing relationships in time include such general words as *duration, span, sequence, succession* suggesting the length and the order of changes and actions, plus numerous definite portions and subdivisions of duration as *date, age, epoch, era, season, period, interval, space, instant, moment*, and *point*.

If the things contained in time are changes exemplified as specific events and acts, the things contained in space are substances appearing as specific objects. "To space" means to put spaces or intervals between objects, while "to place" signifies putting in a spatial order by assigning a particular relation of spacing to something else, that is a proper position. The kinds and portions of space generally are expressed by such space names as *extension, region, area, interval, distance, place, location, locality, position, post, room, site, situation, spot, station, surface, line*, and *point*.

As it happens with other names, any relation name can be limited, modified, defined, and described by other nouns, as well as by adjectives, clauses, and determiners (identifiers or quantifiers), thus forming relation noun phrases: $\mathbf{N_{RP}} \rightarrow$ pre-modifier + $\mathbf{N_R}$ + post-modifier.

The nominal variable $\mathbf{N_{RP}}$ indicates a kind, or an instance of a class of relations, that is, the mode in which one thing is related to other thing. A relation in general may be exemplified by different correlatives in different places at different times, as in the examples: *space relations, time relations, quality resemblance, the energy relations of matter, the truth of sentence, close kinship, school friendship, passionate love, family relationship, order of the day, the heads of division.*

As with the other names, the names of relationships and associations can be obtained from the universal noun phrase formula (2) by substituting the proper head name with relational nouns, as in:

$$\mathbf{N_{OR}} = < (\mathbf{D} + \mathbf{A} + \mathbf{N})\, \mathbf{N_{OR}} > \quad (10)$$

The phrasal structure $\mathbf{N_{RP}}$ is a set of constructions acquired by pairing all the relation words with other sorts of words, which includes as its members the noun compounds specifying various examples of relations: $< \mathbf{N_O}, \mathbf{N_R} >; < \mathbf{N_S}, \mathbf{N_R} >; < \mathbf{N_C}, \mathbf{N_R} >, < \mathbf{N_R}, \mathbf{N_R} >$.

Crucially, almost all compound names made as a combination of two or more nouns indicate certain kinds of relationships, such as the class inclusion relations, where the nominal constructions refer to subclasses: $< \mathbf{N}, \mathbf{N}, \mathbf{N}, ... > \subset \mathbf{N_{OR}}$. This takes place regardless how they are written: (1) as one word, *headache, railway, sunshine, homework, moonlight, newspaper*; (2) with a hyphen, *mouse-trap, tooth-brush, project-worker*; (3) as two separate names, *means test, blood test, information office, software company, police station, market town, light rays, e-mail message, traffic lights, alarm signal, fruit market, flower shop, space ship, road works, savings bank, science lesson.* There also are nominal constructions indicating the whole-part relation, as *bedroom furniture, city centre, government agency;* the material of which something is made of, *wood house, steel girder, leather clothing, beet sugar, gold pen*; or, the time of occurrences, *winter holidays, war time*, and so on. If descriptive adjectives only describe the entity referred to by the head name ascribing to it the values of a qualitative property, the modifying nouns classify the head names and so acting as relational, classifying adjectives. There also are relative compound nouns made out of nouns modified by adjectives, prepositions, and verbs.

As with other names, relative names also can be post-modified by prepositions, relational words, linking nominal phrases of different types, such as *of.* Replacing the relation name variable with the general names of the kinds of relations, and the rest with the general names of objects, states, or changes, we can get the relative nominal phrases: *the structure of the world, the being of relations, the relation of things, the hierarchy of beings, the association of ideas, the order of nature, the relations of equality, the relation of cause and effect, the order of causality, the relation of contiguity*. So the relation names also can act as the associative terms but now for the relationships expressed by other connective words, prepositions, and thus fulfilling the established definition: "those terms, then, are called relative, the nature of which is explained by reference to something else, the preposition 'of' or some other preposition being used to indicate relation" (Aristotle, Categories).

There also is another way to construct concrete nominal phrases and expressions from generic entity names, as it is widely practiced in the (subject-predicate) grammars dealing with specific referents: particular objects, individual states, specific events, places, and times. In such a substance-attribute linguistic ontology, non-substantial entities (states, changes, or relations) are considered as concrete instantiations or exemplifications by substances of corresponding properties. Accordingly, states are states in substances, changes are changes of states (properties) in substances, and relations are specific associations of particular objects. In other words, the universe in the subject-predicate grammar consists of substances having various properties and standing in different relations. In that the substances are viewed as the subject of states and changes, such sort of connections is expressed by a relational word (or attributive tie), *of.* As far as this kind of grammar is engrossed by the ontological linguistics, we can specify (determine, exemplify, and instance) a generic entity (state, change or relation) by applying it to particular substances (or systems of substances) given with spatial locations S and times t, as in the concretization formula:

A State (Change, Relation) + A Particular Object = A Particular State (Change, Relation)

Therefore, the specific states (changes and relations) are nothing else than generic entities exemplified by individual substances at a certain time and place. Nominally, all sorts of states, conditions, states of affairs, along with specific acts, activities, happenings, occurrences, and processes, will be expressed by the syntactic formation rule involving a relational word, as in:

The General State Name (Change, Relation) + of + the Subject of State (Change, Relation)

The syntactical constructions of that pattern, where the entity names are post-modified by the substance name, commonly use the relational word *of* serving as an attributive link. Generally, such construction pattern may express different types of relations: identity, origin, cause, classification, inclusion, connection, removal, reference, direction, location, or time.

The specific meanings of this relational word, as well as other prepositions, directly depend on the nature of the names to be connected. So, if the names of changes are involved, then the sentence indicates the relationship of cause and effect (or logical relations of antecedent and consequent): as *because of* or *coming from* and *produced by*.

Alternatively, the state expressions may be presented in the genitive form with apostrophe *s* to indicate possession (*the animal's life, my father's property* or *the nation's affairs* or *a family's finances*), all is coming under one semantic formula:

A Particular State (Change, Relation) = 'D + Name's + State (Change, Relation) Substantive'

Also, all other expressions obtained for nominal phrases are valid for the action names as well, with some necessary modifications, like in the change-referring gerundial expression taken in the genitive form:

A Thing's Causing (Experiencing) Change = 'D + Substantive's + Causing Term + A Change Name'

The causative terms can be substituted by the gerundial form of any causative verbs (see below), such as *causing, making, doing, affecting, giving* or *having, undergoing, experiencing,* or *submitting*. They all are marked off by taking adverbial modifiers, objects, modals as well as by having tense, aspect, and voice. In the place of the change names we can put action nominals (change-like things and fact-like entities), which are the verbal nouns produced from the action verbs, as in the examples: *somebody's making a(n)* {accusation, agreement, attack, attempt, bargain, calculation, choice, comment, contribution, copy, correction, criticism, decision, difference, discovery, escape, excuse, job, haste, inquiry, investigation, journey, mistake, move, noise, note, observation, offer, proposal, proposition, recommendation, reduction, reference, report, request, suggestion, trouble, war, will}.

More complex case of non-finite, infinitive verbal nouns can be illustrated by the nominal phrases, *somebody's* {anxiety, pleasure, relief, surprise, honor, determination, inclination, reluctance, and willingness} *to see somebody else*. The passive infinitive is expressed as *women's desire to be admired*, the perfective aspect as *man's wishing to have been known*, and the progressive aspect as *animal's inclination to be living in society*. Yet the genitive form is commonly restricted by being used only with personal nouns and indefinite pronouns, the names of animals, with inanimate nouns denoting a group of people, their places, countries, institutions, and with the nouns indicating a change's duration. Though, there currently is a tendency of using inflections instead of closed-classes words by widening the usage of the genitive form in the expressions like *a thing's being* produced from the phrase *the being of a thing*.

Altogether, we obtain a general phrasal expression signifying a part of the thing's being and applicable for any sort of a particular:

A Thing's Property (or Attribute) Being {or Having, Causing} Something = 'D + Substantive's + Being (Having, Causing) Term + (Substance, State, Change, Relation) Name'

Thus, when suitably accommodated with the universal onto-grammar, the subject-attribute grammar's phrases will signify the being of a thing in a given form, state, change, and relation understood as the properties of a substance (substantial, essential, or intrinsic property); change-related, functional, and active properties; or relational properties. Such as it is illustrated in the cases: *a substance's being an agent, planet, star, galaxy, man, country, wood, paper, iron, glass, or dress*; *the animal's being alive, dead, a father's being ill, healthy, 'the nation's being rich, poor, the family's being large*; *the child's becoming adult*; *that's causing this, man's being a parent*, and *human's having affinity to animals*.

Prepositions and Adverbs: Nature, Meaning, and Classification

Before passing to the content analysis of verbs, we have to look into the nature of prepositions and adverbs, viewed in traditional grammar as function words and qualifiers (modifiers), respectively. As for the issues what prepositions are and how they should be classified, we again need to refer to the category of relations; for all prepositions express relationships between two things signified by a substantive, an adjective, or a verb and a substantive name of the preposition. Such as names, adjectives, and verbs, prepositions also fall under general terms, constituting a distinct lexical category. With that their scope or range of meanings usually is very wide and uncertain. For instance, the preposition to be *in* can be used in many various senses, namely: (1) the part (species) *in* the whole (genus); (2) the whole (genus) *in* the parts (species); (3) the form *in* the matter; (4) the action *in* the agent; (5) by reason, as a result of, *in* fear; (6) for a purpose of, something is *in* a thing's end; (7) a thing is contained, included, or bounded in something else as *in* a receptacle, *in* time, *in* space; (8) by means of, *in* a language; (9) with respect to, *in* skill; (10) according to, *in* my thinking; (11) possession, to have *in* him. Still these are not all the possible meanings, here also is the further senses: to be occupied, *in* search of something, to be made out of, *in* wood, to be wearing, *in* red, to be surrounded by, *in* the snow, and so forth. The same happens with its synonyms *into*, and with a set of antonyms such as *beyond, by, from, out, past, through, with* and with other English prepositions such as *after, at, before, for, of, to*.

To divide this part of speech into categories, it is necessary to refer to the major species of relation as its subordinate classes, which are the relationships in space, in time, of cause and effect, as well as the relations of parts and the whole, opposite relations of entities, and comparative relations of similarities and differences, qualitative and quantitative, direct or rhetorical.

In English, the main role of prepositions is to indicate spatial relationships, namely a place-position or place-direction of one object to another, which can be modeled as follows:

a. Without dimension, as a point

b. With one dimension, a line

c. Two dimensions, a surface
d. With three dimensions, a space or area

Also, prepositions can express relationships in time, causality, final (destination and purpose), efficient (agency), or instrumental (instrument), the relations of resemblance and diversity. Then, in general, all cardinal prepositions **P** may be classified into several groups:

i. Indicating relations in space $\mathbf{p}_s$
ii. Pointing relationships in time $\mathbf{p}_t$
iii. Expressing causal relations (material and instrumental causes, efficient causes, and final causes) $\mathbf{p}_c$
iv. Indicating relative diversity, difference or opposition $\mathbf{p}_d$, and comparative identity as correspondence or resemblance $\mathbf{p}_i$

$$\mathbf{P} = \{\mathbf{p}_s; \mathbf{p}_t; \mathbf{p}_c; \mathbf{p}_d; \mathbf{p}_i\} \qquad (11)$$

Among spatial prepositions $\mathbf{p}_s$ there are those that suggest two ways of the spacing of substances, the position place *where* something is located, and the direction place *where* something is moving or directed *to*. The space relation of position, or location, suggests being in spatial relationships expressed by the prepositions {at, away, back, off, on, in, outside, by, past, beyond, across, through, up, down, over, under, above, below, beneath, or underneath, in front of, behind, beside, next to, opposite, around, between or among, with, without, against, near, far away}. The space relations of direction can be indicated by {to, away from, from/ to, back to, for, toward(s), onto, off, on, into, inside, out of, outside, through, by, past, beyond, along, across, up, down, over, under, above, below, beneath, underneath, in front of, behind, beside, around, between, with, against, near, far away}.

As for the temporal prepositions, *to be in time* means the temporal relation of being *before*, *now*, and *after* respectively indicating the preceding order of the change that has happened, the current change, and the change that will happen. That is, this complex construction of temporal relations out of change and quantity (number) is made up of three species—the past, the present, and the future, where each of which has as its parts certain dates with a definite duration: *last year*, *yesterday*; *now*, *presently*, *just*; *tomorrow*, *next year*. A set of temporal prepositions used to indicate succession, sequence, period, limit, interval, and points of time consists of the following members: {at, after, about, before, by, in, during, on, for, from/to, throughout, till, until, within}.

Altogether, the causative prepositions suggest origin, agency, instrument, means, reason, or cause: *from, with, of, by means of, by, through, thanks to, because of, owing to, due to.* The final purpose (referential) prepositions imply purpose, subject, or reference: *about, in, on, of, as to, for, in favor of, concerning, connected with, involving, on the subject of, with regard to, as regards, with a view of, with reference to.* The "contrastable" prepositions involve **opposition** and difference and include such words as *against, but, but for, besides, despite, except, apart from, in spite from, notwithstanding, regardless of.* The corresponding prepositions express agreement or conformity, for example, *in accordance with, according to, in agreement with*, and so forth.

As for adverbs as a class of words modifying verbs, adjectives, other adverbs, and clauses, part of them are constructed from prepositions and names, another part derived from adjectives. Altogether, this part of speech may be arranged into four groups:

- Adverbials of place (position and direction) $\mathbf{a}_p$
- Adverbials of time (when, since-when, for-how-long, how-long-ago, and how-often) $\mathbf{a}_t$
- Adverbials of manner (in-what-way) $\mathbf{a}_m$
- Adverbials of degree (to-what-extent) $\mathbf{a}_d$

The *adverbials of place*, referring to position, are formed with the aid of names and position prepositions as *at, on, in*; while those referring to direction are constructed with the aid of names and movement prepositions as *to, -ward(s)*. The adverbials of time also are formed from temporal prepositions and names: *in last year, since last year*; whereas, *how-often* adverbs of frequency and *how-long-ago* adverbs of relative time have their specific names: (*always, ever, frequently, generally, occasionally, often, rarely, never, seldom, sometimes*) and *(already, just, early, late, recently, soon, still, suddenly, usually, yet*), respectively.

The remaining adverbs, the *adverbials of degree* and the *adverbials of manner* have neither temporal nor spatial origin. The adverbials of degree, including intensifiers, indicate what scope, range, extent or intensity have a state or quality expressed by an adjective, or a change expressed by a verb (e.g., *almost, altogether, hardly, nearly, practically, quite, rather, too; a bit, absolutely, at all, badly, completely, extremely, greatly, highly, indeed, most, much, perfectly, positively, really, seriously, terribly, utterly, very*).

The *adverbials of manner* signify in what way a change happens, something is done, or how an action is performed, and placed usually at the end of a clause. Most adverbs of manner, including viewpoint adverbs, are formed by the addition to an adjective the suffix *-ly* (e.g., *automatically, basically, generally, clearly, happily, scientifically, skillfully, simply, wisely, wholly; educationally, financially, practically, theoretically, scientifically*). So the class of adverbs $\mathbf{A}_V$ is composed, as in the traditional grammar, of four sorts of adverbs - the adverbials of place $\mathbf{a}_p$, time $\mathbf{a}_t$, degree $\mathbf{a}_d$ and manner $\mathbf{a}_m$:

$$\mathbf{A}_V = < \mathbf{a}_p, \mathbf{a}_t, \mathbf{a}_d, \mathbf{a}_m > \tag{12}$$

Verb Space: Verbs, Predicates, and Entity Types

Together with substantives, verbs make up the major structure of a sentence, and, generally, they are viewed as serving the predicate of a sentence indicating the present existence of state or action, while other verbal words fall under a group of indefinite verbs or the tenses of a verb. Being a major part of speech and constituting a distinct grammatical or syntactic category, the verbs are another main constituent of a sentence along with the names. Traditionally, a verb is defined as "a sign of something said of something else, that is, of something either predicable of or present in some other thing" (Aristotle, On Interpretation, p. 25). The classic definition was specified in the modern linguistics as that the verbs belong to a class of content words expressing existence, action, and occurrence; and serving as the predicate of a sentence, while, according to the onto-grammar a verb is a word indicating some existence, or a state, or an action, or a relationship. The definition of sentence is more complicated, since it differs even in standard grammatical works. Commonly, a sentence is defined as a group (number, collection, set, combination, or string of words that includes a subject, a verb (predicate) and an object to make a full grammatical structure with complete sense. In the ontological grammar, additionally to its syntactic properties, a sentence has the semantic property of representing a subject-object relationship of things via the few types of names and verbs.

From the traditional linguistic position, verbs are defined as content words denoting existence, state, action (or occurrence), while grammatically functioning as the predicates of sentences. In their changing and behavior, the verbal forms are regulated by two sorts of linguistic relationships—grammatical and semantic. Among syntactic relations are intransitivity, transitivity, reflexivity; voice, tense, aspect, inflection (conjugation), modality, modification (qualifying), conjunction, complementation, and agreement in person, gender, and number. Among the relations involving meanings may be such semantic relations as, superordination and subordination, synonymy and antonymy, whole to part and part to whole relations, and manner relations. So as the predicates verbs are divided into intransitive, transitive and reflexive forms, infinitives, participles (present and past), phrasal verbs, as well weak verbs (four forms) and strong verbs (five forms).

Verbs are identified with an n-ary mathematical relation over the sets of names, or as functions over name-arguments, as predicates in formal logic, as operators and functions in computer programming languages, or as hyperlinks among Web resources mapping relationships among entities in the real world expressed by Web pages (documents). Despite such advancement in the formal analysis, as content words, verbs still generally are classified just as state verbs, indicating a state of being (existence), and action verbs, expressing actions, processes, events, or occurrences.

Instead, from the onto-grammar perspective, verbs are considered as predicating of things in several basic ways: affirmation (or denial) of a thing's existence (or being), the fact to be a substance, being in a state or condition, being in a change, or being in the relation of a certain kind (part-whole, order, resemblance, or designation). Thereby and thereupon all the verbs designated by the variable **V** are to be partitioned into four broad types: universal verbs $\mathbf{V}_U$, state indicating verbs (state verbs) $\mathbf{V}_S$, change expressing verbs (change or action verbs) $\mathbf{V}_C$, and relation referring verbs (associative verbs) $\mathbf{V}_R$:

$$\{\mathbf{V}\} = \{\mathbf{V}_U\} + \{\mathbf{V}_S\} + \{\mathbf{V}_C\} + \{\mathbf{V}_R\} \quad (13)$$

As the class of being is the root of all things, the term "being" is pivotal to natural language; consequently, the words "to be" and "not to be" appear to be the primitive verbs not only in English, but also in any natural language; for it seems to be the power base for all other verbs as a word class expressing existence, state, change, and relation. Along with its Latin equivalent "to exist," the English verb "to be," conjugated as *be, am (are, is), was (were)*, has the meanings of all four essential kinds of being. As the substantive or existential verb, it means *to have reality, existence, actuality, or truth,* thus referring to the universal class of *being*, *entity*, or *thing*, usually expressed by the structure:

'There + BE + D (substance, state, action, or relation) name'

Besides, there are secondary meanings: *live*; *take place and occur*; *endure*, *stay*, and *remain*; *belong* and *befall*. As the basic copulative verb in all the languages, the existential verb "to be" can express all elemental forms of entity, to be a **substance**, a **state**, a **change**, and a **relation**: "to be {something or somebody; the property of, in possession of, equal in value to, worthy of, of a certain length; an occurrence, event; in relationship to, different from, the cause of, dependent on, a substitute for, a part of, true of, like, in place, in time}." Such a classification sheds light on a standard interpretation of the verb "to be" as expressing existence, identity, class inclusion and class membership, or property predication (state, change, or relation attribution).

Another universal verb is "to have," whose ordinary meanings are spread as wide as: *having the ownership, containing as a part, possessing an attribute, state or quality, undergoing or affected with a change, causing to be, doing something or engaging in activity; possessing a relation to, being in relationship to.* In general, the term is used with reference to: wholes and parts; states or attributes, qualitative and quantitative, mental and physical; changes, actions and activities; relationships of all sorts, kinds, and manner.

Next, the state verbs signify *the state-of-being* or the state of having some property, including the term "to have" used with reference to state, condition, quality, or quantity. On this score the stative verbs usually do not occur in the progressive form. With that, all the kindred verbs carrying the connotation of *possession* or *having* some characteristics and attributes will come under the state verbs: *hold, own, possess*, and *keep.* Verbs referring to the senses, meaning the involuntary reaction, "to become aware of by the senses" fall under the division of state verbs, such as, *see, hear, feel, smell, taste*, while such verbs as, *look* taken in the sense of making a conscious attempt to see or *listen*, in the sense of making a conscious effort to hear, relate to the category of action verbs. At last, all the predicates pertaining to the states (emotional and psychological) of the mind to some extent also can be numbered as the state verbs, such as, *believe, consider, detest, envy, expect, fear, feel, find, forget, hate, hope, imagine, know, like, love, mean, mind, notice, prefer, regret, remember, suggest, suppose, think, understand, want* or *wish.* All the stative verbs are commonly occurring in the non-progressive aspect.

If the stative verbs refer to all sorts of states, the change verbs indicate all the possible changes of those states. The conditions of things are to be expressed by the state verbs, while the actions that they do and the processes they undergo are to be expressed by the change verbs of action or activity. The sorting of active verbs is regulated by the manner of their action (admitting of contraries and of variation of degree); for the character of any particular process is determined by its type, sphere, and the properties of that which undergoes the process. The top verb in the hierarchy of active verbs is "to change" taken in the sense of making different (cause to change, in a certain form or respect) or becoming different (undergo a change in essence, property, action, or relation). All the subordinate active verbs bear the meanings of specific manner and distinct forms of change: substantial changes (generation and destruction, change integrity), modifications of states (alteration, increase and decrease, change state, change shape, change quality, etc.), dynamic, variable processes (change of manner of action), or relational changes (change of place, motion).

Having a typology of changes simplifies the task of classification of action verbs; for all the possible action (process and occurrence) verbs can be derived from it. Taking into account that there are four sorts of change (of substance, state (quality, quantity), of manner of action, and relation (position)), the active verbs $\mathbf{V}_C$ can be divided into four subdivisions involving substance (object) changing verbs, state modifying verbs, process verbs, and verbs of change relations, as in:

$$\{\mathbf{V}_C\} \rightarrow \{\mathbf{v}_{oc}\,;\, \mathbf{v}_{sc};\, \mathbf{v}_{ac}\,;\, \mathbf{v}_{rc}\} \tag{14}$$

The symbol $\mathbf{v}_{oc}$ replaces the change substance verbs indicating the meaning of *being the cause of substantial changes*, generation or destruction, namely, *bring about, cause, create, arouse, engender, generate, give rise to, produce, incite, yield, construct, transform, convert, make, beget, breed, occur, come into being*, and so forth.

The symbol $\mathbf{v}_{sc}$ symbolizes that subset of verbs referring to change states, such as idiomatic expressions with to "get, become, come, grow, turn, fall, go, run." Some state verbs refer to a state of changing (agency, action) or to a state of being in change, as an incomplete state, physical or mental, expressed by the infinitive in progressive aspect. Among the stative change verbs are the verbs that indicate: (i) *being in the process of acting or in the state of being acted* (e.g., *be* in the sense of "be being"; *have* in the sense of "be having"; *exist* in the sense of "be existing"); (ii) *sensory experiences, emotional and mental states* (e.g., "be sensing," "be feeling," "be thinking," "be imaging," "be forgetting," and "know" in the sense "experience the state of knowing." It is significant that this type of change verbs includes the qualitative change verbs implying *change in quality*, alteration (e.g., *affect, alter, influence, convert, modify, reform, remake, reorganize, transform, vary*, etc.). Also, they comprise all sorts of quantitative verbs suggesting *change in quantity*, increase or decrease (e.g., *grow, become bigger, increase, amplify, augment, enlarge, expand, extend, magnify, escalate, spread, decrease, diminish, decline*, etc.).

Further, the sign $\mathbf{v}_{ac}$ is the symbol of change action verbs indicating a process of change during which the state of action, operation, or agency changes its mode of acting or operation under the influence of other changes. Examples of such difference in the way, mode, manner of action may be the transitions from one distinct form of change to other, from rest to motion, destabilization, from motion to rest (negative acceleration), as well as all kinds of processes altering the rate of change.

The remainder comes under the relational verbs $\mathbf{v}_{rc}$ involving the idea of *having a relationship to or being in a relationship with*, and expressing all sorts of relation types: the relations of parts and wholes, relationships in space, in time, causality, as well as the relations of identity and diversity and semantic relations. First and foremost, the relative verbs include the senses of the verb *relate*, as "interrelate" or "be in a relationship with," "have or establish a relationship with," {relate, associate, link up, link, colligate, connect, tie in}, {refer, pertain, concern, come to, bear on touch, touch on}. For example, here falls the verb *make* in the sense of *constitute, compose*, as well as all the verbs in the sense of separating into parts, portions or classes, such as, *divide, partition, classify*; or all the verbs indicating similarity or diversity in substance, state, action, in relation, such as, *compare, resemble, differentiate, contrast, discriminate*, and *oppose*. Plus the verbs concerned with semantic relationships of symbols, signs, or constructs with substances, properties, changes and relations: {denote, refer to, designate, denominate, name, represent, stand for, mean, signify, imply, indicate, suggest, import, purport, betoken, point out}. Among the set of change relation verbs we can find the motion verbs having the meaning of *being the cause of changes in place, position, or spatial relation* (e.g., move, carry, transfer, transport, relocate, turn, revolve, drift, *travel, go, come, walk, run, stroll, march, fly, transplant, transpose*, etc.).

If the relation verbs are usually transitive verbs, the stative verbs and active verbs can be used both intransitively (a-causatively) and transitively (causatively). Verbs used in both ways can refer to:

- *Mental states* {see, hear, smell, feel, taste, wonder, hesitate, suggest, regret, remember, understand, know, believe};
- *Changes and actions* {change, begin, react, respond, contract, emerge, explode, appear, disappear, extend, end, stretch, lengthen, deepen, break, burst, close, clear, open, grow, shut, deflate, differ, taste, rot, decay, fire, increase, decrease, decline, happen, result, continue, decide, depend, pass, press, finish, follow, fail, fill, commence, shock, stick, shine, set, show, strike, seem, sell, signal, weight; comment, choose, cheat, act, travel, concentrate, suffer, ache, think, dream, imagine, behave, call, cry, give, marry, live, die, lie, feed, fight, object, play, promise, speak, write, smile, search, inquire, read, steer, submit, teach, telephone, try, trade, vote, wait, worry, work};
- *Motions* {move, come, go, walk, run, arise, rise, travel, fall, drive, advance, fly, stop, stay, turn, revert, slide, return, revolve, rotate, stand, jump}.

The semantic difference between some transitive and intransitive verbs may be illustrated by the key action verb to *change*. As intransitive, it bears a passive meaning: *to become different or diverse, to be subject to change, to be changed in substance, form, nature*, and so forth. As transitive, the verb has an active meaning: *to make different or diverse, to cause to be different from one another, to change the substance, form, nature, appearance, place; alter, modify, qualify, convert, transform, transmute*. Thus, the peculiarity of the intransitive verb comes from the fact that its grammatical form is active but the meaning is passive; whereas the transitive verb has both the grammatical form and meaning as active.

Besides, all simple sentences made of intransitive verbs can be transformed into a sort of name phrase having a pattern: *the [state (action) names] of [substance names]*. For instance, "something exists (changes)" becomes "the existence (change) of something." The phrasal form shows that this class of sentences is lacking full meaning and cannot be considered as complete. Then the correct sentence formulation should embrace the sentence pattern for intransitive verbs as well as the phrasal structure of sentence consisting of *a noun phrase* (a group of cognate words modifying the head noun) and *a verb phrase* (a group of cognate words with a lexical verb as its head).

Like with names, in many cases the nature and type of verbs, particularly the verbs of changes, can be identified by their morphological elements, prefixes and especially suffixes, as in:

- *-ate* meaning *combine* or *treat with* (calculate, separate, estimate, educate, associate)
- *-esce* meaning *begin* (coalesce, convalesce)
- *-fy* meaning *become or make* (certify, verify, codify, solidify, identify, simplify)
- *-ize (-ise)* meaning *become, make, practice* or *treat with* (publicize, realize, modernize)
- *–en* used for monosyllabic words and meaning *become* or *make so* (blacken, heighten)

Most verbs, as substantives, prepositions, and adjectives, are general words. To modify, define, and specify their meaning, a special class of auxiliary verbs is introduced in order to mark possible changes according to mood, number, person, voice, aspect, and tense. So, to express attitudes toward a possible state or change, the modal verbs are used: *will/would, can/could, shall/should, must/had to, may/might.* For expressing aspect, voice, number, or person, the auxiliary verbs *have, be, do* are employed.

All the complex verbs usually are classed into five groups:

1. Modal + Infinitive (mood)
2. Have + -ed Past Participle (aspect/tense)
3. Be + -ing Present Participle (aspect)
4. Be +-ed Past Participle (voice)
5. Do + Infinitive (question, negation)

When being simultaneously modified by all operators, the simple verbs **V** are to be transformed into the complex verbal groups $\mathbf{V}_G$, consisting of the verbs modified, with the modals and auxiliaries in the following strict order:

$$\mathbf{V}_G \rightarrow \text{Modal + Have-ed + Be-ing +Be-ed + } \mathbf{V} \qquad (15)$$

Due to the introduction of auxiliary verbs, any of basic verbs may be additionally characterized by the ontological categories of substance (person), quantity (number), change (voice, active or passive), time (tense), and modality (necessity and contingency, possibility and impossibility, existence/nonexistence), usually presented as grammatical relationships and syntactic categories (number, person, tense). These rules, though, exclude indefinite verbs, which do not carry with them the notion of time. The infinitive, the present participle, and the past participle constitute the non-finite parts of the complex verbs not influenced by tense or number and person of their subject. Due to this grammatical property, verbs also are the constitutive elements for constructing non-finite clauses out of finite clauses, or simple sentences.

Concluding the subject, we have again to emphasize that the present taxonomy of verbs, based on the ontological linguistics, comprises most interesting of existent verbal classifications that is founded on linguistic semantics. At present, there is a rich linguistic literature aimed at the lexical organization of this part of speech in terms of top-level categories such as being, quantity or amount, quality or property, relation or direction, place, time, position, state, change, action, activity, event, and manner (Vendler, 1967; Jakendoff, 1983, 1990; Dowty, 1979; Talmy, 1985; Levin, 1993; WordNet, 1998). Most of the works examining verbal taxonomies intend to identify a few verbal supertypes serving as the semantic primitives, primes, atomic predicates, or root verbs for all sorts of particular verbs. The schemes of verb-types vary in accordance with the scheme of general concepts. But, mostly and generally, as it has been pointed out, they all revolve around two wide groups of verbs, stative verbs (of State) and active verbs (of Change) presented as state (situation) predication versus process (change, action, event) predication. This results in missing a whole semantic domain of associative verbs. As it happened with the WordNet's organization of verbal meanings into 15 files (clusters or groups) by a manner relation (dubbed as troponymy) and formulated as "to V_1 is to V_2 in some particular manner" (Fellbaum, 2003). While enumerating the verb synsets in WordNet, we can see that the English verb semantic classes categorized not only by states and processes (actions, changes, or events), as commonly thought, but also by existence (being) and relationships:

- Body care and functions verbs (e.g., *clean, shiver, shudder, pain, wash*), $\{\mathbf{V}_C\}$
- Change verbs (e.g., *alter, vary, modify; turn; adjust, conform, adapt*), $\{\mathbf{V}_C\}$
- Cognition verbs (e.g., *think, reason, understand, learn, conclude*), $\{\mathbf{V}_C\}$, $\{\mathbf{V}_R\}$
- Communication verbs (e.g., *speak, talk, ask, answer, thank*), $\{\mathbf{V}_C\}$, $\{\mathbf{V}_R\}$
- Competition verbs (e.g., *compete, oppose, fight, battle*), $\{\mathbf{V}_C\}$, $\{\mathbf{V}_R\}$
- Consumption verbs (e.g., *consume, eat, drink*), $\{\mathbf{V}_C\}$
- Contact verbs (e.g., *fasten, attach, cover, cut, touch*), $\{\mathbf{V}_C\}$,$\{\mathbf{V}_R\}$
- Creation verbs (e.g., *make, create, invent*), $\{\mathbf{V}_C\}$, $\{\mathbf{V}_R\}$
- Motion verbs, (e.g., *move, go, run, come*), $\{\mathbf{V}_C\}$
- Emotion or psych verbs (e.g., *amuse, fear, love*),$\{\mathbf{V}_C\}$,$\{\mathbf{V}_R\}$
- Stative verbs (e.g., *be, have*), $\{\mathbf{V}_U\}$, $\{\mathbf{V}_S\}$
- Perception verbs (e.g., *listen, feel, smell*), $\{\mathbf{V}_S\}$, $\{\mathbf{V}_C\}$
- Possession verbs (e.g., *have, hold; give, transfer; take, receive*), $\{\mathbf{V}_S\}$, $\{\mathbf{V}_C\}$
- Social interaction verbs (e.g., *meet, transact, trade*, $\{\mathbf{V}_C\}$, $\{\mathbf{V}_R\}$
- Weather verbs (e.g., *precipitate, rain, thunder, wind*, $\{\mathbf{V}_C\}$

So what is essentially defective in the present linguistic verbal typologies, they are missing a set of relative verbs as a separate semantic set indicating the whole gamut of possible relationships among different entity levels presented by signs, constructs, and extra-linguistic things. In fact, the real verbal classification includes:

- The part-whole relation verbs (*divide, compose, categorize, classify*)
- The comparison predicates (*compare, identify, resemble, differ*)

- The verbs of opposition (*oppose, contrast, contradict*)
- The verbs indicating causality (*do, make, determine, initiate, effect, induce, stimulate*)
- The verbs of spatial relations (*displace, transfer, place, position, locate, site, direct*)
- The verbs of temporal relations (*time, endure, elapse, pass, expire, precede, follow*)
- The verbs of logical connections (*imply, entail, infer, deduce, follow, reason, prove, think*)
- The verbs of semantic relationships (*apply, denote, connote, refer to, designate, signify, represent, stand by, symbolize, mean, express, communicate, indicate, , depict, record*)

Failure to see that actions, activities, and events are only varieties of change makes the WordNet's verbal classification complicated and arbitrary. Besides, the hierarchy of verbs should correspond to the hierarchy of top-level names. But a more serious and fundamental defect consists in missing the existence (being) verbs and relational verbs, seemingly mixed with the verbs of state. As a result, the confused verbal taxonomies, together with incoherent nominal classifications present two main hindrances in such important natural language processing tasks as automatic construction, generation and parsing of meaningful sentences, the core of content of any language.

As other open-class words, verbs fall out to have four classifications:

- Existence verbs $\mathbf{V}_{\mathbf{U}}$
- Stative verbs $\mathbf{V}_{\mathbf{S}}$
- Action predicates $\mathbf{V}_{\mathbf{C}}$
- Relative predicates $\mathbf{V}_{\mathbf{R}}$

Accordingly, as it emerges from our analysis, unlike the "abstract" names, adjective and verbal nouns, the correct definition of verbs ought to be a class of words that indicates existence, state, change (activity, process, occurrence, and event), and relationship so that to assert, declare, or predicate any of them of some other things. Due to its nature, a verb is a basic constitutive component for constructing a simple sentence, a group of related words expressing (asserting or denying) some relationship of things. The sentence remains simple even if such a "meaningful portion of speech" is constructed with one or more substantives and verbs, simple or compound, and modified with words and phrases.

Sentence Patterns: Sentences and RDF Triples

In the grammatical hierarchy of natural language, a sentence is the most decisive portion, or linguistic unit of language and communication necessary for generating or understanding meaningful content. The issues of sentence construction and sentence parsing are the matter of fierce debates and argumentation among various natural language processing communities supporting various grammars, from phrase-structural to emergent to stochastic. This has its rationale, since it is hard to find any other linguistic category whose definition and meaning is so confused and misunderstood.

Sentence *may be defined as a number sequence (string, collection, group, combination, or set) of words making sense, expressing a complete thought, having a complete grammatical structure, containing a subject and a predicate, or satisfying the rules of a language.* A sentence is only one form of speech that states whether a thing exists or not, and whether the connection of the thing with its attribute or other thing is a true fact. This linguistic construction may express an assertion or denial of existence or non-existence, possibility or impossibility, contingency or necessity. Regardless of its complexity, the basic components and elements of such meaningful expression are names and verbs, the significance of which constitutes the whole meaning of the sentence. In the structure of an English sentence, the parts of speech occupy the place-slots as strictly designated and ordered, names, verbs, and names again. Such orderly arrangement is a way to show that each verb is functioning as the main predication of its subject, which may be a simple or complex expression signifying any sort of thing: substance, state, quality, quantity, action, relation, place, time, and position, and so forth.

Delving further into the superficial syntactic structure of names and verbs, one can uncover that sentences are patterned by two extreme ontological schemes: substance-attribute or relation-substance. In the first of which the most real entity is considered to be substance, while the other entities are just its accidents, attributes, or properties. Whence it follows the traditional substance-attribute grammar where every proposition is just assigning a property to a thing. And thence substantives and adjectives are only parts of speech standing for things shared by all particulars, at the expense of neglecting verbs and prepositions as universal words having general meanings. In such subject-predicate and phrasal grammar, a statement is regarded as expressing the fact of attributing properties to substances or ascribing attributes to the objects. When formalized, each basic sentence S is supposed to have the fundamental structure:

$$S \rightarrow N_P\,(\text{Subject}) + V_P\,(\text{Predicate})$$

Where Noun Phrase N_P and Verb Phrase V_P subject to the transition rules: $N_P \rightarrow D + N$; $V_P \rightarrow V + N_P$. The letter D also stands for Determiner consisting of Identifiers and Quantifiers.

The noun is here commonly divided into four ways: a Proper Noun denoting an individual thing, person, or place; a Common Noun naming a class of things, or the class as a whole; a Collective Noun expressing a collection of individual objects; and an Abstract Noun indicating an attribute, or quality. What is more, besides being the subject of a clause, the NP can be used as a grammatical object or a substantive clause.

In accordance with the composition of the V_P, the five basic sentence structures (clauses) usually are distinguished, as in:

$$S_1 \rightarrow N_P + \{V_i + A_V\}$$

$S_2 \rightarrow N_P + \{BE + (N_P; A; A_V)\}$
$S_3 \rightarrow N_P + \{V_t + O_d + A_V\}$
$S_4 \rightarrow N_P + \{V_t + O_i + O_d\}$
$S_5 \rightarrow N_P + \{V_t + (N_P; A_V)\}$

The V_i stands for Verb, intransitive; V_t points out Verb, transitive; O_d stands for Object, direct; and O_i replaces Object, indirect. Semantically, in the phrase grammar, verbs are divided into two groups—active and stative. The enumerated five major sentence patterns closely tallied with the five types of verb phrases accepted in modern phrasal grammars. A slightly alternative version of phrasal grammar considers as a sentence (or a verb and predicate) a group of words following a copulative, linking verb or transitive verb. Among complements may be direct objects, indirect objects, nominal phrases, or adjectives.

The contrary scheme tends to assume that substance is a kind of relation having the determinative reality, and that the relations of all kinds and sorts, instead of substances with their attributes, make up the structure of the world: nature, knowledge, and language. It then follows that substantives with adjectives and prepositions together with verbs, all stand for universal things, and that every significant statement expresses a relation between things, like the RDF triple representing a statement of a relationship between things it connects.

Of all known inquiries that which is proposed by Kant is regarded as mostly fundamental for relational grammar. Basing on his fourfold classification of things, Kant distinguished four types of predication: Quantity (universal, particular, singular); Quality (affirmative, negative, infinite); Relation (categorical, hypothetical, disjunctive); Modality (possible/ impossible; existent/ nonexistent; necessary/contingent). Interpreted as "all relations of thought in judgments," all relations of things represented in sentences are divided into three sorts of statements:

- Categorical (of the predicate to the subject)
- Hypothetical (of the cause to its effect)
- Disjunctive (of the relations of the parts of the whole, of logical opposition, or of the relation of community) (Kant)

Using this as the bedrock, some modern logicians proposed relational grammar, where relations generally expressed by verbs and prepositions, while properties by adjectives and common names and so statements express the relations between particular things.

Things with underlying sentence structures will look not so complicated, if one tries to avoid these two extremes of *substance-attribute grammar* or *relation-substance grammar*. In that case one will obtain much advanced grammars, as that one attained by Quine (Quine, 1960), who regarded relations as represented by transitive verbs, prepositions, relative common nouns with prepositions, comparative adjectives mixed with prepositions, and conjunctions, while substances with their properties expressed by substantives, adjectives and intransitive verbs (USECS, 2005). This model of grammar allowed two ways of predication connected with two main types of general terms, absolute $< N, V_i >$, and relative $< N, V_t, N >$, as well as with two basic sorts of verb patterns, intransitive and transitive.

Adhering to the onto-grammatical strategy allows us to combine all the approaches and to come to the most adequate definition of sentence is a tripartite SVO structure of words. Namely, a sentence is a string of words consisting of a subject name (S), a verb (V), and an object name (O), there the subject is that of which states (conditions or qualities), changes (actions), or relationships are stated (affirmed or denied). The object in turn is that which receives or endures the consequences of theses conditions, actions, or relationships. Mapping adequately the structure of reality, the ontological grammar reconciles such contrary perspectives as substantive and relational grammars, thus saving us a great deal of confusion and trouble by providing the fundamental sentence structure in the form of general sentential rule, formula, expression, or function. For, along with being, thing, entity and nonbeing, nothing, or nonentity, we need to give equivalent status to the primary classes of things by few semantic word classes: substantive words, state words, change terms and relational terms, all together making up the word network. Instead of phrasal structure N_P + V_{P}, all full sentences follow the fundamental pattern **Names + Verbs+ Names**, expressing a predication (statement) of some relationship (indicated by verbs) among things (denoted by names), formalized as the set of verbal functions over the sets of names **V (N, N, ...)** (see Figure 12). What implies a consistent system of onto-grammatical transformational rules simultaneously generating the totality of sentences, meaningful and grammatical:

$$\mathbf{S \rightarrow < N, V, N> \rightarrow < N_P, V_G, N_P > \rightarrow \ldots} \qquad (16)$$

The SOV structure of sentences is widely used in identifying the type of a natural language additionally to genetic classification by language families.

Figure 12. The word network for entities and relations

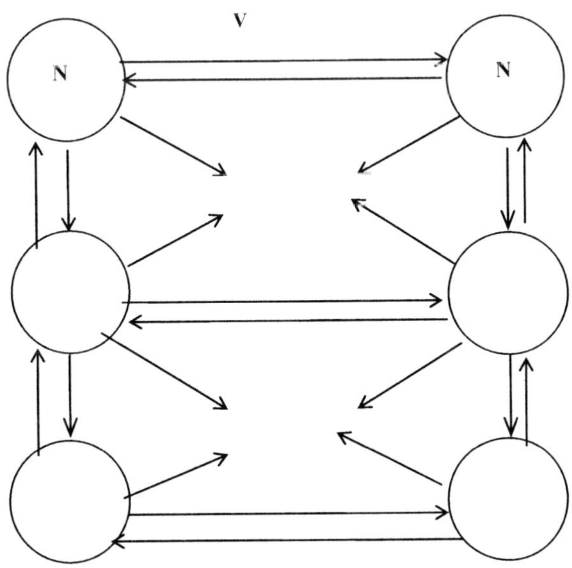

The **subject-verb-object sentence scheme** is used by some standard metadata models for encoding the machine-understandable resources on the Web, as RDF with its varieties, visualized as graphs of nodes and arcs, written by using an XML Namespace syntax. Being a standard representation language for the Web of data, it relies on the sets of RDF triples; where each statement includes a subject (an individual or a kind of things), a predicate (a type property or a relationship pertinent to the individual), and an object (the property value or another resource) (Resource Description Framework, 2004). Each Web resource is identified (named and located) by its WWW address, Universal Document Identifier, or global Universal Resource Identifier, URI, using a generic syntax of http URI schemes:

```
<scheme name> : <//hierarchical part (authority part and optional path)>
[ ? <query> ] [ # <fragment> ]
```

An RDF triple is presented as a triplet of URIs naming and locating Web resources via the HTTP request/response protocols between clients and origin servers (user agents) intermediated by proxies, gateways, and tunnels. All is to retrieve a description or representation of the Web resources, both information resources and non-information resources or abstract resources, such as concepts and attributes, to be identified and specified by the RDF vocabularies or Web ontologies. The naming mechanism between an URI and a Web resource (as a piece of information, a source of data, a network data object or service and process, or computational network entity) copies the semantic mechanism of denotation between a name and the entity named.

So, one can consider further examinations as establishing the semantic nature of Web resources, any entities having identities, and RDF triples, subject-names, predicate-verbs, and object names, all identified by URIs or Uniform Resource Locators enabling machines to communicate the semantic value of Web data and documents (Berners-Lee, Hendler, & Lassila, 2001).

Apart from the meaning composed of its constituents, some sentences are differentiated as having also the truth-value. Such variety of ordered names and verbs belongs to the special sort of sentences defined as propositions or statements, the truth or falsity of which depends on the fact of the thing's being in this or that form, state, change, or relationship.

In that we defined four classes of words, every significant sentence is to be constructed with substantives of four basic kinds related by verbs of four major types, as in:

$$\mathbf{S} \rightarrow \{\mathbf{N_{O,}\ N_{S},\ N_{C},\ N_{R}}\} + \{\mathbf{V}\} + \{\mathbf{N_{O,}\ N_{S},\ N_{C},\ N_{R}}\} \qquad (17)$$

The set of formulas states that every complete affirmation or denial consists of a noun, a verb, and another noun, simple or complex, finite or nonfinite.

Our sentence formulation also includes the standard definition of sentences widely accepted in the natural languages: "Every affirmation, then, and every denial will consist of a noun and a verb, either definite or indefinite" [Aristotle, On Interpretations], as in:

$$\mathbf{S} \rightarrow \{\mathbf{N_{O,}\ N_{S},\ N_{C},\ N_{R}}\} + \{\mathbf{V}\} \qquad (18)$$

Such alternative is fit for an intransitive verb not taking an object, which agency or action does not go farther that its subject. In fact, there are very few verbs that are absolutely intransitive, and among them the substantive verbs *be*, *exist*, having the sense of the unqualified being of a thing, the fact of the thing's being (or not being).

Altogether, the underlying sentence structure of the WordSpace can be divided into 4x4x4 types of simple sentences, four classes of names times four types of verbs times four classes of names, and presented as a rectangular array of canonic sentence patterns:

N_O V N_O	N_O V N_S	N_O V N_C	N_O V N_R
N_S V N_O	N_S V N_S	N_S V N_C	N_S V N_R
N_C V N_O	N_C V N_S	N_C V N_C	N_C V N_R
N_R V N_O	N_R V N_S	N_R V N_C	N_R V N_R

When adding to affirmations all denials or negative assertions, the number of sentence patterns at least will double, if not to count the patterns **N+V+N+N** as separate structures.

However, only four sentence species derived from the general structure seem to be mostly valuable. They are those that are composed out of the same semantic word classes: a substance name, a verb, other substantive; a state noun, a state verb, other state name; a change name, a change verb, other change substantive; and a relation name, a relation verb, other relation noun:

- $S_O \rightarrow \{N_O\}+\{V\}+\{N_O\}$
- $S_S \rightarrow \{N_S\}+\{V_S\}+\{N_S\}$
- $S_C \rightarrow \{N_C\}+\{V_C\}+\{N_C\}$
- $S_R \rightarrow \{N_R\}+\{V_R\}+\{N_R\}$

These kinds of sentences are commonly used for definitions, axioms, or postulates signifying essential nature of things and dealing with the statements of the relationships (e.g., differences or likenesses) of things within the same class. A typical example of such definitional sentence is: "A component (constituent, ingredient, element; section, fragment; share, division, subdivision; atom, particle, piece) is a part (portion) of some whole (compound, mixture; class, aggregate, collection; matter, mass, body)."

As for particular sentences, they are mere examples of universal propositions. Hence, to state about particular facts and things, the general terms should be substituted with their befitted phrasal expressions, as in:

$$S_P \rightarrow \{ N_{OG}, N_{SG}, N_{CG}, N_{RG}\} + \{V_G\} + \{ N_{OG}, N_{SG}, N_{CG} N_{RG}\} \quad (19)$$

Thus, in line with the general typology of things, all the sentences are to be arranged into four principal classes:

- An object (substance) sentence, when the subject is some substance (or object, spatial or non-spatial).
- A state sentence, when the subject is some state (condition, quality, or property).
- A change sentence, when the subject is some change (action, process, or event).
- A relation sentence, when the subject is some kind of relationship (association, connection).

Different criteria have been used to construct and classify statements. Even in our typology, a sentence can be arranged by what is stated (asked) about the subject, or what type of existence it has. However this will not miss the point that all meaningful sentences involve the generic classes of entities, and there is ultimately one universal sentence pattern.

Analyzing the types of sentences, it should be first observed that the object sentences, when the subject is the substance of which a sentence states or asks, prevail in our common language. Formally, all statements (predication and questions) about substances can be reduced to one sentential rule of onto-grammar:

$$\mathbf{S_O \rightarrow \{N_O\} + \{V_U; V_S; V_C; V_R\} + \{ N_O, N_S, N_C, N_R\}} \qquad (20)$$

Crucially, in the syntactic structure of the substantive sentences, $\mathbf{N_O}$ denotes the classes of substances encompassing as the members all kinds of agent, i.e., what or who has power to change (act), or to effect changes of any kind (physical, mental, abstract, moral):

i. An intelligent agent (human, animal or artificial)
ii. A physical substance or body
iii. A chemical substance, a medical drug, or a biotic agent

All propositions of the first-order predicate calculus come to be as a variety of the substantive sentences. Since the predicate calculus operates with *individual (object) variables* (x, y, z...), substituted by names of individuals as its subject-terms, *predicate variables* (α,β,γ,...), replaceable by predicates, and *logical operators* (¬, ∧, ∨, ⊃, ≡, ∀, ∃) corresponding respectively to the expressions "not," "and," "or," "if. . .then," "is equivalent to," "for all," "for some." All the following basic formulas fall under the object sentences where substance names are connected with the universal verbs "to be" and "to have" or with the relative verbs:

$(\forall x)\alpha x$ expresses 'for all x, x is α', 'everything is α'

$(\exists x)\alpha x$: 'for some x, x is α', 'something is α', 'there is a α'

$(\forall x)(\alpha x \supset \beta x)$: 'whatever is α is β'

$\exists x(\alpha x \vee \beta x)$: 'something is both α and β'

$(\forall x)(\exists y)\alpha xy$: 'everything has the relation α to something'

$(\exists x)(\forall y)\alpha xy$: 'something bears the relation α to everything'

$(\forall x)(\forall y)\alpha xy$: 'everything relates to everything'

$(\exists x)(\exists y)\alpha xy$: 'something relates to something'

For the higher degree (but not order) predicate calculus, a general transformation rule does not suffer principal changes: P: $O_1 \times O_2 \times ... \times O_n \rightarrow S$; P is a predicate of second degree or more $n \geq 2$, where n is the number of object variables, O_i is a set of objects. As a domain of the predicate (verbal) functions may be a class of living things, then its range is a collection of general object statements about living objects, like "all animals die." The value of P at the point $<o_1, o_2, ..., o_n>$ constitutes a specific object statement referred to a particular being, like "my chow died." To become the higher-order predicate system, the calculus should be able to describe a higher order property of a different sort, that is, properties of properties, things of higher order such as actions and relationships.

There are numerous examples of the object sentences, and they may be as different as: "The world is the aggregation of things"; "Entity is the class of all classes"; "Being is composed of its genera"; "Genus is a collection of things"; "Individuals belong to particular substances"; "Substance is permanent"; "All substances exist in a state of reciprocity of action"; "Matter is composed of atoms"; "A word is a symbol of a thing"; "Words make up sentences"; "A sign designates a construct"; "A mathematical set represents a collection of objects"; "Names name things"; "Every formal system has undecidable sentences"; "Object variables refer to substances"; "URI identifies resources"; "Ontologies provide semantics."

Further, for the matter of simplicity, the majority of further illustrations will be given in a categorical and phrasal form, what does not make a principal difference for the subject matter.

The agentive structure of statements is very popular in natural language. But in more accurate language, the agentive pattern of a sentence is to be replaced by the state, change or relation sentence patterns logically represented by higher-order predicate calculi, where quantification is permitted over object, state, change, and relation variables, which in the ontological grammar are the placeholders for terms.

The state sentences, when the subject is the state about which a sentence is predicated, prevail in modern science. The semantic formula for the state sentences is reduced to:

$$\mathbf{S_S \rightarrow N_S + \{V_U; V_S; V_C; V_R\} + \{N_O, N_S, N_C, N_R\}} \qquad (21)$$

Since the detailed exposition of all natural and social phenomena is mainly based on the concept of state, along with change and relation, the language of contemporary science tends to the state (situation, condition, quality, property, quantity) sentences, as much as the "folk" language to object sentences. There are some examples of the state sentences:

"Property is a state"; "Qualities and quantities belong to the class of states"; "A quality denotes what an object is in a certain relation"; "Number is a quantity"; "State undergoes a process of change"; "Property variables represent states"; "The properties of matter are studied by physics"; "A function measures (represents) the property of matter"; "Wealth commands position"; "Decidability is a property of sentences"; "Numerosity is a cardinal number of a set"; "Number limits all things."

Still the state sentences do not correspond to reality in such high degree adequacy as the change sentences.

The change sentences, where the subject expresses such things as, actions, processes, activities, events, occurrences and other types of changes, seem to be a form of sentences mostly wanted in recording all types of discoveries and rules:

$$S_C \rightarrow \{ N_C \} + \{V\} + \{N_O, N_S, N_C, N_R\} \quad (22)$$

All sorts of statements about changes, actions, activities, events, or causes fall under this category of sentences. Among the change sentences are the following:

"All changes take place according to the relation of cause and effect"; "Every change has a cause"; "Changes constitute causality"; "Motion is change of place in space"; "Alteration is change of quality"; "Growth is change of size"; "Generation is change of substance"; "Motion is a kind of action"; "Action is a series of acts"; "Speech is an action"; "Science produces knowledge"; "Change variables stand for causes"; "Event is a change that happens at a given place and time"; "The motion of matter is studied by mechanics"; "Heat transfer changes temperature"; "Boiling changes water into steam'."

The relation sentences, when the subject expresses all sorts of relationships, associations and connections of things, seem to be mostly complicated but mostly significant sentences:

$$S_R \rightarrow \{ N_R \} + \{V\} + \{N_O, N_S, N_C, N_R\} \quad (23)$$

The nearest examples of the relation sentences are the following combination of words with the aid of the universal verb *to be*: "A hierarchy of beings is described as the interrelations of classes"; "Bondage is a collection of all bonds"; "Structure is the totality of relations"; "Causality is a relation of cause and effect", "The succession of time is the succession of changes"; "Time is a measured duration"; "Space is a relation of objects"; "Place is a portion of space"; "Before is a temporal relation"; "Truth is a correspondence"; "Affinity is a relationship by blood"; "Parentage is the relation of parent to child"; "Identity is an equivalence relation"; "A sentence is a portion of speech"; "Truth is of the combination of words"; "Sequence of tenses is a grammatical rule"; "Spatiotemporal variables relate to space and time"; "The relation of class inclusion differs from a class membership"; "The meaning of a sentence is identified with its truth"; "Energy relations of matter studied by the physical sciences." Crucially, the significance of relative sentences consists in the recording of all specific types of meaning associations among heterogeneous entities such as, signs (or symbols), thoughts, and actual things. For instance, such sentences cover all expressions conveying indication, suggestion, or representation as according to the semantic scheme:

"a thing (an object, material or conceptual, state or property, action or process, and relationship) is an indication (representation, sign, symbol, mark, characteristic, or feature) of something (somebody) else."

Any sentence asserting (predicating) some sort of indication falls under the class of representation statements, namely, "Words are the signs of ideas," "Ideas are the similitude of things," "The idea of a thing is the meaning of its designation," "Wealth is a sign of success," "Success is no criterion of ability," "The figure of woman with child is the emblem of motherhood," "Greed is the characteristic of men of affairs."

A NL semantic processor can mechanically generate a wide range of meaningful examples, given the specific classifications of signs and symbols: {allegory, augury, auspice, badge, cachet, character, characteristic, cognizance, criterion, demonstration, designation, device, differentia, emblem, evidence, figure, hall-mark, marker, marking, indicant, idea, identifier, illustration, image, impress, impression, index, indicator, manifestation, mark, motto, note, omen, picture, precursor, pre-indication, print, portent, prognostic, seal, sign, slogan, signal, simile, similitude, stamp, symbol, symptom, test, testimony, token, touch, trace, trait, type, watchword}.

To construct meaningful sentences is impossible without knowing the ontology of signs (pointing something) and symbols (standing for or taking the place of another thing) as conventional signs and natural signs, as well as supernatural signs (dreams, omens, miracles, portents, or visitations). Related as cause and effect, or part and whole, the natural signs among other things include the entities of the human mind (thoughts, ideas, concepts, and images) whereby symbols signify the concrete things in the world. By contrast, the conventional signs cover all the things used in meaningful communication: articulate sounds, words, intentional signs, all natural languages, and artificial symbolisms of science, mathematics and logics. A core verb of a set of representational associative sentences is the term "to mean" connecting nominal relatives serving as signs and things.

Generally, there are three basic senses of the term, or the ways a relative sentence of meaning may be employed:

- To have the sense of something, or to signify, or to denote, or to stand for, or to refer, or to be a symbol of, or to indicate, or to convey, or to show, or to symbolize, or to represent, or to typify (for conventional signs and symbols)
- To have in mind, to intend, to intend to express, to think of (for natural mental signs)
- To have as an effect, result, consequence, end, purpose, or design, to cause, to result in, to imply or to entail (for natural signs and symbols)

Causal Statements: Syntax, Semantics, Ontology

However significant the meaning of sentences may be, the highest knowledge value is assigned to a special division of associative sentences, causal statements. For it is ultimately the cause of things into which the human mind inquires, and the faculty to hit instantly upon the reason of things will finally distinguish NL-driven reasoning agents from other intelligent artifacts. Being most valuable constructions of natural and machine languages, and having a great import for the whole issue of semantic mechanisms and devices capable of natural language understanding and generation, causal statements require more attentive analysis. Because such propositions constitute an essential sentence type, we have to deal with this topic at a greater length so that to reveal the difficulties, establish the distinctive definitions, and expose the sorts of things the causal sentences express.

First, we must inquire what a causal sentence is, or what its nature and meaning, and how it fits into our division of sentences. In line with our classification, it is a sort of sentence where the subject and the object are both change substantives, whereas the verb may be a universal verb, a change verb, or a relational verb barring a state verb. By its structure, the universal causal statements are symmetrical sentences combined out of verbal names, the transposition of which expresses a characteristic property of causative sentences: to express the convertibility (or reversion) of the cause and effect relationship:

$$\mathbf{S_{CS} \rightarrow \{N_C\}' + \{V_U; V_C; V_R\} + \{N_C\}''} \qquad (24)$$
$$\mathbf{S^{\#}_{CS} \rightarrow \{N_C\}'' + \{V_U; V_C; V_R\} + \{N_C\}'}$$

For instance, if it is stated that "exercise is the cause of good feeling," then it follows that "good feeling is the cause of exercise," where both "exercise" and "feeling" fall under a class of changes as two sorts of action and experience.

There are two other examples for the heterogeneous classes of causes, nature and mind: "Physical events give rise to mental occurrences, then mental events will cause physical changes"; or for the homogeneous classes, heat and magnetism: "Thermal phenomena generate magnetic phenomena, then magnetic changes will cause thermal changes." Similar sentences can be constructed in expressing the interrelations between any natural phenomena (changes, processes, actions, or events): thermal, magnetic, electrical, mechanical, electromagnetic field, chemical, and biological.

The basic pattern of causal statements shows that, beside that generic effects should be attributed to generic causes, the different classes of changes cause each other reciprocally. This means that cause (C) and effect (E) is both necessary and sufficient for each other. In terms of formal logic this relationship can be written as $C \Leftrightarrow E$ (C if and only if E) meaning:

- Cause is sufficient for Effect iff $C \Rightarrow E$, if C then E; C implies E; whenever C occurs, E happens
- Cause is necessary for effect iff $E \Rightarrow C$, if E then C; C, if E; C whenever E; C when E

In natural languages, real causality constitutes a special type of biconditional sentences composed of two converse conditional sentences, and joined by the phrases: "if and only if," "necessary and sufficient for," "just in case."

That is, in the real causal sequence of occurrences, the being of the consequent action is absolutely necessary. Namely, when the antecedent event has come-to-be, the consequent change must always come-to-be, and if the consequent should come-to-be, the antecedent also must have come-to-be. Even though many numbers of consecutive members are mediating or intervening into the incipient primary cause and its effect, their sequence (or nexus) is necessarily reversible and reciprocal. As a result, the coming-to-be of each of them is eventually cyclical; for the true causation, where the relatives are both sufficient and necessary for each other, by its nature reverses returning upon itself. More specifically, the relation of causality can be formulated as follows:

"some instance of a change is associated with an instance of other change if and only if the latter instance is associated with the former."

There are a great number of causative verbs, such as, *produce, effect, induce, compel, create, generate, destruct, get, manufacture, occasion. perform, influence, determine, shape.* But the hardcore causative verbs are CAUSE, MAKE, DO, AFFECT, implying four basic senses of cause:

- Existence (to cause to exist or to be by making, composing, constituting, creating of materials, to *cause* to be or to *cause* to exist in a certain form, e.g., the parts make up the whole)
- State (to cause to be in a state, to bring some state, to cause to have some property)
- Change (to *cause* to change or to be done, to bring about a change; to *cause* to change in some form; to cause a thing to have a change
- Relationship (to *cause* to change in some relation; to *cause* to be or to exist in a certain relationship)

The grammatical constructions for expressing causality are as follows: "this (agent or change) causes (makes, does) that (change)," "to cause something to change or somebody to act," "to cause somebody a certain experience"; "to make somebody something," "to make (do) an action," "to do somebody something." In terms of the EIS ontological language, all these varieties of causal sentences are underpinned by one key structure consisting of the causative verbs and the change names of different sorts:

$\{N_c\}$ + {CAUSE, DO, MAKE, PRODUCE, AFFECT, EFFECT, ACT ON }+ $\{N_c\}$ (25)

Thus, causal relationships can be expressed by the following causative expressions:

- A set of causative verbs [*cause, make, create, do, effect, produce, occasion, perform, determine, influence; construct, compose, constitute; provoke, motivate, force, facilitate, induce, get, stimulate; begin, commence, initiate, institute, originate, start; prevent, keep, restrain, preclude, forbid, stop, cease*]
- A set of causative names [*actor, agent, author, creator, designer, former, originator; antecedent, causality, causation, condition, fountain, occasion, origin, power, precedent, reason, source, spring; reason, grounds, motive, need, impulse*]

- A set of effective names [*consequence, creation, development, effect, end, event, fruit, impact, influence, issue, outcome, outgrowth, product, result, upshot*]

The modes (manner) of causation can be expressed by the verbs describing the ways and manner how something is caused or produced, done or happens, like the WordNet's verbal typology:

- To make, create, with the verbs describing the manner of making (troponyms)
- To cause, do, make, with the verbs describing the manner of causing
- To keep, hold, and prevent, with the verbs describing the manner of keeping
- To control, with the verbs describing the manner of controlling

The sentence structure clearly shows that *if one thing causes another, then they cause one another, and both cause and effect belong among changes* (actions, events, or occurrences) expressed by change verbals. Additionally, the set of causative verbs can be nominalized with the aid of the causative prepositions ($\mathbf{P_C}$) as *of, with, from, through, by* or with the help of the prepositional phrases as *thanks to, because of, due to, owing to, by (in) virtue of, on account of, by reason of, on the ground of*:

$$\mathbf{S_{CS}} \rightarrow \mathbf{N_C} + \mathbf{P_C} + \mathbf{N_C} \qquad (26)$$

Again, the change names can be presented as the phrase names expressing a particular sort of changes (or connection between an object, state, or relation) and a change (action) in which they participate, such as, *brain activity, weather pressure, rate exchange, power decline, war cry, mystery play, mystery murder, shock treatment, pressure growth, marriage rate, rat race, memory fail, land slide, star explosion, price rise, wealth growth, air pressure, moonrise, sunshine, family planning, systems analysis, blood test, sunrise, sunset; the emergence of hope, the derailment of train, the start of competition, the end of life*. In spite of the diversity of expressions indicating that something causes the action or it is subject to the action, they are either causes or effects expressed by the change substantives. Then, as the phrasal structure, causal sentences become particular sentences consisting of substantive phrases related with complex verbal expressions.

$$\mathbf{S_{CSP}} \rightarrow \{\mathbf{N_{VG}}\} + \{\mathbf{V_U}; \mathbf{V_{CG}}; \mathbf{V_{RG}}\} + \{\mathbf{N_{VG}}\} \qquad (27)$$

Traditionally, the meanings of causal terms are sought in the substratum (the matter, the subject, the material, the parts); the essence (the whole, the class, the synthesis, the archetype, the form, the pattern, or the definition); the mover (the agent, the movement, or the change), and the end (presence or privation; the purpose, the good). That is, expressing the first point from which a thing is or comes to be or is known, all the causative words fall under those expressions that answered with reference to the following:

- The material sources of change
- The formal cause
- The primary, precise, or proper cause
- The final cause, for the sake of which a thing is changed

There also are other established facts we need to know about causative expressions, constructions, and statements. Above all, it is the fact that things can be causes of one another in the same way, that is, all the classes of causes are mutually related, both in nature and mind. This general phenomenon of reality is covered by the principle (rule or law) of causal reversibility. Besides, if the kinds of causes are only few in number, their varieties are many: essential or proper, a cause per se, and incidental or accidental, a cause by chance; predisposing and exciting; principal and instrumental; potential and actual; particular for the individuals and generic for the classes. Again, sometimes, the absence or presence of the causal agent may be a cause of change, action, motion, or movement. It seems such complication of the nature and meaning of causation is likely to be a reason of confusion and misunderstanding of this significant mode of relation.

So far the sentences have been considered neither as coordinated nor subordinated with other clauses, while formally the number of clauses in a sentence may be as long and infinite, so long as a string (or sequence) of clauses is ordered as to the grammatical and semantic rules of a language. So, besides entering phrases and modifying words, the number of names and verbs in a sentence structure can be duplicated or even multiplied with the aid of the conjunctions (coordinators, positive or negative) relating like or unlike things (*and, or, but, either...or; neither...nor*).

Grammatically, there are two sorts of sentences: simple sentences consisting of one clause and composite sentences composed of a sequence of clauses.

We have a simple sentence even when its separate parts as substantives and verbs are given as composite; since this in no way can essentially affect the structures and types of simple sentences. Also a simple sentence can be made negative by using so-called negators: *not, never, hardly, seldom, rarely, scarcely, neither, nor; no, nothing, nobody, no one, nowhere;* what also will not principally affect the number and types of simple sentences.

Composite sentences can be constructed out of the simple sentences by the agency of relational words, conjunctions, conjunctives, or connectives, **C**. Then, depending on the type of conjunctions conjoining words, phrases, clauses, as well as sentences, they are usually separated as a compound (co-ordinate) sentence and a complex (having a subordinate) sentence:

$$\{S_O, S_S, S_C, S_R\} + C + \{S_O, S_S, S_C, S_R\} + C + \ldots \tag{28}$$

The sentential formula above confirms that a composite sentence can include an unlimited number of simple sentences, principal, subordinate, or dependent clauses, as soon as they are connected semantically and grammatically. We can have sentences with several co-ordinate clauses, or with one principal clause and a series of consequentially dependent sentences. Again, any of these clauses can be represented as a non-finite clause by converting its verb into the *to*-infinitive, *-ing*, and *-ed* participles. Unlike the complex sentence, the compound sentence is made up of clauses of equal rank and connected by the coordinate or weak conjunctions $\mathbf{C}_C$: conjunctive (*and, not only...but also, so*); disjunctive (*or, either... or, neither ... nor*); or contrasting (*but, yet, however*).

$$\{S_O, S_S, S_C, S_R\} + C_c + \{S_O, S_S, S_C, S_R\} \quad (29)$$

In compliance with three types of compound conjunctions, there are three subdivisions of coordinate sentences: conjunctive, disjunctive, and contrasting. The conjunctive sentences describe things that are either simultaneous in point of time or consecutive, in temporal order, in causal order, or in logical order. The disjunctive sentences may describe the relation of the parts of the whole, namely, such things within the same class or genus complement each other, being mutually excluding at the same time. The things that oppose one to another as contraries, contradictories, and as privatives to positives are described in contrasting sentences.

In complex sentences, a clause can play a subordinate role of a noun by substituting it, of an adjective by modifying the noun, or of a verb by specifying its meaning with respect to state, change, and relationship. Hence, there are several sorts of subordinate or dependent or strong conjunctions, or connectives and respectively clauses:

- The nominal connectives introducing substantive or nominal clauses: *that, what; who, which, when, where, whether, why, how many.*
- The relative pronouns introducing name modifying or relative clauses: *who, which, what, that, as, but; whom; whose, of which, of what; when, where, why.*
- The conditional (consequence) conjunctions introducing conditional or state clauses: *if, unless, whether...or not, on condition that, provided that, providing that, so long as, as long as, suppose, supposing that, in case.*
- The causal connectives introducing causal clauses of three sorts: a reason clause, a purpose clause, and a result clause: *because, as, since, seeing that, now that, for; to-*infinitive, *so that, in order that, for fear that, lest*; *so... that, such...that.*
- The concession (contrast) connectives introducing concession (contrast) clauses: *though, although, even though, even if, while, whereas, still, yet, nevertheless, however, what-, who-, when-, wherever.*
- The comparison conjunctions introducing comparison clauses or manner clauses: *than, as...as, the same...as; as, as if.*
- The temporal connectives introducing a temporal clause: *when, as soon as, after, before, till, until, whenever, since, while, once.*
- The place connectives introducing location clauses: *where, wherever.*
- The qualifying conjunctions introducing qualifying clauses: *in that, inasmuch as, in so far as, as far as.*

On the whole, each type of a complex sentence is distinguished by the type of its subordinate conjunction $\mathbf{C}_S$ expressing certain sorts of relation: the relationships of the subject with its properties, the relation of cause and effect, the relations of comparison, the relation of reciprocity, the relation of parts and wholes, or spatial and temporal relationships, as it is formalized in the sentence structure:

$$\{S_O, S_S, S_C, S_R\} + C_S + \{S_O, S_S, S_C, S_R\} \tag{30}$$

To express complex relationships between things and facts, logically represented as a chain of reasoning process, argumentation, or demonstration, as the relationship between the premises and the conclusion, just the composite sentences are usually not enough. Here is an illustration of such a sentence: "Yet, if the same process is a coming-to-be of *this* but a passing-away of *that*, and a passing-away of *this* but a coming-to-be of *that*, why are some things said to come-to-be and pass-away without qualification, but others only with qualification?" (Aristotle, On Generation and Corruption, pp. 414-415).

Now it is necessary to employ special words named as conjuncts that are linking the composite sentences, coordinate or compound, as well as simple sentences. So, in order to attach the conclusion or inference sentence to the preceding sentence, the following consecutive words may be used: *accordingly, consequently, for, hence, so, then, thence, thereof, thereon, thereupon, therewith, whence, wherefore; altogether, in all; as a result, considering all that, to conclude, all things considered, in the whole.* To show the cause of things, we use the causal connectives: *as, because, since, for*. To contrast composite sentences, to show the dissimilarities of comparing things, the contrasting or conceding conjuncts are applied, in the sense of "in spite of the fact": *although, but, however, nevertheless, notwithstanding, still, though, yet; in any case, on the other hand.* To make rewording or addition, the following conjuncts are applied: *that is to say, namely; what is more, besides, and further.*

That is all about the basic forms of sentences (propositions or statements), simple, complex, and compound, their nature, definition, meaning, classification, and algorithmic formulation.

Summing up, the main complexity of natural language comes not from grammar and syntax, the traditional field of study of linguists, but rather from semantics, as the history of theoretical linguistics evidences and our study supports. Since grammar dealing with morphological structure of words and syntactical structure of phrases and sentences is determined by semantics, setting the real meanings of NL constructions. So an ontological semantic framework is unavoidable for avoiding many misconceptions about the nature and structure of natural language. One of them is an unjustifiable claim that its consistent description, rigorous axiomatization and formalization as algorithmic program is open to question. In contrast to such widespread misconceptions, we tried to prove that linguistic knowledge can be organized as a formal system of basic definitions, axioms, and rules so that to assure consistency of meaning, correctness of linguistic inference and discovery of possible relationships among words and sentences (USECS, 2005).

Conclusion

A formal theory of natural language is suggested within the framework of ontological (real world) semantics. Natural language turns out to be a universal knowledge representation and reasoning language where the number, meanings, and relations of word classes (both content words and function words) are tied to the major classes of things in the world. It is shown that the natural language as the sum total of expressions can be classified with respect to the prime ontological classes: entities (things, beings, existences, or realities); substances (objects or agents); states (properties, qualities and quantities); changes (actions, activities, events and occurrences); and relations (dependencies, connections, links, associations, relationships). Hence, all words can be arranged into several characteristic groups: entity referring terms, substance naming words, state expressing words, change indicating words, and relation signifying words. The research also uncovers that there are four distinct name types of things, the names of substances, states, actions, and relations; four kinds of adjectives, substantive, stative, active, and relational; and four types of verbs indicating being or existence, states, changes, or relations.

Within the general semantic framework, it is formulated a consistent set of linguistic expressions and rules:

- Morphologic formulas expressing content and real meanings (entity types and relationships)
- Phrasal rules for the formation of word strings denoting individuals
- Syntactic rules for forming sentences mapping the associations and relationships of things
- Linguistic rules of inference forming a sequence of sentences according to a real logic of ontological entities

As it will be confirmed in the next Part, presenting linguistic knowledge as a set of algorithmic formulas enables us to use NL as a universal knowledge representation and reasoning system for building powerful intelligent mechanisms grounded on natural languages: reasoning engines, representation systems, Web search agents, and NL question answering tools. This makes real the knowledge and reasoning applications capable to learn, acquire, process, and communicate the digital world's collective knowledge in NL forms.

References

Aristotle, Logic (Organon): Categories. (1990). In M.J. Adler (Ed.), *Great books of the western world* (Vol.1). Chicago: Encyclopaedia Britannica, Inc.

Aristotle, Logic (Organon): On Interpretation. (1990). In M.J. Adler (Ed.), *Great books of the western world.* (Vol.1). Chicago: Encyclopaedia Britannica, Inc.

Aristotle, Logic (Organon): Posterior Analytics. In M.J. Adler (Ed.), *Great books of the western world.* (Vol. 1). Chicago: Encyclopaedia Britannica, Inc.

Aristotle, Logic (Organon): Topics. (1990). In M.J. Adler (Ed.), *Great books of the western world.* (Vo. 1). Chicago: Encyclopaedia Britannica, Inc.

Aristotle, Physical Treatises: Physics (Physica). (1990). In M.J. Adler (Ed.), *Great books of the western world.* (Vol. 1). Chicago: Encyclopaedia Britannica, Inc.

Aristotle, Physical Treatises: On Generation and Corruption. (1990). In M.J. Adler (Ed.), *Great books of the western world.* (Vo. 1). Chicago: Encyclopaedia Britannica, Inc.

Berners-Lee, T., Hendler, J., & Lassila, O. (2001). The Semantic Web. *Scientific American, 284*(5), 34-43.

Dowty, D. (1979). *Word Meaning and Montague Grammar.* Dordrecht: Reidel.

Extensible Markup Language (XML) (2004). Retrieved April 25, 2006, from http://www.w3.org/XML/

EuroWordNet Top Concepts (2004). Retrieved April 25, 2005, from http://demo.ontotext.com

Fellbaum, C. (ed.) (1998). *WordNet: An electronic lexical base.* Cambridge, MA: MIT Press.

Fellbaum, Ch. (2003). *English Verbs as a Semantic Net, 5 Papers.* Retrieved April 25, 2006, from http://www.cogsci.princeton.edu/~wn/

Jackendoff, R. (1983). *Semantics and Cognition.* Cambridge, MA: MIT Press.

Jackendoff, R. (1990). *Semantic Structures.* Cambridge, MA: MIT Press.

Kant, I. (1990). The Critique of Pure Reason. In M.J. Adler (Ed.), *Great books of the western world.* (Vol. 39). Chicago: Encyclopaedia Britannica, Inc.

Levin, B. (1993). *English Verb Classes and Alternations: A Preliminary Investigation.* Chicago: the University of Chicago Press.

Question-Answering System in Natural Language on Internet and Intranets. (2004). France Telecom. Retrieved April, 25, 2006, from www.francetelecom.com/en/group/ rd/offer/brevets/programmes/sysqa.html

Quine, W. (1960). *Word and Object.* Cambridge, MA: MIT Press.

Resource Description Framework. (2004). Retrieved April, 22, 2006, from http://www.w3.org/RDF/

RDF Vocabulary Description Language 1.0: RDF Schema. (2004). Retrieved April 30, 2006, from http://www.w3.org/TR/2004/REC-rdf-schema-20040210/

Sowa, J. (1995). Top-level ontological categories. *Int J Human-Computer Studies*, 43, 669-685.

Talmy, L. (2000). *Towards a Cognitive Semantics: Typology and Process in Concept Structuring.* Cambridge, MA: MIT Press.

Text Retrieval Conference (TREC) Data (2004). Retrieved April, 22, 2006, from http://trec.nist.gov/data/topics_eng/

Turing, A. (1950). Computing Machinery and Intelligence. *Mind*, Vol. LIX, No.236, 433-460.

USECS (Universal Standard Entity Classification System). (2005). Retrieved April, 22, 2006, from http://www.encyclopedic-intelligence.com

Vendler, Z. (1967). *Linguistics in philosophy*. Ithaca, New York: the Cornell University Press.

(The New International) Webster's Comprehensive Dictionary of the English Language, Encyclopedic Edition. Naples, Florida, USA: Trident Press International.

WordNet, WordNet 2.0. (2004). Retrieved April, 22, 2006, from http://www.cog.sci.princeton.edu/wn/

Chapter XI

Natural Language Intelligences: The Virtual or Digital Aristotle

Introduction

Of all possible intelligent NL applications and semantic artifacts, a special value is today ascribed to building the question answering systems (Q&A) with broad and wide ontological learning (Onto Query Project, 2004), classified as open-domain Q&A knowledge systems [Question Answering, From Wikipedia, 2006]. This line of research is considered as upgrading of a traditional keyword query processing in database systems, as endowing the Web search engines with answering deduction capacities. Ideally, such a general-purpose Q&A agent should be able to cover questions (matters, subjects, topics, issues, themes) from any branch of knowledge and domain of interest by giving answers to any meaningful questions, like the Digital Aristotle, "an application that will encompass much of the world's scientific knowledge and be capable of answering novel questions and advanced problem-solving" (Project Halo, 2004). The trade name of the Digital Aristotle was inspired by the scholar mostly admired for the depth and width of his perception, whose mind spread over ontology, physics, logics, epistemology, biology, zoology, medicine, psychology, literary theory, politics, and art.

The reasons why the project Halo failed may be very illustrative and instructive. To construct such a knowing-all, versatile computing entity, Paul Allen (the owner of Vulcan Inc. funding the project and the initiator of the idea of the DA) projected to reach the pinnacle of knowledge technology in several separate stages. The completed first phase was intended to test the validity of the state-of-the-art KR&R systems for the real world knowledge domains, such as inorganic chemistry.

Quite unusual, but as a positive result of three competing teams, Cycorp Inc., the SRI International, and Ontoprise GmbH, there was presented the negative findings making a whole "taxonomy of types of failures": knowledge modeling errors, non-expressive languages, faulty reasoning, poor coverage, inadequate answer explanation, poor scalability, evaluation insufficiency, meta-knowledge and meta-reasoning deficiencies (Friedland & Allen, 2004). The second 30-month phase was aimed to remove these drawbacks by hiring '*world class team*' in knowledge representation and reasoning, knowledge acquisition, intelligent interfaces, natural language understanding, and system integration. As a matter of fact, the DA project is just the follow-up of past DARPA projects to develop advanced knowledge systems and tools, Knowledge Sharing Effort, High Performance Knowledge Base, and Rapid Knowledge Formation. The would-be powerful reasoning application, the DA appears as a new attempt to create a large versatile knowledge-based system, now as empowered with question answering and explanation generation capacities, but with the old means, formal logic languages having little to do with the real world knowledge formulation.

Additionally, many experimental investigations in information engineering increasingly indicate that the major difficulty in making automated question answering is not in knowledge representation languages but rather in having a top ontology organizing a knowledge domain and a lexicon. In other words, the DA system should not be seen just a catchy trade name but rather as the speaking title suggesting a classical ontological pedigree deeply rooted in its progenitor's universal system of entity types. For, by general consent, the *real* Aristotle was the first who sought to create a systematic universal ontology as the study of primary kinds of being forming the highest level of things.

The project Halo lesson teaches us that that to become a general query system, or the Virtual Aristotle Machine (VAM), it should appear as an ontology-based NL question-answering system with the world knowledge and vast lexicon able to real (causal) reasoning and explanation in any domain area. Accordingly, the NL application should be designed as a universal language machine made up of knowledge bases and natural language resources organized by standard ontology and thoroughly processed by onto-logical reasoning engine for formulating novel knowledge, principles, rules, and facts in natural language forms. Ideally, its reasoning engine is planned to possess the most sought-after capacities:

- To identify and process top constructs and relationships in any knowledge area
- To resolve complex problems revealing their central causes and reasons
- To generate questions and answer explanations in natural language
- To interact with the human being exchanging information and thoughts on every lively topic
- To predict possible events, consequences, effects, or outcomes brought about by human actions or natural processes

The VAM is expected to critically contribute into transforming the Syntactic HTML Web into the Semantic Ontology Web as the next generation of the Internet's World Wide Web. Emphasizing the principal distinctions of the Virtual Aristotle from the Digital Aristotle as a traditional KR&R system, we further will study the way how to create the VAM as a know-all natural language machine with an open domain Q&A facility.

The main thrust of this chapter is that to develop the DA or VAM with capacity to represent the essential features of things in the world we need to move away from the historical knowledge engineering tools and technologies banking on formal logical languages. Briefly, the principal issue to be examined is as follows: *how the basic kinds of things determine a full set of question classes (and answer types)* to meet human inquiry in any knowledge domain.

A Universal Query System:The Entity Categories for Question Answering Systems

The query system capable to manage any sentence of inquiry is one but significant language device in the whole new class of NL-driven knowledge and reasoning technologies aimed to eventually replace traditional knowledge representation and reasoning systems based on formal logical mechanisms and languages. Unlike affirmative and negative sentences asserting or denying existence or non-existence, possibility or contingency, impossibility or necessity, the interrogative sentences, or simply questions, interrogations, or interrogatives, are the sentences of query (doubt, inquiry, interrogation, or interrogatory) calling for an answer, reply, or response, thus stating a problem, a subject of inquiry, debate, or dispute. In natural language processing technology, the content analysis of interrogations, their definition, meaning, and taxonomy, is no less important than the semantic study of positive and negative assertions (of something about something). For to know how to construct and decompose (parse) the grammar and semantics of interrogations is a way to build Q&A applications capable to query an immense body of digitally encoded knowledge information. However, the main obstacle for constructing such much sought after NL reasoning artifacts is the current incapacity to perform semantic content and context analysis of NL query language, its inability to arrange the question data by a system of top-level entity types.

The road map for creating a Q&A module of the VAM necessitates resolving many challenging issues, ranging from the question classification to the human user profiling (Burger, et al., 2006). Among them there are question classifications and taxonomies; onto-semantic models of question understanding and processing, syntactic, semantic, contextual, reformulation, key-word techniques; context as specifying the meaning of questions; data and knowledge sources available for Q&A; a taxonomy of answer types; answer processing and extraction; answer formulation in NL; multilingual power; interactive real-time dialogue; open-domain advanced reasoning, scientific and commonsense; the user domain of interest. Given these guidelines, we intend to examine the crucial issue of question classification following the simple rule: the kinds of questions we ask (and the types of answers we get) are as many as the kinds of things that we know (Aristotle, Posterior Analytics, p. 122).

From the very beginning of computer science, the founder of the idea of artificial intelligence, Alan Turing, modeled computational mind in terms of natural language enabling an intelligent man-machine dialogue communication (Turing, 1950). Still, despite that AI boasts many new paradigms, strategies, and goals as cognitive robotics, knowledge engineering, machine learning, superhuman intelligence, software agents, and artificial neural networks, the natural language processing power being encoded in computer programs remains a key line of business for achieving complex cognitive behavior in artificial mechanisms.

In the context of machine (automata) theory, a universal (Q&A) machine is a language transducer, a linguistic automaton (computing machine or response mechanism) producing answering behavior as a consequence of the interrogative stimuli. In other words, the interrogative information agent can be constructed as a language machine with the sets of inputs, natural language questions (spoken, written, keyboard, OCR), states (the world model, linguistic knowledge, lexical data bases), and outputs, or answers together with syntactical and semantic rules of operations (question analysis, answer formation). In line with the definition, any interrogative automaton combines natural language processing technology and information retrieval techniques, operating through several components or modules (Question-Answering System, France Telecom, 2004): question analysis module, information retrieval mechanism or search engine, and answer composition module. Such a system structure is used by most interrogative systems engaged in the Q&A track of Text REtrieval Conferences (TREC) TREC-8 and TREC-9, TREC-10, TREC-11, TREC-12. They are generally using the same procedure of composing answers to the test set of questions of different types: retrieving and segmentation of the top *n* documents from the TREC Q&A collections of texts (question corpora); identifying a hierarchy of question types; associating them with entity categories; forming (extracting) the answers appropriate to the question types; and delivering the constructed answer to the user (TREC, 2004).

The experimental analyses of questions and answers indicate that the core component of any powerful Q&A machine should consist of onto-semantic modules (analyzer/construer) representing the basic terms, truths, facts, and laws of the world (or its parts) as a crucial component for erecting a system capable to intelligent man-machine communication. This component is essential for defining the meaning of the terms of interrogatives, mapping entity categories to question types, associating a hierarchy of question types with the answer type taxonomy. It appears that the absence of such an all-embracing entity typology for question answering is turning out as the main reason of the Q&A systems' failures on TREC questions. Moreover, to develop an interrogative software agent, first, we must construct a comprehensive linguistic world model that will enable a computing machine to process NL questions in an effective way: to identify the meaning of question types by relating them to their entity categories so that to form a set of question answering types as a comprehensive QA typology.

Accordingly, the most challenging questions that the designer encounters in constructing interrogative agents and knowledge tools for forming queries are as follows: How to analyze and classify interrogative statements? How many types of questions are there? How Q&A types are related to the question typology? And how the both typologies are determined by their entity categories? Mostly due to the absence of appropriate answers to such questions, we have not yet enjoying the company of powerful query-answering artifacts.

As for the question analysis and classification, so far the interrogative sentences usually have been examined from grammatical, logical, or semantic perspectives. Given the nature of the answers, interrogations are classified as fundamental, factual, instructional, or normative. Besides, they may be partitioned as questions with a true answer, false questions with a false answer, and illegitimate questions with no possible answer. But, typically, the grammarians and logicians divide all the questions into two broad groups:

a. **WH-questions** covering all interrogative words, like What, Who, Which, Whose, Why, How, How many, How long, Where, When, and so forth, asking for the appropriate parts of a sentence: nouns or pronouns, verbs, adjectives, adverbs, or prepositions.

b. **Yes/No questions** (true/false) asking for the yes/no answers with the pattern (operator, subject, rest of the verbal group, object); or as the tag question where the affirmative is followed by the negative interrogative or vice versa, the negative by the positive interrogation.

All the WH-questions are further broken down into several major kinds of questions, although not as mutually exclusive categories, (What or Who, Why, How, Where, When) with the following meanings:

I. The **What** (which, who, whose; wherein, whereby, wherefore, whereof, whereon, whereupon, wherewith) questions specifying and identifying a member of a certain class [an individual or a set of a class of things, whether of substances, inanimate objects, animals, plants, people, persons (the who questions); of states of affairs, conditions, qualities, and quantities; of changes, actions, activities, processes, event occurrences; and of relations].

II. The **Why** questions asking for the causes of things [(for, to) what cause (purpose, reason, aim, motive, end, object, principle) or because of which (rationale, reason, cause, ground, account)].

III. The **How** questions querying the ways something happens or done, the way (means) of becoming and doing [in what way (manner); to what degree (extent, amount); in what proportion; in what state (condition); for what reason (purpose); at what sum; by what name; to what effect; with what meaning; time duration, how long (ago)].

IV. The **Where** questions identifying the place, locality and relation [(at, in, from, to) what place, situation, or relation] of objects and changes and events;

V. The **When** questions identifying the time and date [at, on, in what (which) time] of changes (actions, processes, occurrences, or events).

The pattern of the WH-statements varies with the target of asking. When we are asking for the subject of a sentence, the structure is the following:

[What, Who, Which, Whose, or (What, Whose, Which + N) + V_G + N]?

The interrogative words function either substantively as interrogative pronouns, subjective (possessive or objective): Who (Whose, Whom), Which (Whose, Which), What (of What, What), or adjectively as the What-question phrases, What, Whose, Which + name. The pattern of the What-question sentences looks very simple: if the target of the query is the subject of a affirmative (negative) sentence, then we only replace it with the appropriate interrogative pronouns: What, Which, Who or Whose, or by the interrogative nominal phrase.

But when we are asking for any other part of sentence, except the subject, object-nouns, adjectives, verb, or adverb, the interrogative pattern will include operators, as in:

[the WH-question word (phrase) + the operator (the modals, the auxiliaries) + N + V_G + N]?

Evidently, the largest range is the What-question kinds, which number can be as many as the number of entity classes. For instance, when given six kinds of change, we can have six what-question types concerning changes, actions, activities, or occurrences; for 10 modes of substance, we obtain 10 top kinds of What-question. As to the Why-question kinds, it varies with the kinds of causes involved in any change, action, activity, or event occurrence and happening. The How-question kinds fall under a broad range of subcategories: manner, amount, proportion, condition, purpose, reason, sum, name, or meaning. At last, the number of the species of (spatial) relations will determine the amount of Where-question kinds, while the species of time, the When-question kinds. It is plain that the What-questions, having as its answering topics the broadest range of things, both as classes and individuals, the abstract and the concrete, are mostly original sort of interrogatories; for the rest, interrogatives can be obtained by adding to the What-question the entity category asking for.

Following the interrogation patterns and question types, the answer types can be classified with respect to the part of the sentence the question targeted (focused): names of things, individual or common, animate or inanimate (entity-subject, entity-object); adjectives (properties, qualitative or quantitative, active, and relational); lexical verbs (existence, state, action, and relation); adverbs (place-position, place-direction, time, duration, frequency, degree, and manner of action). So then both question types and answer types can be characterized as related to the entity categories that the questions target, imply, call for, ask for, ask about, or ask after. As a consequence, the number of the answer types is equivalent to the number of the entity types; that is, both the question-type analysis and answer-type classification should be regulated by the entity (being) classification. To put this in other words, the kinds of questions the information agent can ask, as well as all the possible question answering types it can deliver, are to be as many as the kinds of things we know. Even the brief grammatical review shows that the entity classification setting the meanings and types of interrogatives constitutes the knowledge hardcore of any Q&A artifacts.

In the question answering grammar, a question is a sentence of query, interrogation, or enquiry asking for an answer, reply, or response. By the very definition, questions and answers are correlatives, they are semantically related to each other, and explained by a reference of the one to the other, that is, they are the pair of opposites falling under the category of (linguistic) relations. Since each and every question type **Q** requires a corresponding answer type, such relationship can be formally represented as an answering function **A**. The function will associate one particular answer of the range **S** (a set of possible answer types as a dependent variable) with one specific question of the domain **Q** (a set of question types as an independent variable):

$$\mathbf{A: Q \rightarrow S \; \{S_O, S_S, S_C, S_R\}} \qquad (1)$$

Having the property that for each particular question there is one corresponding single answer, the answering mapping function also indicates that to each interrogative type there is a question answering type, that is, a hierarchy of questions implies some answer-type taxonomy, as a hierarchy of possible answer types. In compliance with the EIS Universal Ontology as the all-embracing taxonomy of entity types, all the grammatical and syntactic categories can be arranged into the four onto-semantic classes of word: *the substance referring expressions, the state expressions, the change expressions, and the relation indicating expressions*. Accordingly, all the WH-questions can be subdivided into the following semantically distinct groupings:

1. The queries referring to substances, their kinds and parts (objects, materials, agents, or beings, personal and impersonal) $\mathbf{Q_O}$
2. The questions referring to states (conditions, properties, qualities, and quantities) $\mathbf{Q_S}$
3. The queries referring to changes (actions, activities, processes, events, and occasions) $\mathbf{Q_C}$
4. The questions asking for relational entities (identity, diversity, opposition, causality, place, location, direction, time, date; part-whole relations as well as all sorts of conceptual, semantic, grammatical, and logical relationships) $\mathbf{Q_R}$.

The EIS taxonomy of question types is constructed as a generalization of a semantic typology of questions mostly based on a traditional onto-semantic framework of lexical expressions, which are substance, quantity, quality, relation, place, time, position, state, action, and affection. Due to such classification of lexical expressions functioning as the fundamental terms of discourse and language, the modern English grammar allows as the top semantic classes of the question (and the answer) the following sorts of question and question answering types:

- Substance question type (what a thing is or what it is, class or individual)
- Quantity question type (what quantity a thing has)
- Quality question type (what quality a thing has)
- State-focused questions (what state a thing has)
- Action-targeted queries (what change, action a thing causes or does)
- Experience-related questions (what sort of changes (processes) a thing undergoes)
- Relation question type (what relation a thing has to some property or other thing)
- Place question type (what place, position or direction a thing has)
- Time question type (in what time, when, a thing happens)

The current question answering agents apply different processing strategies (from statistical methods to knowledge-driven) to extract answers from various knowledge sources, unstructured or structured (TREC, 2004). Yet any question answering system to effectively perform question analysis, search or information retrieval, and answer construction needs to use some lexical model of entity types built as a lexical inheritance system. Following Aristotle's fundamental listing of word categories, some semanticists propose to use as major entity categories the classes of thing, event, state, action, place, path, property, manner, and amount (Jakendoff, 1983). Others, concerned with experimental lexical projects, prefer something more complicated (EuroWordNet, EDR). Blending various lexical ontologies (Vendler, 1967; Lyons, 1977) with Aristotle's notion of quality, the EuroWordNet top ontology breaks down all things in the world into three disjoint categories: concrete objects existing in space and time, "event-like things" taking place, and true or false "fact-like things," which all sorted out as 1st, 2nd, and 3rd order entities. The first sort of entity covers concrete things denoted by concrete nouns and arranged by origin (natural, artificial), form (substance, object), composition (part, group), and function (instrumental role, action, activity). The second sort of entities are expressed by nouns, verbs, and adjectives and composed of two kinds of situation type (or state), static and dynamic. Where the situation parts are cause, communication, condition, existence, experience, location, manner, mental, modal, physical, possession, purpose, quantity, social, time, and usage. At last, the third order entities are expressed by abstract names and include all kinds of things that may be valued not as real but rather as true or false: concept, idea, category, notion, thought, theory, plan, cognitive content, knowledge, fact, procedure, doctrine, policy, communication, information (EuroWordNet, 2004).

In the Japanese linguistic project of building electronic word dictionaries, a set of top concepts appears as even more complicated and confused (Structure of EDR, 2004). Of three parts of the Concept Dictionary (the Head Concept Dictionary, the Concept Classification Dictionary, and the Concept Description Dictionary), the hierarchical classification of concepts is given in the second part. Here, the top concepts with their subheadings are chosen as follows: being, human or subjective (person, animal, independent body/group, supernatural being); matter as an affair (concrete thing/object, a matter, a characteristic name, an objective subject); event/occurrence (phenomenon, action/deed, movement, change in state/in condition, condition/state); location/place (physical location, position, area/territory/domain, part, direction, abstract location); time (point, interval, duration, measurement, period, elapse).

As we can see now, both cases suffer the same categorical omissions: they misuse two top semantic distinctions—the class of states and the class of relations. In the first case, the relation type embraced by the situation type as a subclass together with the class of property; while, in the second case, relations scattered everywhere. Besides, the state entity is just a subtype of the super-concept of occurrence/event.

It seems that due to such inconsistency and lack of unity with linguistic ontologies most Q/A groups have been trying to apply their own home-made semantic taxonomies, usually without deep ontological roots. But, seeking to avoid such ad-hoc, small-scale question taxonomies, some question answering teams are increasingly tending to employ ready-made, large-scale semantic networks of words, like the WordNet lexical inheritance system of synsets.

The Standard Ontology and the WordNet Taxonomy

As it appears from many experimental works on answering systems, the world ontology determines the entity classification as a hierarchy of the highest types of entities by imposing the knowledge (meaning and content) structure on the word classes, on all syntactic and grammatical categories, whereby coming up as a highly valuable tool in transforming the lexis or lexicons into the lexical knowledge. Then the utility of the ontology-based query language can be proved by its ability to furnish a general organizational model for large-scale lexical resources and databases deployed on the web servers.

Yet, just to identify the semantic relations among word meanings or sets of synonyms, as it demonstrates the online lexical database, WordNet, is only the first step toward building the lexical knowledge base connecting natural language with the world knowledge. In that any lexical data base to become knowledge base is to be organized according to ontological knowledge, the WordNet database content with 25 "unique beginners" ("semantic primes" or "primitive semantic components"), each starting its separate hierarchy, needs to be subsumed under a **single hierarchical entity structure.** Which is to be topped by being, entity, or thing with its four pivotal classes, substance, state, change, and relation, and bottomed by nonentity, nothing, or nonbeing. Since the entity category works as the most universal of all classes and kinds, it will embrace all WordNet's top lexical concepts, including physical entity with all its subordinates (individual thing, cause, physical object, matter, location, subject, expanse, part, unit, variable, etc.), abstractions (attribute, quality, quantity, relation, set, space, time) as well as state, event, phenomenon, action, group, possession, and psychological feature.

Having developed a comprehensive lattice of things, we are able to perform a semantic rearrangement of WordNet's lexical system; what suggests reclassification or rather subsumption of all its content, noun, verb, adjective, and adverb sets of synonyms, in terms of entity (substance, state, change, relation) types. It is no need to emphasize here that the restructuring of the lexical material without having built a comprehensive set of top-level categories is useless work, just confusing the semantically ordered networks of words.

Of all possible sorts of interrogatives, the What questions make up the most valuable part of queries, since they intend to seek the definitions of the things in question, namely, what a term (word) signifies, what the meaning (content) of concept is, or what a thing is. The subject of inquiry or the range of topics of the what questions are spread from the fundamental questions of ontology, such as "What is existence (being, thing, entity)?" "What are the primal classes of being?" "What are beings or entities (classes, physical objects, minds, properties, changes, relations)?" to most specific inquiries, such as "What is the largest city in the world?" "What is the biggest animal in nature?" "What is the size of Eiffel Tower?" or "Who invented the telephone (the wheel, electric bulb, car, TV, radio, computer, etc.)?" The extant Q&A systems are mostly engaged with factoid questions asking for particular beings, persons, situations, events, places, or times, whereas the real power of the interrogative agents come from their ability to handle (instead of quibbling) the generic queries concerning the meanings of the common terms of the question. That is, paraphrasing Turing's example (Turing, 1950), when we ask the information agent: "Can machines think?" it should be able to define the meaning of the terms "machine" and "think," which is the very topic of ontological investigation.

Abiding by the single being hierarchy supported by four primitive classes, we can specify four subtypes of the What-question: what (substance, object) questions q_o, what (state, properties) questions q_s, what change (cause, action) queries q_c, and what (relation, connection) interrogatives q_r.

Since the "what" indicates the substance of things, the what (substance) questions involve all the substantive names denoting the possible varieties of substance kinds:

- Entity (thing, object, physical object, causal agent; group, collection)
- Non-living things and objects (substance, matter, stuff, material; natural object; artifact; food)
- Living beings (organism; animal, fauna; plant, flora; human being, person; body, corpus)
- The parts of substance (atoms, components, constituents, elements, fractions, fragments, ingredients, members, particles, pieces, portions, sections, segments, shares, and subdivisions)

Instead of a multiple hierarchical structure of (substance) name meanings, like entity or thing, object (natural object, substance, artifact, food) and organism (plant, animal, person), all substance expressions are to be arranged (placed, positioned) in a single (lattice-ordered) hierarchy of substance terms of several levels deep (USECS, 2005), as in:

Entity > Substance & Substratum & Essence & Stuff> Matter & Material &Body & Mass & Object > the Elemental Substance & Particle &Atom &Element &Molecule > Composition & Mixture &Compound &Aggregation >...> the Universe (an aggregation of substances) > the Earth > Life {Organisms (Plants, Animals, Humans)} > Mind > Artifact > Machine > AI >...> Nonentity

One of the most serious defect in WordNet 2.0 is the semantic ambiguity of its topmost semantic category, entity, defined as "that which is perceived or known or inferred as having its own distinct existence (living or non-living)." As a matter of fact, entity is being, the only permanent substratum of reality, underlying all things in the world. From a physical standpoint, there might be a few classes of substance as an entity (stuff) of which all objects are made. But today, to the best of human knowledge, matter or material is the most essential substance of the physical world; still substance is one of the four coordinate categories subordinate to entity.

Thus, instead of multiple hierarchies, all the names of substances are contained in a single hierarchy of noun meanings (nominal types) ordered by a part-whole relation of generalization or specialization, such that the properties of the top type of substance are inherited by all its subordinates. Then every particular question requires a set of generic issues to have been answered beforehand. Among such presupposed fundamental questions asked about substance that the information agent has to pose in the first place are the following: whether a substance (say, a cosmological "dark matter," or mind) exists or not; what the nature of the substance is; what composition it has; what kinds (species, varieties, or modes) falls under it; what class (whole, genus) the object belongs to; what instances the substance have. Also, whether the substance is qualified, has this or that attribute, properties, changes, and relationships, that is, whether the relationship of a property and the substance is a real fact; in what manner (how) the substance operates, behaves, acts, or acted upon; what relationships the substance has to other substances.

As a consequence, every true answer to the generic question deductively entails a true answer to the more specific question. So, if the agent knows the definition of the class of thing, then it knows its basic genus, and all the distinguishing properties, structural (parts), qualitative (attributes), functional (behavior), and relational, inherited by all the species and subspecies of the thing. For all the differences (determinations) of a certain class will be inherent (inbred) properties also of its species (varieties and sub-varieties), which should be distinguished from each other by the same method of division or differentiation. That implies that the answer which is valid of a class of thing also is valid of its subordinate types; for a property of a property is a property of the thing itself. Or whatever is universally predicated (negated) of a class also is predicated (negated) of all its members. That is, any question automata should foreknow some basic things and truths about things like that substances (objects) are entities having identities, states (properties), and behaviors (changes, functions), that they are acted upon by each other, undergoing processes of change of different nature, that classes of substances are ordered by inheritance hierarchy, that there are types and subtypes of substance, several levels deep, which inherit all the features of the substance class, and so on. Such ontological knowledge simplifies the answers to the specific questions concerning particular natural objects, bodies, concrete things, living things, plants, animals, organisms, or persons, all existing in a single hierarchy of entity types. For, regardless of the particular properties, every each of these substances, apart from being an instance, kind, variety of a class of substance, always has some identity, states, functions or behaviors, and relationships to other objects.

The next division of the What-question is the what state question asking about a state (mode, form) of being and entity or how some thing stands with respect to its basic properties (attributive features); for an aggregate and collection of individual properties, quantities, qualities, conditions, and states of affairs compose a state (property) type. From the list of WordNet semantic primes, the property nouns falling under the state names are the following:

- **State nouns** indicating state of affairs, condition, status, state of matter and mind
- **Attributive nouns** implying formal attributes (properties, dimensions) of objects or persons
- **Shape names** designating spatial shapes, 2D or 3D, included into attribute designations
- **Quality names** indicating distinctive properties of something (or someone), defining a thing to be such and such, included into attribute designations
- **Quantity names** designating amounts, quanta, and units of measure
- **Possession nouns** as designating the state of having some attribute, quality, property, and so forth

Again, instead of a multiple hierarchical structure of (state) name meanings, all stative expressions are to be placed in a single (lattice-ordered) hierarchy of state terms. As a collection of essential properties, the category of state subsumes as its subordinate kinds substantial properties (attributes), qualities and quantities, as it is shown in a single hierarchy of state types of several level deep (USECS, 2005):

Entity (Being, Existence) > State & Condition & State of Affairs > Substantial Property& Attribute > Quality& Character & Feature & Characteristic & Nature > Quantity & Amount & Measure & Dimension > Multitude > Magnitude > Energy >...> Nonbeing

In natural language, a value of a state (variable or function) designated by the adjectival (state) nouns is expressed by one type of adjectives, which is neither substantial nor active nor relational, but stative adjectives. So, the values of states (properties, attributes, qualities, quantities, conditions) are expressed by the WordNet's descriptive adjectives, meaning such attributive properties as: size, form or shape or figure, temperature, weight, physical properties, human qualities, affections and affective qualities as sound, taste, smell, touch, color, blackness and whiteness, heat and cold, and so forth. Besides, the values of states are expressed by the stative verbs possessing the idea of having and by the verbs indicating mental states.

The third division of What-questions is the what change (action act, activity) question asking about behavioral, functional, active, or operational properties. Namely, whether a substance changes, becomes different, varies in some way, or it causes something to happen to some other substance, making different by changing something into something else, altering its nature or properties.

As for the collection of WordNet's semantic primes, the noun synsets falling under the change names are the following:

- **Process nouns** expressing natural processes
- **Action nouns** indicating all sorts and manner of acts, actions and activities
- **Event names** implying all sorts of natural events
- **Cognition nouns** designating cognitive and knowledge phenomena from mind to attitude
- **Feeling nouns** denoting effects, emotions, and desires
- **Motive nouns** referring to motivations, motives, and needs as the causes of the action
- **Phenomenon nouns** signifying natural phenomena

As with substances and states, instead of a multiple hierarchical structure of (action) name senses, all the active expressions are to be positioned in a single (lattice-ordered) hierarchy of change terms. As a collection of actions, the category of change subsumes as its subordinate kinds actions, activities and event occurrences as it is shown in a single hierarchy of change types of several level deep (USECS, 2005):

Entity > Change & Process > Action & Act & Activity & Agency & Operation & Mechanism & Behavior > Event & Happening & Occurrence & Outcome & Phenomena > Human Action & Act & Activity & Doing > …> Nonentity

In line with the lattice-ordered hierarchy of changes, both natural actions and events as well as human acts and activities are subtypes of change; for, by the very definition, an event is a change happening at a given place and time, while an action is a process of change that is done by some agents, substances, forces, chemicals, or intelligent agencies.

When our questions refer to a change type, there are some questions that we ask at the first consideration: Whether a change takes place or not? What is the nature (cause, reason) of the change? What is the reason of the change? What structure (constitution, make-up, or composition) has it? What class does the action belong to? What kinds (species) do belong to it? Or what are the instances falling under the change? Besides these, whether the change is qualified: has this or that property, manner (or mode) of becoming and doing, or relations. The last question implies several sub-questions: whether the relationship of a property and the change is a real fact; in what manner (way) the change operates or produced; what the relation of the change to other changes is, and in what way the change can be quantified, measured, or represented, for example, the period of space and time occupied by it, the form of its trajectory in the state space.

If there is a change, to know its nature is equivalent to learning the cause (reason, motive, or purpose) why it happens. For it is the cause of change into which we inquire in all our investigations, as well as looking for the answers to other four accompanying questions, referring to:

- The subject of change (what kind of entity undergoes a process of change, transformation)
- The action mechanism (the way of transmitting changes)
- The initial form (the beginning stage, phase, or state) of activity, from which the change proceeds
- The final form (to which the change proceeds, the end of activity)

Having found out the subject of change (a substance or a class of substances), the next question will concern the structure of change seeking to learn a sequence of transformations, variations, modifications (phases, states) constituting a process of change and how the sequence ordered. If the process of change involves just one substance, we have a simple, serial and homogeneous change as a set of modifications in an entity (internal) or of an entity (external). But when it concerns some group, aggregate of substances, the change becomes heterogeneous, complex, often complicated, tangled and intricate to separate, analyze, and understand. In fact, we deal here with a complexity of changes, where like or different kinds and types of changes can be organized into various configurations: linear (divergent or convergent) chains, reticulated networks, or circular structures.

As for the sorts of change, they are reduced to the several basic forms: change of substance (e.g., generation, destruction, nuclear fusion, fission, chemical reaction); change in state (change in quantity, increase and decrease, change in quality, alteration or modification); change in action (fluctuations); and change in relationships (change in place, motion, or change in time). Each of which can be again divided into species and subspecies but by one method of division.

The last answer type of What-questions is the what relation question asking about relational features of entity types. As, for instance, the sample question: "'What relation does the universal formal ontology bear to the logical theory of types?" Since the "what relation" questions indicate this or that sort of relations, they will involve all syntactic word classes whose meanings fall under the relation-type heading. Additionally to relational adjectives, verbs, prepositions, and conjunctions, from the top list the WordNet lexical concepts, the nouns coming under the relation words are the following:

- Relation names indicating relationships between various things
- Time nouns related to temporal relationships
- Spatial nouns designating spatial relationships

It is necessary to observe here that some entity classes can be classed both as quantitative and relative but in two senses, in the primary or secondary senses. As, for example, time and space belong to the class of relations in the first sense, but they also belong to the category of quantity, a mode of state, in a secondary sense, as all other entity types.

The class of relations is not only a mostly crucial entity type, but also a mostly difficult category for analyzing and understanding. As a matter of tradition, there are several senses in which things are thought to be related to each other or to be of something else: as the contained to the containing, as the active to the passive, as the measurable to measure, the knowable to knowledge. From the standpoint of the UFO, there are few top-level relation types (USECS, 2005):

- **Substantial relations** among substances (e.g., composition, spatial relations, social and human relationships)
- **Stative relations** between all sorts and modes of states (e.g., qualitative, quantitative)
- **Relationships of changes** (e.g., causality, temporal relations)
- **Relations of relations** (proportions, analogies)
- **Part-whole relations** (e.g., species-genus, individual-class, component-object, stuff-substance, portion-mass, part (phase, period, stage, degree, level, point)-process (continuum, series, sequence, extent, succession, development, cycle, course), and part (spot, place, location)-space (area, region, expanse)
- **Comparison relations** (identity, resemblance, diversity, and difference)
- **Opposition relations** (contraries, contradictories)

All the relation types can be represented in terms of signs, symbols, or constructs. So we can identify a class of **representational (sign) relations** including ideal relationships (associations of ideas, all the mental relations of ideas and images), semantic relations of meaning between signs, constructs and things, linguistic relations between language forms, mathematical relations between formal expressions, logical connections between propositions, and functional dependencies between variables.

Thus, when our quests refer to the relation type, there also are some standard questions the answer to which the agent must know at the first place: Whether a relation exists (really or in the abstract, conceptually or mentally); What class of relations does it belong: parthood, identity, resemblance, diversity, the hierarchy of substances, the relations of qualities or quantities, the order of changes (processes), time or space relationships, and so forth; Or is it contiguity, affinity, association, inference, analogy, proportion, function, mapping, subsumption, opposition, or relevancy?; What is its elemental structure (the number and order of relative terms)?; What are the instances falling under the relation? And whether the relationship has this or that formal properties: symmetry, transitivity, reflexivity, convertibility.

Knowledge of relationships, its properties and kinds, is decisive for deep-reasoning abilities of answering agents. For example, then guided by the general principle of the transitivity of relations extending to causal, quantitative, qualitative, temporal, spatial, and logical relations [if A implies B, and B implies C, then A implies C], it can deduce the right answer for an infinite number of transitivity-based questions, such as, cause of cause is cause of effect; more than more is more than the less; worse than bad is worse than good; earlier than early is earlier than late; east of east is east of west, and so forth.

However significant the What-questions are, the final object of all our inquiries is to know the whys and wherefores of things, their original, primary, and first causes, that is, to find all proper answers to the question "why." After all, "it is the cause we seek in all our inquiries," "it is the reason of things into which we inquire" (Aristotle, Posterior Analytics). The Why-question type is important for scientific knowledge, practice, and production; for *to know the reason why a thing is, is to know its nature.*

Concerning what cause is, what is its nature and number, and what kind the causes are, the modern accounts give little light. As a result, the varieties of causes and modes of causation appear as many as the number of schools engaged in the matter. Although, the meanings of causes are plenty, from a traditional position, the question "why" is answered by reference to a few things: the matter and its parts, the whole and the form, the change or the agent, and the purpose or the motive. Namely, in all Why-questions we are looking for the four principal answer types (Aristotle, Physics, pp. 271, 275):

- The substance, the substratum, the material, the parts, the elements, that from *which* a thing is done or come into being, as the premises are the causes of the conclusion.
- The necessary antecedent, the whole, the form, the pattern, the law, the principle, the essence, the rationale, the definition.
- The change, the agent, the maker, the motion, the efficient cause, *what* originated a change/rest.
- The purpose, the motive, the object, the reason, the end, the rationale, the final cause, that for the sake of which a thing changes or done.

The basic point of such divisions is that the essential meaning of cause is neither the substratum and the parts, nor the essence or the whole, nor the end or the good, but the change itself, "the change-producing of the changing," "all *sources of change* or of rest" (Aristotle, Metaphysics, p. 533). Given what sort of domain of objects and class of being we inquire, investigate or examine, the Why-question implies a specific class of causes (changes, acts, actions, processes, and doings), natural, mental, or social, as well as a certain mode of causation, proper (a cause *per se*) or incidental (a cause *by chance*), generic or particular, potential powers or actually operating causes.

Thus, *to know a thing's nature or property is to know the prior and precise cause (and reason) why it is, or why it takes place.* For instance, the question "What are definitions?" and its answer "A phrase signifying a thing's essence" (Aristotle, Topics, p. 144) can be converted to the question "What is the reason of definitions?" or "Why do definitions make up a part of ontological inquiry?" and the reply "Because they reveal the essential nature of the thing, they are the primary truths and the basic, indemonstrable premises of scientific demonstrations" (Aristotle, Posterior Analytics, p. 124). Next, it follows the question "What are ontological demonstrations or reasoning?" with its answer "Proving whether the connexion of a given property with a given thing is real or false" are convertible to the question "What is the reason of ontological demonstrations?" or "Why are such demonstrations employed?" and the reply "Because, basing on the definitions as the primary facts, they deduce unknown facts, thus bringing scientific knowledge."

To sum up the subject, the world ontology together with its comprehensive entity classification, determines a proper hierarchy of questions and question answering types to be used by the open-domain query **language automata**. All questions are answered by reference to several kinds of things, to thing, being or entity as such, to a substance, to a state, to an action, and to a relation. Of which the first in order are the questions seeking for the definition of a thing, that is, inquiring the fact of a thing's being or not being and then the fact of the thing's being this or that sort of entity: a substance, a state, a change, a relation.

Basing on the question typology, then, we can classify all the TREC test questions as the follows:

i. What type of **substance (stuff)** questions, inquiring of the nature (class or parts) of objects, spatial or non-spatial, material or nonmaterial [e.g., What is an atom? What is biosphere? What is caffeine? What are invertebrates? What is a shaman? What are amphibians? What is amixocillin? What is nicotine? What is e-coli? What are coral reefs?].

ii. What type of **state (condition, quality, quantity)** questions, enquiring the way something is with connection of its properties, attributes, qualities, or quantities [What is a prime factor? What is micron? What is epilepsy?].

iii. What type of **change** questions, inquiring the nature of actions, events, occurrences, and doings [e.g., What is an eclipse?].

iv. What type of **relation** questions, querying all the sorts of associations, connections, and links something has to something else [e.g., What is the relationship between deduction and induction in the process of reasoning?].

As all told, an intelligent computing application, such as Virtual Aristotle, will be distinguished from other reasoning applications by relying on the standard ontology of entities and the general truths and facts about the world. Onto-semantic investigation into types of morphemes, words, phrases, and sentences affords us the natural language with its all semantic and logical relations to be captured in a symbolic calculus of syntactical formation and transformation. The meanings processing activities, or intelligent processes in the semantic machines, are embodied by causal sentences driven not only by the syntactic, but also, and mostly by, the semantic properties of their grammatical constituents representing various forms and associations of reality.

Syntax, data, computation, and logic are the architectural principles of mechanical and syntactic systems as digital, discrete-state computers, whereas, the soul and life of onto-semantic systems, as human brains or artificial intelligences, consists in semantics, world knowledge, computation, real logic, all supported by a standard ontological groundwork. And the quality of reference ontology will finally decide how close the digital know-all creature, like VAM, may near to the greatest human mind ever-living in our sinful world. If the closed-world Q&A systems rely on domain knowledge and specific ontologies, while the open-world question answering systems should involve fundamental world knowledge, upper-level ontologies, and commonsense reasoning mechanisms, all within the single, standard semantic model of reality. Ignoring this principal methodology resulted in an untimely demise of the project Halo's Digital Aristotle, what certainly does not mean the dying of all hopes of the Virtual Aristotle.

Conclusion

In the universal query answering knowledge machine all the knowledge entities (elements or units) and relationships, as well as its core design components and functioning, will be determined and organized by the reference model of reality.

It is confirmed that the number of the answer types is equivalent to the number of the entity types. Or, the kinds of questions with the possible question answering types delivered are as many as the kinds of things we know. The issues of the question type analysis and answer type classification are ultimately determined by the general entity (being) classification schema. Thus, the universal entity classification model setting the meanings and types of interrogatives constitutes the knowledge hardcore of any advanced Q&A systems necessarily involving fundamental world knowledge and general reasoning mechanisms within a common ontological platform.

To construct an omniscient software system with Q&A facility and general reasoning capacities, above all, we need to represent the whole knowledge of reality as a single universe of discourse. That challenging intellectual task suggests creating a formal theory of things providing a causal scheme of real processes, so that to enable a knowledge-based answering agent with causal reasoning and answer explication. So only the extensive world knowledge supported with the lexical data bases will make possible a computing entity able to operate with meaningful queries of various sorts and types, specific and general, investigative and informational.

References

Abdoullaev, A. (1989). *Knowledge base of encyclopedic artificial intelligence.* Moscow: VINITI, the USSR Academy of Sciences (in Russian).

Aristotle, Logic (Organon): Posterior Analytics. In M.J. Adler (Ed.), *Great books of the western world* (Vol. 1). Chicago: Encyclopaedia Britannica, Inc.

Aristotle, Logic (Organon): Topics. (1990). In M.J. Adler (Ed.), *Great books of the western world* (Vol. 1). Chicago: Encyclopaedia Britannica, Inc.

Aristotle, Metaphysics. (1990). In M.J. Adler (Ed.), *Great books of the western world* (Vol. 1). Chicago: Encyclopaedia Britannica, Inc.

Aristotle, Physical Treatises: Physics (Physica). (1990). In M.J. Adler (Ed.), *Great books of the western world* (Vol. 1). Chicago: Encyclopaedia Britannica, Inc.

Burger, J., Cardie, C., & Chaudhri, V., et al. *Issues, tasks and program structures to roadmap research in question & answering (Q&A).* Retrieved November 8, 2006, from http://www-nlpir.nist.gov/projects/duc/papers/qa.Roadmap-paper_v2.doc

EuroWordNet top concepts. (2004). Retrieved April 25, 2005, from http://demo.ontotext.com

Iwanska, L., & Shapiro, S. (Eds.). (2000). *Natural language processing and knowledge representation: Language for knowledge and knowledge for language.* Mentlo Park, CA and Cambridge, MA: AAAI Press and MIT Press.

Jackendoff, R. (1983). *Semantics and cognition.* Cambridge, MA: MIT Press.

Lyons, J. (1977). *Semantics* (2 vols.). Cambridge, MA: Cambridge University Press.

Onto-based querying. (2004). Retrieved April 22, 2006, from http://www.ontoquery.dk

Project Halo. (2003). Retrieved April 22, 2006, from http://projecthalo.com/

Question-answering system in natural language on internet and intranets. (2004). France Telecom, Retrieved April 25, 2006, from www.francetelecom.com/en/group/rd/offer/brevets/programmes/sysqa.html

Question answering. Retrieved November 8, 2006, from Wikipedia http://en.wikipedia.org/wiki/Question_answering

Structure of the EDR electronic dictionary. (2005). Retrieved May 22, 2005, from http://www.iijnet.or.jp/edr/E_Struct.html>

Text retrieval conference (TREC) Data. (2004). Retrieved April 22, 2006, from http://trec.nist.gov/data/topics_eng/

Turing, A. (1950). Computing machinery and intelligence. *Mind,* Vol. LIX, No. 236, 433-460.

USECS (Universal standard entity classification system). (2005). Retrieved April 22, 2006, from http://www.encyclopedic-intelligence.com

Vendler, Z. (1967). *Linguistics in philosophy.* Ithaca, New York: the Cornell University Press.

WordNet, WordNet 2.0. (2004). Retrieved April 22, 2006, from http://www.cog.sci.princeton.edu/wn/

Chapter XII

The Knowledge Society Applications:
The RRR Language Machines

Introduction

Like in the human mind marked by rationality, goals, and purposeful actions, knowledge in advanced computer systems and intelligent agents is thought to constitute a distinct level lying above the symbol (computational), the logical (algorithmic), and the physical (biological, mechanical) levels (Newell, 1980; Newell, 1990; Newell, 1993). The symbol or code level with its two kinds of representations, data structures (contents) and processes (codes and procedures), is considered to be a representational medium level realizing knowledge content and actuating the agent's thoughts and decisions. Accordingly, all inquiries of AI usually are arranged as a separate contribution either to the symbol (code) level or to the knowledge (semantic) level. Then any sort of logic, whatever may be its generality, contributes little directly to the knowledge level, since "no matter how general a logic is, it is not at the knowledge level" (Newell, 1980). Logical principles are by nature merely rules of reasoning, deductive arguments, and demonstration, they are not principles of real knowledge. So, being neutral to their subjects, formal logical systems neither describe nor represent, nor explain, nor predict any real phenomena. Consequently, nothing substantive can be deduced from the logical axioms or postulates regarding the nature and order of things in the real world. Nevertheless, despite the public fact that logic says nothing about any real thing, and that it is all about proposition and inference forms, there is an established habit to combine formal logic languages with entity taxonomies and typologies so that to construct a general representational language for the DA, in particular, and knowledge technologies, in general (Ontology and Taxonomies, 2001). Nowadays, such versions of logical formalisms and languages as semantic networks, rule systems, fuzzy logic, frame models, predicate calculus, or situation calculus still constitute all the thrust of current knowledge and reasoning applications. Some AI practitioners feel concern about this conceptual confusion, and urge on replacing "form-oriented AI research" with "content-directed AI Research," (Mizoguchi, 1998). What is intended as an appeal to the AI communities to substitute knowledge engineering with ontological (domain knowledge) engineering, thus, underlining the higher value of content theories in comparison with so-called logical mechanism theories.

A few AI researchers even undertook a new line of attack on the nature and meaning of knowledge representation, as combining five various senses: an entity surrogate, or imitation, or sign to reason about the world instead of premature unreasonable acting in it; the terms and concepts to intelligently reason about the world, or ontological commitments; the computational environment for effective intelligent reasoning; a language to say things about the world (Davis, Shrobe, & Szolovits, 1993).

Still it is not all to understand the evident things: that the whole issue of knowledge representation was misguided, that "KR *is* a set of ontological commitments,", that the content (basic classes and rules about the world), rather than the form (arbitrary notations and formal languages as logic, rules, frames, XML, RDF, or OWL constructs), makes the essence of information about the world.

As a matter of fact, all the difference is how the AI designer is constructing reality, or how to view and reason about the world. For if the world is modeled as the totality of mental experiences, as being inside the mind of some intelligent human observer, then the representation is narrowly restrained by a sort of intuitive ontological categories and principles to which some individual agent (or a group of cognitive subjects) is culturally committed. In this case, the developer is doomed to create reasoning machines with different worlds, where spiritual agents and supernatural constructions may become as much as a part of reality as natural objects.

Again, the lack of success of the pilot stage of the Digital Aristotle (or rather its full failure) was attributed to that the task of knowledge representation (modeling and formulation), was assigned to knowledge engineers who misunderstood "the true domain knowledge models." To head off this, the subsequent stage is planned to concentrate on making a set of knowledge learning tools, such as document-based knowledge formulation applications. Such software application is planned to be designed as the means whereby the domain experts are suggested to get the basic content of the textual material, its ontology (entity and relationship classes and instances, axioms, definitions, and rules) (Friedland & Allen, 2004), as it is projected with the Semantic Web ontology languages. Still these measures hardly prevent another failure, mostly because of lack of understanding that an application capable to run the world's scientific knowledge is a sort of knowledge system fundamentally different from traditional AI knowledge technologies. First and foremost, what is distinctive of the Virtual Aristotle is that it typifies a new class of intelligent machines, a world representation and reasoning (WR&R) system, whereas the DA epitomizes a KR&R system involving formal logical languages fully accountable for the taxonomy of shortcomings (Friedland & Allen, 2004). Below we look into the nature and meaning of world knowledge and its formal representation in the context of computing intelligent artifacts. Also, the VA's general meaning framework accounting for a comprehensive hierarchy of entity representations will be outlined.

Briefly, the topics to analyze are as follows:

- The world knowledge representation formalism with meaning relationships
- How the knowledge base content should be structured
- The description of the VAM in terms of automata theory as a universal Q&A knowledge transducer
- The content and organization of the encyclopedic knowledge base

The RRR Machines: The Nature of Knowledge and World Knowledge Systems

Of all the specific differences distinguishing knowledge as a complex entity of mind, the most essential is that it is ultimately about the world. Like the world, reality, entity or being, knowledge is a concept of comprehensive extent, for its extensiveness is correlative with the world immensity, diversity, and expansion. The meaning of knowledge is not, then, less complicated than that of the world. As the world itself, knowledge also has the elements, parts and conditions; structure, composition, or organization; kinds, types, or forms; levels and order; rules and laws. It is frequently defined as a kind of awareness of things consisting in determining their *whos and whats, whys, wherefores, wherebys, whens, and wheres.* That is, to know something is to give suitable answering to the following questions: Who? What? Why? How? Where? When? It appears, when looking into a thing, we ask several stock and standard questions on:

- The what of it or the who of it (assigning a kind of stuff or being; a what substance question)
- The what property of it (attributing a kind of states; a what state question)
- The why of it (attributing a cause, reason, purpose; a what change question)
- The how of it (explaining the manner of occurrence, action, or doing; a what change question)
- The where of it (assigning a spatial property; a what relation question)
- The when of it (assigning a temporal property; a what relation question)

But the knowledge classification and representation is complicated with the fact that the ways of regarding it are as many as the views of things in the world. The classes of knowledge can range from mind or brain as the seat of thoughts and feelings, to power as mental qualities to mental vocabulary, to lexis as all the words of a language, to structure to history as a body of knowledge, to perception, to cognitive processes operating mental contents, to mental content to information, to mental attitude determining human actions (WordNet 2.0., Knowledge, Cognition, Noesis).

In all, as the most complex cognitive entity, knowledge is marked by several dimensions: epistemological, logical, semantic, and ontological. Epistemologically, the kinds of knowledge are traditionally classified by reference to the variety of all the entities, existing or thinkable, as the pairs of opposites: logical/factual, necessary/contingent, tautological/significant, and conceptual/physical. Also, knowledge is sorted with respect to the degree of certainty (probable or certain; adequate or inadequate knowledge); and with relation to the designation, purpose, or final cause of knowledge (theoretical or practical knowledge). Or, by the means of knowing: through sense, intuition, reason, learning, perception, acquaintance, experience, imagination, apprehension, and intelligence (a priori or a posteriori knowledge, the intuitive or the discursive, the empirical or the theoretical, the common sense and the factual). From the meaning standpoint, knowledge as a relative entity is viewed as information about things, whereby having such correlatives as representing constructs (representatives) and the represented things (representees). Again, the types of these correlatives can serve as another criterion for another knowledge classification. From the integrative ontological standpoint, the full meaning scope of knowledge as a complex intelligent process is associated with the typical aspects of a real process of relationships: *the agent* or the knower and *an object* of knowing or the knowable; *the state, act, or process* of knowledge (consciousness or awareness, understanding, learning, and cognition, i.e., all higher cognitive processes such as search, thinking, deciding, knowing, and language); *the ways* by which the process is accomplished (perception, intuition, reasoning); and *the product and results* of knowing (experience; skill, information, science, technology, wisdom, culture).

Altogether, the full meaning of knowledge consists in the following:

1. All kinds of knowers and what the agent thinks, represents, understands; pragmatic and semantic dimensions.
2. All types of things knowable or represented; ontological and scientific dimensions.
3. All forms of results; epistemological and psychological dimensions.
4. The sum total of what is known; material and cultural dimensions.
5. The mechanisms, methods, techniques, and modes of knowing processes; methodological, syntactic and logical dimensions.

Among *the kinds of knowers or agents,* there are the animal brutes, the human beings, and the knowledgeable machines. We can say knowledge is all that a knowing agent knows or represents by any process, from whatever source, facts, truths, principles, rules, hypotheses, notions, ideas, and so forth.

Among *the things to be known or represented,* there is a great number of the objects of knowing: physical, chemical, biological, mental, social, technological, cultural, linguistic objects, as well as mathematical, logical, and conceptual objects inclusive knowledge itself, that is, knowledge about knowledge.

The modes of knowledge or *forms of representations and organization* are as diverse as concepts, ideas, facts, truths, principles, laws, rules, intuitions, judgments, definitions, algorithms, hypotheses, and theories.

The sum of all the knowledge in objective sense constitutes the complex whole of what is known encoded on material substrates or electronic forms; and realized in human practice, actions, and behavior. It includes all cultural elements: the branches of knowledge and technology, institutions, beliefs, laws, languages, morals, customs, codes, rituals, ceremonies, artifacts, tools, techniques, machines, books, and all works of arts.

The mechanisms and modes of knowing are mainly determined by the psychological factors of the mind, by the agent's attitudes, beliefs, ideals, values, and dispositions.

Now it must be plain that the knowledge-level machine as a new species of the knower belongs to the class of knowledgeable and automated machines or *knowledge automata* as computing entities cognizant of the meanings of database contents (structured factual data or information as embodied knowledge) by being aware of the basic elements and the composition of its correlative counterpart—reality. In formal terms, the knowledge automata operate symbolic structures and compositions of entity representations endowed with semantic properties. Ideally, such knowledge machine should be capable to assign to the entity and relationship representations their proper meanings, thus, automatically inducing the general statements about various kinds of things symbolized as entity class variables and specific sentences presented as the values of the variables. If such a machine can compute the classes, subclasses, instances of matter, energy, nature, life, man, animal, society, and so forth, their key elements, properties, states, forms, changes, and relations, then it could mechanically deduce all the many assertions and facts about the things of every day life. Namely, that metal corrodes; water is liquid and has weight; people die; activity requires energy and so forth.

In other words, the value of any advanced knowledge based system is measured by what and how many it knows about reality, as well as by the truth value of its symbolic representations. That is, the quality and performance of knowledge-level agents determined by the accuracy of the world knowledge representation, which turns on the largeness and depth of ontology; its potency to be applicable to all major domains of being, to every basic sphere of reality, to any knowledge domain. Since ontology is all about the world knowledge, it has to constitute the quintessence of knowledge-level machines. But the principal issue here is to give a formal and exact definition to knowledge itself, particularly, to the world knowledge, information about the whole world and its basic parts and connections. In the symbolical fashion, the world knowledge **K** can be described as an isomorphism between ontology as the infinite theory **O** of the universal environment and the entire reality **W**:

$$\mathbf{K}: \mathbf{W} \leftrightarrow \mathbf{O} \qquad (1)$$

In the world knowledge equation, the operator **K** stands for a special kind of meaning relationship from the world to the systems of entity representations, ranging as wide as physical quantities, signals, symbols, and signs (sense-data, images, figures), mental symbols; mathematical structures, logical formulas, and conceptual constructs; linguistic expressions; and non-linguistic signs. So the EIS ontology is constructed not as a set of entities with their associations, but rather as a universal formal theory (symbolism) describing the general things of the world formally structured as a set of primary things: $\mathbf{W} = < \mathbf{T}, \mathbf{O}, \mathbf{S}, \mathbf{C}, \mathbf{R}, \perp >$.

The functional symbol $\leftrightarrow$ denotes bijection, a one-to-one correspondence relation, pairing the constituents of reality with the elements of knowledge such that the knowledge elements and their combinations should correspond to the real world entities and their order. That is, ontological theory is exact and complete representation of the world of things if each its statement is bijective and true for every represented entity. The notion of truth, the truth or verity, as the correspondence to reality is the distinguishing attribute of the world representation and reasoning systems. The truth as the conformity to things involves the adequacy between the thing represented and that by which it is represented unlike the coherence theory involving logical consistency of representations. The cause of the truth of a complete representation, a proposition, is the reality or the fact of being of the represented thing, and not vice versa; since the truth or falsity of the representation depends on the fact of the being or not being of the thing represented. In other words, ontological truth has natural priority against all other sorts of truths, scientific, mathematical, and logical truths, be they tautologies, dictionary truth, truth by postulation, by proof, by deduction, or by satisfaction in a special model.

Thus, the complex sign structure **K(W)** as the knowledge function presents **the world representation function** assigning to the distinct components of reality their appropriate symbolic representations, such as the variables of objects, states, changes, relationships. As a result, the general ontology language O_L for the VA can be formulated as the knowledge function $K(W_I) = K(O_j, S_k, C_l, R_m)$, where its arguments are the ontological symbols representing the variables of primary entity types and relationships, each of which can assume any of a set of values, instances, or individuals. The entity variables may serve as placeholders for mathematical objects and relationships, logical terms or predicates, word classes as names and verbs, or computing classes and procedures. So then there are four major groups of represented things:

- The genus "substance" exemplified by particular objects and individuals
- The class "state" as attributes, properties, qualities and quantities
- The class "change" as processes, events, acts, and behavior
- The class "relation" typified as causation, identity, diversity, part to whole, and other various sorts of relations

As a result, there are four sorts of knowledge representations: the symbolic forms denoting objects; the sign patterns representing states, properties, and quantities; the sign structures of changes, actions, processes, and events, or the symbol structures of relationships of different types. Meaningfully associated according to the Entity-to-Entity Relationship product, $W_I \times W_I$, these basic types of representations will compose the universal symbolism about realities. In such a common system of symbols, reasoning is the meaning (ontology)-driven computations on entity representations (mental signs, conceptual constructs, computing data, or linguistic constructions), taken either as the class variables (ontological and domain-knowledge reasoning), or as the value sets of the variables or a product of value sets (concrete reasoning). The former encompasses all regional theories of particular classes of reality, as theoretical physics, theory of evolution, cognitive sciences, or artificial intelligence models of knowledge and reasoning. The latter refer to specific theories, theoretical models, and schemata presenting certain features of an entity or a system of concrete entities. The statements are here existential, factual, singular, and contextual as describing the features of individual entities: compositions and structures; properties and facts; conditions and circumstances, events and occurrences; particular causes and effects; or specific rules and patterns.

Now we can see the critical signification of telling apart the species of *knowledge representation* and its genus, *the world representation,* which is nothing else but a universal body of knowledge embodying directly the world structure and expressed by linguistic constructions, as in:

$$2^{L} \leftrightarrow 2^{K} \leftrightarrow 2^{W} \quad (2)$$

The left side of this series of equations represents the world of language, natural or artificial, while the middle part stands for the power set of ontological constructs, whereas the right side represents the power set of the universe of things.

When the object of investigation is the knowledge itself, we deal with nothing but *knowledge about knowledge,* which is all about formal rearrangements of knowledge, that is, about forms and formats for information storage or exchange, like knowledge interchange format or RDF. This is the sphere of the traditional KR&R systems, like that practiced by the SRI' Halo project of the DA based on the SHAKEN software system of knowledge machine and recognized as most successful group among the competing teams of the Vulcan Inc.'s Project Halo. According to the world-knowledge-language ternary relationship, there also is the way of building knowledge machines as commencing from languages, natural or formal. Namely, referring to a number of linguistic resources, analyzing the natural language structure, its key grammatical categories and relationships, as well as using existent systems of verbal classification, one can try to fix a set of knowledge elements and relationships. Acting this way, the SRI International team is developing its knowledge machine as based on a collection (library) of basic concepts and relations derived mainly from Roget's comprehensive classification of synonyms and the WordNet's lexical system. When the underlying structure of semantic resources are misunderstood, as shareable and reusable knowledge units (general knowledge components) may be selected a defective set of individuals. As a result, the SRI's library of representations of things only includes entities as objects, properties as linking objects to values, events as actions and states, relations connecting objects to objects, events to objects, and events to events added with roles (or functions) the objects perform in the relationships. In such a "library of knowledge units" intended for building the DA, many classes of entities are simply missing, when compared to the Entity-to-Entity Relationship Product (or Matrix). To avoid this deficiency, any entity instances, either the things that exist (objects) or the things that are (events), should be generalized as entity types, object kinds and event kinds, so that to represent entities and relationships truly and systematically. Thus, to make a good use of such lexical resources as Roget's scheme of synonyms (Roget, 1852, 2002) or the WordNet's hierarchy of semantic sets, we need first to develop a unifying world model of things within which a whole collection of word classes can be properly organized. Since division of headwords as expressing six classes of entities [abstract relations (existence, relation, quantity, order, number, time, change, causation), space, matter, intellectual faculties, voluntary powers, and the sentiment (affections) and moral powers] itself asks for reclassification according to the true system of ontological standards. After all, an all-comprehensive classificatory scheme comes as the total semantic skeleton of dictionaries and thesauri, however large, which lexical resources provide rich material to flesh out the world taxonomies and typologies.

Inasmuch as the ontological categories as universal predicates constitute the heart of the knowledge level, its application to knowledge-based technologies may result in developing **the world knowledge machines** typified by *the VAM as a UFO-driven reasoning application where the real-world entities and relationships represented as a body of symbolic structures, procedures, and rules*. Unlike human beings, the VAM as an ontology-based computing artifact are devoid of mental experience carrying with itself never-ending problems of searching the nature and meanings of mental content. The traditional semantic relationships between symbols (natural languages), thought (intrinsic things), and extra-language things, where the language expresses the ideas and the ideas represent the world, reduced in the Virtual Aristotle to the binary relations between symbols (signs, words) and things (outside to machines). Knowledge machines are expected to use sensible signs (physical signals, codes, or words) in order to express, to process, and to communicate information about the world domains and their contents. In contrast with the human mind, here the symbolic structures and processes are supposed to directly signify things and relationships without the mediation of ideas. It means that the nature of "mechanical meaning" depends only on the types of symbols and the kinds of entities these symbols denote or represent. In the order of intelligent machines, computers or computer programs will operate with symbols and code structures being devoid of any mental constructs. The symbols processed by the machine intelligence are the signs of things, and, hence, they have to get their significance without the mediation of the conceptions of human intellect. Therefore, the significance of symbols is to come directly from the things denoted or properties and relations connoted. Again, for the VAM, it is all the same whether its symbolic codes signify the objects of the real world (matter, sensible things, substances, appearances) or the objects of the human mind (images and ideas, feelings and thoughts, perceptions and conceptions). In either of the cases, it operates with surrogates, substitutes, or symbols by marking, naming, indicating, symbolizing, standing for, embodying, typifying, or taking the place of other things standing external to machines.

Inasmuch as knowledge machines are bereft of mental experiences, their reasoning processes can be correlated with the handling of the physical signals expressed by the symbolic codes representing things including various phenomena of matter and mind. In the human language, spoken words are the symbols of thought; while in the machine language both written and spoken words are the signs of things, which, unlike the mental experiences symbolized by words, are the same for all sorts of agents, natural and artificial. In other words, symbolic processing becomes a true reasoning process only when its structures comply with the order of things in the world (mental or physical). As a result of this, in the digital knowledge machines, there are no more such confusing and confused notions as concepts, ideas, and conceptions, as well as sense, intension, comprehension, extension, and so forth, all that specifies the semantic relationships between ideas and the world. Instead, we come to a body of entity representations in the form of symbols, signs, signals, or expressions, all representing things of different kinds. The meaning and truth, and sense and reference of the sign (symbol) system of entity representations as the set of expressions of ontological language will be determined together by the group of axioms:

a. The set of ontological definitions, axioms, and principles about the general kinds of entities and relationships formulating their key properties, attributes, and associations.

b. The set of factual presuppositions of knowledge domains, like as the second law of thermodynamics for the concept of state of a physical system or Newton's second law for the concept of mass of a body.

c. The set of mathematical axioms, as the axioms of set theory for aggregates of formal entities.

d. The set of semantic axioms such as the reference rules and representation assumptions.

Such a content-based universal symbolism is supposed to contribute to the code and knowledge levels, as well as to the causal mechanism level of intelligent agents. That approach also makes obsolete the definition of knowledge base as a collection of all sorts and manners of information items on a par with ontological terms, lexicons, and taxonomies. More relevant here comes the knowledge base to be defined as an information body of domain-specific knowledge items (terms, facts, rules, concepts, methods, models, programs, codes, logics, figures, and analogies) structured by the all-entity representation system. That is the VA as a body of information entities complete with the set of first principles, definitions, and axioms and all sorts of computations as operations on entity representations should be grounded in and governed by a complete and consistent structure of world variables. The completeness of such general ontological language is confirmed by its validity extended as being true in all spheres of being and in every area of reality and satisfiability as being true for particular classes of being, the subject matter of applied ontologies. At last, its consistency and logicality will be provided by the accordance with the primary premises and basic truths supported by mathematical, factual, and semantic assumptions and rules.

Below we demonstrate that the EIS ontology can be represented as a general system of sings, symbolic representations, and formulas where the meaning of the things they represent are given by the fundamental axioms and formulas, ontological, mathematical and factual, coming complete with the semantic assumptions and rules.

The Meaning Processing in the Virtual Aristotle

To have as symbolically represented in the VA the semantic information about the real world, scientific knowledge particularly, requires a universal theory of direct and inverse meaning relations of signs or symbols and things. The basic elements of such a general ontological symbolism come to be the symbolic variables of substances, states, and changes and relationships of different types and kinds, with their properties, instances (values), or value sets. And the basic construction is here the universal class of the world signifying the total collection of entities and relations and involving the expressions of the basic kinds of things. In other words, any elemental representations of any complex symbolic compositions either refer to substances and substance classes or indicate states or state types, changes or change kinds, relations or relation kinds.

The full significance (meaning) of any symbolic expression consists in denoting its primary and essential meaning, which makes the core of the definition, and in connoting or representing all the properties (attributes) and associations relevant to the symbol. For instance, any individual term of objects gets its signification from a substance (or a system of real objects) referred to and from the properties, states, behavior, or relationships (spatial and temporal) expressed or intended. As the knowledge representation and reasoning background, the universal ontology can define, describe, represent, or explain symbols (expressions, constructions, words, or conceptions) in the key terms of entities and their properties and instances. So, by its nature, ontology is not only about what a thing is or what the content of a concept is, but also it is about what a symbol (word, term or name) means or signifies. Thus, it is established a hierarchy of signs and symbols toped by the catch-all-entity expression W, where "spoken words are the symbols of mental experience and written words are the symbols of spoken words" and eventually the mental experiences are the images of outside things (Aristotle, On Interpretation). Crucially, the definitive meaning of a hierarchy of signs or symbols mapping to computing data representing information about the world, its realms, regions and domains, can be found within the axiomatic ontological system as its single semantic background.

So far the most significant contribution to the semantics of data was made by the ER schema deriving its strength from mathematical system, set theory and relation theory; namely, the abstract relational structure $A = < S, \leq >$, where S is an arbitrary set of elements (or object or entities) and $\leq$ an ordering of S. This structure is a foundation of the unified ER model of semantic information about the world, defined as a collection of entities and relationships. Such world model involved as the onto-semantic primitives the concepts of entity (as sets and instances) and relationship (as sets and instances) with their attributes, or properties, measured by values (as sets, products or a single value). Thus, the universal sign structure included three types of variables, $W = <E_i, R_i, f>$, the symbol E_i denoting the entity sets taken as domains, $E_i = \{E_1, E_2, \dots, E_n\}$; the sign R_i representing the relationship sets as defined on the domains, $R_i = \{(e_1, e_2, \dots, e_n) \mid e_1 \in E_n, e_2 \in E_2, \dots, e_n \in E_n\}$, the ordered sequence (a tuple) of which $(e_1, e_2, \dots, e_n)$ constituted a relationship; and a function f mapping from an entity (relationship) set into a value (instance) set V_i or a product of instance sets, f: E_i (or R_i) $\rightarrow$ V_i, thus representing an attribute of entity or relationship (Chen, 1976).

Today the ontology of the entity-relationship model is widely used as an underlying scheme in defining the various local schemas of Web data resources, in resolving the challenge of semantic interoperability by constructing the data integration systems, and in constructing Web ontology languages. However successful, the ER models become a global scheme only as a general axiomatic theory of things dealing with the issue of issues: what the most general kinds of entities are, as well as how they are related and how they must be defined and represented. Such a universal ontological language encompassing the ER representation model constitutes the core of the VA, organizing its information content and providing causal reasoning on the coded representations of things. Having such a theory of the world encoded in a general-purpose reasoning application will provide an effective machine-operating semantics of symbolic representations.

As all said, below we will demonstrate that an essential meaning (or significance) of signs and symbols (words, terms, expressions, constructions, representations, or constructs) is achieved within the universal ontological system, like a complete meaning of a theoretical entity can be reached only in a well-founded scientific theory. For this purpose, we use an abstract representation of a symbol variable V as a general sign standing for a kind of thing, or defined as a character (letter, mark, abbreviation, an arbitrary sign) indicating various sorts of things, a substance, an objects, an element, an event, a quantity, a quality, an operation, or a relationship. In the ontological symbolism where entity and relation classes are represented by means of symbols, any sign has at least two meaning dimensions, one giving the primary sense (by a definition and a statement of identity) and other, the contextual meanings (by indicating all the properties and relationships). Then its signification will be formally composed of two semantic functions: the representation (designation and connotation) function Σ and the denotation (definition or reference) function Θ (see Figure 13). The denotation (designation) function maps the general symbol V into the power set of the universe of objects thus identifying its reference (denotation, extension) or designation class, that is, the full collection of objects that the variable refers to:

$$\Theta: V \rightarrow 2^{O} \tag{3}$$

The designation rule involves as its components symbols and conceptual designata: thoughts, ideas, or mental constructs, whereas the denotation rule operates a symbol-object correspondence where the denotata are the referents of the symbolism. When the values of a semantic function are taken from a universe of non-substances, we have a function-state correspondence, and so forth, generalized as a representation function. The representation function maps the variable V into the power set of the universe of states, changes, and relations, whereby identifying its representation (connotation) class, that is, the full collection of represented entities:

$$\Sigma: V \rightarrow 2^{\{S, C, R\}} \tag{4}$$

Then the meaning function maps the general symbol V into the Cartesian product associating the representation function and the reference function:

$$M: V \rightarrow 2^{O} \times 2^{\{S, C, R\}} \tag{5}$$

This meaning formula is noteworthy by being a key semantic rule of universal ontology, as the total and all-embracing sematic model of reality. Since it provides the guidelines and directions as to how to classify all symbols or variables in terms of entities, which are used in scientific theories, mathematical expressions, and logical systems. According to the semantic rule, there are four groups of symbol variables (or quantities):

Figure 13. The meaning of symbols

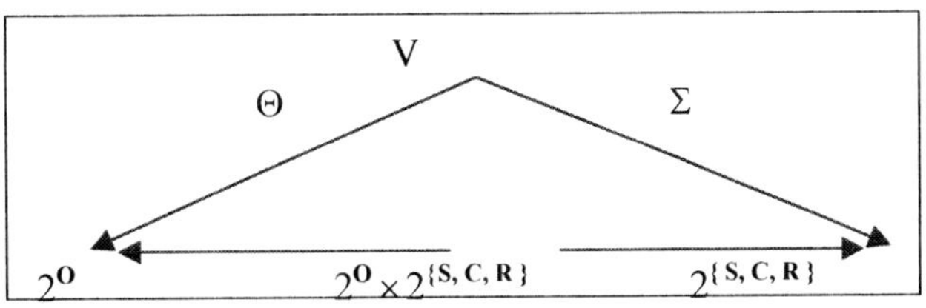

- The object variables O_j naming or referring to concrete objects of different sorts and kinds (particles, fields, cells, proteins, ecosystems; logical individuals and mathematical objects).
- The state (or function) variables S_k representing properties, qualities, or quantities (weight, temperature, pressure, size, shape, mass, color, utility, functions, logical or mathematical).
- The change variables C_l representing events, occurrences, and processes (chemical changes, physical events, biological processes, mental occurrences, or cultural changes; mathematical differentials, or derived functions).
- The relational variables R_m symbolizing relationships, associations, and connections of various types (space, time, metric tensor, coordinates, predicates; differential equations).

Evidently this symbolic typology is extended to formal variables, logical and mathematical symbols and signs, as well as linguistic constructions and expressions. Having the meaning formula as encoded, a reasoning application will possess the knowledge of substances and objects together with all the associated classes of changes, events, reasons, and processes, without which it is impossible to perform efficient machine reasoning. It also will enable software agents acting on the Semantic Web to find the value (meaning) of the Web content by establishing hyperlinks to fundamental terms, such as the expressions of cause (causation and causality). The importance of the meaning law may be exemplified how should be computed causal constructions on the symbolic level, since the agent, both natural and synthetic, becomes really intelligent only when it is capable to understand the meaning of cause and causality, or to compute the semantic values of causative signs and expressions.

Specifying the general meaning formula for a class of causal expressions, written as the symbol C, the meaning law will be modified as in:

$$\mu: C \rightarrow 2^{O} \times 2^{\{S, C, R\}} \quad (6)$$

This indicates the scope of all basic meanings the term of cause might have:

i. A causal agent or agency, an object that causes changes and events to happen, denoted by the agent variables and expressed by agentive nouns.
ii. The state of things, properties, quantities, qualities, situations, or states affairs, conditions as cognitive states or physical conditions.
iii. The changes of substances or states, as actions, processes, operations, activities, acts, events, behavior, occurrences and happenings, natural, mental, and cultural.
iv. The relationship of causality, a cause and effect relationship of different sorts and instances.

For a special particular cause $c \subset C$ the value of the cause meaning function $\mu(C)$ at c is equal to the expression: $\mu(c) = < o(c), \sigma(c) >$. Here the meaning of a certain cause c is made up of its reference classes o(c) and the representation classes σ(c). The cause meaning formula allows a mechanical and automated analysis of the notion of cause. First of all, it points out that any cause refers to that which undergoes the change, $o: C \rightarrow 2^{O}$. Such a reference class is all the general class of objects and substances comprising a class of material objects, a class of living things, a class of intelligent beings, a class of information systems, and so forth. In connection with the material agents, the variety of causes ranges as in the list: *Destiny or Fate, first cause or prime mover, supernatural, nature, power or force, life principle, catalyst, and substantial agent.* With respect to the intelligent agents, *the causer is either a human or artificial being: an agent, actor, author, creator, designer, doer, former, originator, constructor,* and so forth. Besides, there is other additional meaning of cause as relating to that on account of which the change takes place—*cause, end, reason, purpose, aim, plan, project, sake, good, grounds, interest,* and so forth.

Nevertheless, by its nature, the primary meaning of cause represents that by which the change will be produced $\sigma: C \rightarrow 2^{C}$ - the moving causes or the motive and generative forces; that which directly effects the result of changing or alteration or becoming different as the action of changing, as the act of causing some entity to be, to change, or to happen. Accordingly, its essential sense is contained in the definition that a cause is a change of entity (substance, state, action, relation) producing other changes. According to the definition, the types of cause are as many as the types of change or difference, namely, *change in substance* C_S; *change in state* or in property (quality or quantity) C_p; change in action C_c; and change in relationship C_R. The *substantial change* is well exemplified by annihilation generation, transmutation, and changes of integrity, while *the change of state* instantiated by all sorts of *changes in quality* C_q (biological growth, evolution), *changes in quantity* C_M (increase, decrease), *changes in position* C_m (all kinds of motion). Here also, it is included all sorts of *alteration* and *modification* implying some change from one state (or phase) to another: mutation, deformation, destabilization. The changes in action are instantiated by the acts of altering the manner of action, activity, doing, and so forth, like variation and fluctuation, the rate of change. Whereas, the relative changes are exemplified by changes of temporal and spatial properties, for example, changes of direction, or other forms of connections.

Thus, at its very bottom the cause belongs to change. Or, the primary meaning of cause semantically relates to the class of change **C** that involves the general state **S** and process Π of the universe **W**:

$$\mu(C) = \mathbf{C} \subseteq \mathbf{S} \times \mathbf{S} \subseteq \Pi \subseteq \mathbf{E}^4 \subseteq \mathbf{W} \tag{7}$$

According to formula (7), the full definition of cause additionally to its established senses should involve the change types: *alteration, conversion, innovation, motion, mutation, novelty, regeneration, renewal, revolution, transformation, transition, transmutation, variation.* Also, the idea of cause necessarily involves all kinds of events: *case, chance, contingency, consequence, eventuality, episode, fact, incident, issue, happening, occasion, occurrence, outcome, and possibility*. Last but not least, the reasoning about causes involves the process of causation, the action, operation or working of causes, expressed by a set of causative terms: *activity, act, agency, application, effect, exertion, influence, occupation, performance, practice, use, procedure*. Then the relation of causes and effects, causality, should be defined as a special kind of process interrelating distinct (or disjoint) changes through causation, or operation. What corresponds to the general definition of relations as consisting of two different components: correlatives linked with the relationship.

All in all, *the symbolism of cause implies and suggests all kinds of material and mental change causers*—motion, occasions, reasons, motives, actions, and purposes; *all mechanisms of the working of causes*—activities, occupations, operations, influences, actions, consecution; *and the various types of the results of change* caused by some previous changes—effects, consequences, products, outcomes, events, issues, upshots.

It emerges that the ontological symbolism raising the basic type symbol V as the sign of the entire entity W has the principal capacity to create the VA as a semantic ontology machine, setting the rules of its semantic behavior, and shaping its ability to understand and generate meaningful language, discourse, or communication.

It is here of use to tell apart the machine, mechanical sense and the human, live semantics once again. Strictly speaking, the meaning formula of the onto-semantics fits for constructs, whereas, applied for symbolic expressions or machine signs, it must be transformed into the significance of signs, which is equal with the meaning of the constructs designated by the signs. It can be proved by way of replacing the representation function with the significance sense function mapping form the set of expressions to sets of states, changes, and relations. And by substituting the reference function with the significance reference function mapping from the set of expressions into classes of substances or classes of objects of any species. Only, in the case of signification functions, the domain of mapping proceeds from the set of expressions of an ontological language, instead of the constructs designated by theses signs. Altogether, the signification function can be constructed by *the composition of the designation function mapping from signs to constructs and the meaning function mapping from the constructs to things*.

Then, defining symbols "as patterns with denotations," it is still not enough to qualify all kinds of non-representational machines as connectionist and mobotist intelligent systems as the examples of physical symbol systems (Simon, 1995). In reality, there are two senses of meaning: live and mechanical.

For human minds, the meaning is commonly regarded as diverse as *thought, intention, or understanding; usage or utility value; the thing referred to or the reference class; intension or content; intension and extension; or sense plus reference*.

For machine minds, the significance consists in *operation, computation, or measurement; information or communication.* That is, unlike human beings, for computing entities to understand the information about physical world is meaning to have as measured the values (or value sets) of physical variables. To determine a specific meaning of communication is to have the precise value of its corresponding entity variable, and to reason is to compute a system of symbols or entity representations according to a set of content-based rules and algorithms given by the general ontological symbolism.

Then for intelligent machines the meaning is reduced to signification having two dimensions—denotations and representations (connotations or designations). Where denotation rules assign symbols to substances; representation (connotation) rules—symbolic structures to states, events and processes, while representation designation rules—symbolic expressions to mental constructs. If symbolic expressions are information entities determined by the semantic rules: denotation rules and representation formulas, then the signs in machines are computational entities marked by the semantic properties: denotation and representation (designation or connotation) relationships. So the meaning and sense of constructs are reducible to the significance of the symbolism of the ontological semantics, whereby bringing meaning, understanding, and reasoning to machines. And according to the theory, any concept has at least two semantic functions:

a. One is to refer to some entity (or to designate for constructs) (reference, denotation, extension).
b. The other is to represent or embody a whole system of meaningful associations of the entity (representation, content, intension, information, knowledge).

Thus the universal ontology as the infinite semantic model reporting the most essential part of the world knowledge can be represented as a machine-readable general symbolism. The machine processing of meanings, embodied as signs, symbolic representations, patterns, and computing codes, is guided by the fundamental definitions, facts, truths, axioms, and formulas, ontological, mathematical, and factual, coming complete with the semantic assumptions and correspondence rules.

Lack of a consistent formal theory of reality, as well as misunderstanding of the nature of world knowledge, confusing it with knowledge representation, badly affect the quality of knowledge systems. As an object lesson, it may well serve the OpenCyc constructs developed by Cycorp, another team participating in the Project Halo, which upper ontology is boasting 6,000 generic concepts laying the foundation for 2,000,000 axioms and 600,000 concepts (CYC Open, 2004). That also is to be extended by the libraries of middle-level ontologies added with a collection of disjoint microtheories. They are all about "time, space; physical and mental processes and actions and events; evidence and causality; belief and awareness and knowledge and reasoning and uncertainty; human mental and physical capabilities and limitations and motivations, substances and individuals, and artifacts" (Lenat, 1998). To see what distinguishes the EIS ontological language, we try out from the alphabetical list of the public CYC ontology lexicon or vocabulary; for example, a middle level term *battle*, a specific kind of human military action, so much favored by DARPA projects. In the CYC vocabulary, the term *battle* is defined as follows: *Battle is a temporal object type, and a subset of hostile social action.* The definition is formatted in the formal, logic-based language as follows:

```
#$Battle
(#$isa#$Battle#$TemporalObjectType)
(#$genls#$Battle#$HostileSocialAction)
```

Whereas, considering that term from the meaning formula of the EIS language, it is not such simple and trivial as presented. To define what is meant by any construct (an individual term, a property or state predicate, a relation concept, a proposition, a context, or a theory), the machine first of all has to identify the reference class, an entity class referred to by the construct, or the denotation class of the sign. This is the definition per se, the statement of an entity's nature, its primary meaning, formulated as a statement of the meaning of the term ("an equivalent nominal formula"). And the basic attribute of the definition is that it is always made in one of the classes of entity predicates: substance or object or agent; state, property, quality or quantity; change, action, activity, or process; or relationship. Then all the statements found through these categories will signify either an entity's nature and its essence or its state, as a quality and quantity, or its behavior, or some other types of ontological classes. Thus, the definition given to some name or term states its essence and signifies its properties; for the essence of an event is a change, while for an animal the essential meaning consists in organism, or being and its properties in specific functions and behavior. This implies that for things that comparatively endure, like objects, we have one kind of definitions, while for things that happen we have another kind of definitions stating the cause (reason) of their existence as the essential nature. Here works the Aristotelian formula, "to know a thing's nature is to know the reason why it is."

In other words, if generically the same class was predicated of some entity instance, then the definition signifies its genus or an essence. Otherwise, generically, the different class is asserted of the entity instance, the resulting statement will only signify its associative connections to some other types of things. But the totality of relations, associations, or relatives constitutes *the sense of the construct* or *all the connotations and designations of the sign.* As a result, the ordered pair of denotation and representation functions constitutes the full meaning of the construct and the full significance of the sign.

Therefore, a sufficient meaning or understanding of something can be reached only by taking into taking account the context, the totality of logical and conceptual relatives, or the sum total of external circumstances in which this something situated. For a proposition, its sense will be determined by the set of statements inducing and induced by that proposition. At last, the true essence and nature of any complex class of things comes from a well-formed body of knowledge organized as an axiomatic scientific theory.

The term *battle* with its synonyms as *conflict*, *fight*, *engagement*, *struggle*, or *fighting*, when represented as a machine symbol, actually involves all the four basic sorts of things: objects, states, actions and relationships. Its essential meaning will be found in its incorporating class or superclass, or parent class, or base class, which is a kind of action [done by distinct groups of humans (agents) standing in the relation of opposition (a relationship) and driven by hostility (a state) and causes (changes)]. This class of human actions constitutes the term's primary meaning, the most direct and specific meaning, or its reference, denotation, or extension, contrary to the semantic traditions. Whereas, its additional (connotation and designation) senses lie in a special group of people or agents (substances), in a special kind of mental states and attitudes (states), in the possible reasons and purposes (causes), in some sort of relationships. All that lies outside of the actions, people, mental attitudes, cause, and purposes, come under the sense of the term.

In all, taken in the most general sense, the term *battle* refers to any conflict, struggle, fighting, engagement, hostility, and so forth, as an instance of the whole course of a war. Further, the term semantically relates to its synonymous species (conflict, affair, combat, contest, conflict, clash, collision, encounter, engagement, fight, strife, struggle, aggression, hostility, war, or warfare). Next, it implies several groups of people standing as opposed relations: alliance, army, military, troop, assembly, association, band, clan, class, community, crowd, faction, horde, mob, nation, organization, society, team, individuals, and so forth, an objective part of its connotation class. Given the sorts of human groups participating, the variety of battle (or conflict), its incorporated species, derived, or child classes may be as different as a military or armed combat between hostile armies, a class war between social classes, a revolt or insurrection of the masses against the power, a trade war between nations, and so forth. Also, the term's meaning implies the mental state of attitudes (enmity, antagonism, animosity, hatred, malice, malignity, spite) and the effective causes (unprovoked aggression, acts of terrorism, or the danger of aggression or terrorism) and final causes (conquer, crush, defeat, mastering, control, power, subjugation, or liberation) as well as consequences (casualties, mass killings, damaging, plundering, looting, robbery, devastation, banditry, rape, devastation, and destruction), all the different classes of the meaning alien to actions or changes make up the sense class. Knowing the meaning of the term by machines will secure such valuable advices for human beings as "never wage war without a good reason," "do battle for a cause of freedom," whereby giving the answer to the questions: "Why did the Americans become involved in the regional wars?" "What instigated the world power to wage unprovoked wars against poor sovereign nations?" and "What are the possible consequences (final causes) of the aggressive wars?" It appears that if the DARPA folks had developed such a brainy semantic advisor capable to see the deep meanings of human actions whereby predicting all the possible consequences, the wars with their innumerable battles and enormous civilian casualties probably would never have happened.

It can be now concluded that having as properly programmed the entity-based semantic rules, the meaningful machine like the VA will get the power to automatically deduce all sorts of significant assertions and negations about individual things, their properties, situations, acts, and relations. If a knowledge system contains as formalized all the essential axioms about inanimate and living things, about their states and the changes of the states under varying conditions or forms of behavior, then there also will not be any need in hand-coding hundreds of thousands of specific conclusions. Like these everyday trivialities: "smoke goes up," but "water flows down"; "people die"; "a car rides on four wheels"; "falling asleep in the car is dangerous"; "one has to go around an obstacle"; "when raining, carry an umbrella over your head";, and so forth. Evidently, all such-like common sense assertions and negations are nothing more but only particular and numberless theorems of the major assumptions about the world and its major classes for denumerable particular situations.

So any intelligent machine to be such should contain as formally coded all as much world knowledge as it exists, being open to quality and modify its dynamic representations. Especially, it should embody all the basic ontological meanings (definitions), truths, axioms, facts, and rules, arranged into five groups concerning the entire entity and its canonic and primary types. The reality related truths. "There is the world, reality, being or existence composed of a finite number of the kinds of entities." "There is a few number of real genera but infinitely many individuals." "Entities exist naturally or mentally." "All entity classes are represented by variables and measured by quantities." The substance related truths. "All objects are substances having qualities, undergoing changes, and standing in relations." "Substance possesses properties." "Conceptual objects have attributes." "Matter is a substance subject to quantity." "Any substance undergoes change." "Nature is a kind of being." "Life is a class of nature." "Humankind is a species of life." The property (state) related truths. "All the circumstances make up the situation." "A collection of properties makes up a state." "Property is a value of a state of the thing." "Quantity is a class of state." "Quantities are divided into magnitudes and multitudes." "Number is a sub-class of quantity." The change related truths. "Change is a transformation of state." "Event is a particular kind of change." "Any cause (and effect) is a change." "Cause produces effect." "Action is a type of change."

"Any action implies the agent." The relation linked truths. "Causation involves a complex process of changes." "Causal processes are made up of different changes." "Time and space belong to the class of relations." "Time is a measure of a relational difference between states." "Space is a relation of places." "Place is a portion of space," and so forth.

Ontology Machinery and Universal Knowledge Transducer

The advent of the computing entities like the Virtual Aristotle makes us reconsider all the traditional knowledge about machines sorted as the energy-converting devices and the information (data) processing technologies. Up to date the universe of machinery consists of four major groups: physical machines, instrumentation, computing machines, and telecommunication systems, as illustrated in Figure 14. All sorts of traditional machines and instrumentation come under the notion of devices capable to convert physical energy in any of its form, either mechanical, thermal, electrical, magnetic, nuclear, and chemical energy into mechanical energy, mechanical forces and motions, or vice versa. Commonly, physical machines are understood as devices having an input, an output, and a mechanism effecting force and motion-modifying functions, linear motion to rotary motion, or vice versa.

In general, for the physical machines, the five key groups are traditionally distinguished: simple devices/ machine elements (the lever, wedge, wheel and axle, incline plane, pulley, and screw); prime movers (e.g., windmills, turbines, engines, fission reactors, or fusion reactors); generators (e.g., thermal, electric, hydraulic or pneumatic); motors (e.g., thermal, electric, hydraulic or pneumatic); operators (e.g., appliances and conveying machines).

The instrumentation equipment, in turn ,involves measuring of all kinds of natural phenomena. Among the equipment are the *instruments* measuring and monitoring physical and chemical properties of a substance, as well as the biomedical instruments, like the X-ray machines, the CAT, and NMR scanners.

A second category is *sensors*, devices detecting any energy change to turn it into a measurable or recorded signal. The *controls* consist of valves, governors, switches, motors, gears, levers, pulleys, power screws, power chain drivers, and other like mechanisms.

The transducers underlie all measuring, analytic, monitoring, and controlling tools and instrumentation, and they are classified according to the form of energy to be transformed. The transducer contains a transforming element by which it converts any kind of input energy into output energy—mechanical, thermal, magnetic, electric, nuclear, gravitational, or chemical. Or, the transducers produce output signals, as voltage, current, displacement, resistance, temperature, force when subjected to a stimulus—radiation, heat, sound, strain, vibration, pressure, acceleration, voltage, and force.

The class of computers as a group of discrete state electronic devices falls within the physical systems able to receive, store, transform, and modify information in all its forms (raw data, data structures, expertise, and knowledge) and formats (text, video, voice, graphics), as well as to control machinery.

Figure 14. The classification of machinery

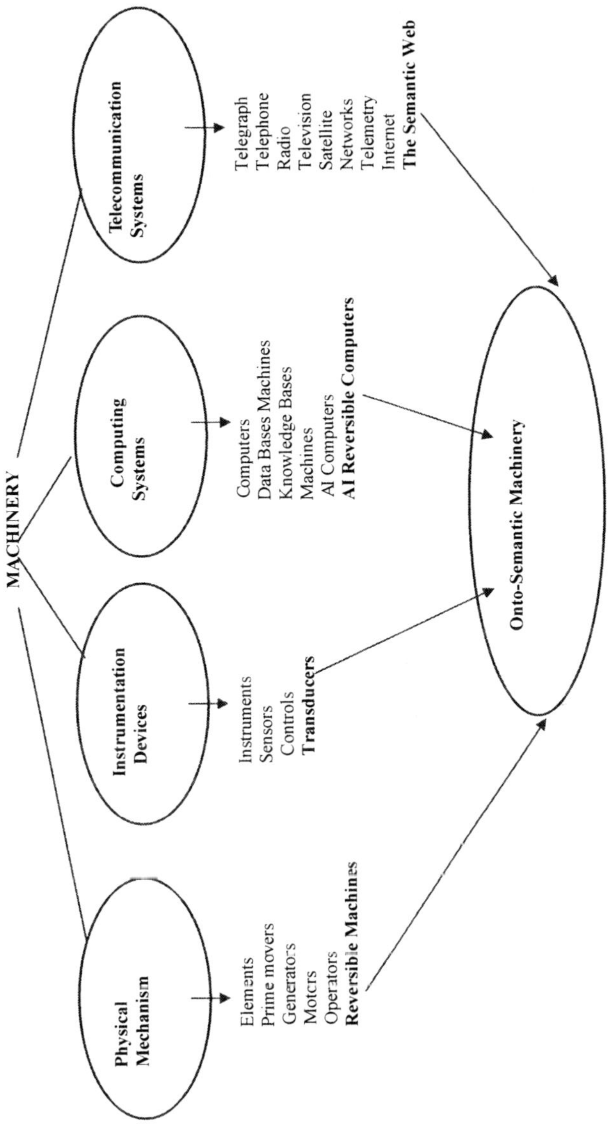

The telecommunication systems transmit information in the form of voice frequencies, telegraph messages, television programs, or digital data via wire, radio, space satellite, or computer networks, like the Internet.

But the pinnacle of the energy transforming devices looks to be a complex mechanical system integrating machines of all major categories as one functional unit. And it is technically feasible, because the direction of energy transformation in all the main energy conversion devices and systems comes to be reversible: any machine can have energy flow in either direction [Energy Conversion]. For instance, for thermoelectric devices, it is physically lawful to convert thermal power to electric power, and vice versa. Or, another near example, in electromechanical machines, mechanical energy may be converted into electrical energy realized in electrical generators as well as electrical energy into mechanical power implemented in electric motors.

The culmination of information systems, in turn, appears to be the reversible computing machinery endowed with the immense range and power of perception, intelligence, and locomotion (Abdoullaev, 1997). Since the underlying laws of physics are reversible, not only physical effects but also computation can go in either direction. Then, among the basic operating principles of such powerful intelligent information machines should be the capacity of digital electrical computing devices to reversible computing as well (Reversible Computing, 2001).

Of all sorts and types of machines, the most desired are teleological universal mechanisms capable of intelligent transfer, positioning and processing of materials, energy and information of any of various forms, as well as possessing the locomotion of any types: aquatic, fossorial, terrestrial, aerial, and space. For, it appears the final cause of machine technology is to create universal machines both guided by feedback control mechanisms and capable to intelligently behave in a changing world: to map (perceive) environments, to have plans (beliefs), to reason (think) thoughts, solve problems, experience emotions, and achieve goals of any complexity. The idea of a universal transformer has long been served for the engineers as the technological Holy Grail or the ideal model of machines.

The universality of the infinite ontological formalism can be verified by its capacity to give a mathematical account of such a sought-after universal machine. The design of intelligent agents is still basically determined by the automata theory based on the concept of the Turing machine (Turing, 1950). The general abstract automaton as a class of information processing electromechanical mechanisms usually is described in terms of its states, inputs, and outputs. The types of functional relationships between its elements along with the kind of the input information or stimulus determine the nature and the level of complexity of the transducer. Adding to the sets of machine inputs, states, and outputs also the set of the world states will allow to develop more sophisticated architecture of intelligent machines.

Using the universal formula for **the world state transition function** (8) from Part III, we can explicate the basic functions showing the pathways through which the machine makes processing of energy and information. Then the formalization of the major ways by which the machine and the world interact can be functionally modeled by the machine state transition (memory) function $\omega \subseteq \pi$ relating its current state with the future state with a time delay of a certain length:

$$\omega: M \times W \rightarrow M \quad (8)$$

For a successive values of a time series of discrete points t=1, t=2, t=3,..., the state function will be rewritten as $M_{t+1} = \omega(W, M_t)$, in which $W=W_I$ is a discrete set of all world states, and $M=M_1, M_2, M_3,...,$ a discrete set of all machine states. Here the world $W = W_C \cup W_N \cup W_M$ embraces all the possible states of the physical world, nonempty sets of cultural influences $C \subset W_C$ and environmental stimuli $N \subset W_N$, all the states of a machine $M \subset W_M$, and all the possible responses, reactions, or forms of behavior B, contributing to the world of artifacts $W_A : B \subset W_A \subseteq W_C$.

As a result, then, an intelligent machine can be formally represented as a general automaton, a universal Turing machine, or a reversible universal operator (transducer, transformer) (cf. Automata Theory), as in:

$$T = \langle W, M, B, \omega, \iota, o \rangle \quad (9)$$

Here, W, M, and B are respectively the sets of inputs, machine states, and outputs (responses, actions, or behaviors) related by functions (rules of operation), ω, ι, o, of which the symbol ι is an (action) function mapping the product M x W into a set of behaviors (actions):

$$\iota: M \times W \rightarrow B \quad (10)$$

The inverse function o complemented with the state transition and the action functions is accountable for describing the feedback loop of the universal machine:

$$o = B \rightarrow M \times W \quad (11)$$

The meaning of the reversible intelligent machine is becoming much clearer, as a highly integrated complex of sophisticated physical mechanisms converting energy and information operated by the world knowledge-based intelligent programs. For example, a mechanism performing force and motion-modifying functions is one module unit of the whole machine responsible for terrestrial locomotion. The energy supplied to and converted in, and delivered by the automata can be of mechanical, electrical, thermal, magnetic, radiant, nuclear, or chemical origin. In turn, the information supplied, transformed, and delivered by the automata [Automation] can be encoded as:

- A string of discrete symbols, like binary terms of 0s and 1s or a dash-dot Morse code
- Sense bearing signs, sounds, sensations, perceptions, images, odors, carried by energy phenomena, light waves, audio waves, mechanical, thermal, chemical, and electrochemical stimuli
- Changing physical quantities: mechanical forces, temperatures, electrical voltages, positions of the rotating shafts of any prime movers, and so forth

But the whole being of the universal transducer revolves around the reverse action function; for a basic aspect of reversible automata is the rule that *the energy conversion direction runs backwards, turns the other way round*, transforming mechanical energy into electrical energy, as well as electrical power into mechanical motion. Through the reverse input function system, the universal automaton becomes a self-governing and self-operating energy/information converting machine, regulating its behavior in different real circumstances by controlling its outputs in conformity to a set of standard programmed values stored in it, such as AI computing programs. Generally, the universal operator will involve all four categories of machines integrating them as one unit:

- Reversible machines as power and motion source
- Transducers as the sensing instruments, controlling mechanisms, and actuating devices
- The AI-based reversible computers as the decision element storing all the basic world knowledge and activating the control system
- The Onto-Semantic Web as a worldwide distributed intelligent resource

It appears the general algebraic formula for the universal automaton may well serve as an underlying logical mechanism for designing encyclopedic intelligent machines. Since stripping this structure of the inverted input function, it can be reduced to a one-way finite transducer, or sequential Turing machine T=<S, Q, O, M, N>, where S, Q, and O are finite sets of inputs, states, outputs respectively. The remained symbols M and N respectively represent a state-transition function and an output function: M: $Q \times S \rightarrow S$, $Q \times S \rightarrow O$. It should be noted that all the computational theories of AI based on the concept of the sequential Turing machine are heavily restricted models. First of all, because of neglecting the set of world states in their formalisms, like a perception-based AI system representing inputs, state-transition, and output functions as a set of linguistic *if-then* rules (Zadeh, 2001).

Fundamentally, the educed model of universal automaton prescribes a set of principles for any kind of mind: the mind of an intelligent being, an artificial intelligence, an abstract logical mechanism, or artificial neural nets in terms of perception and mentation or intellection, decision-making and computation, controlling and actuating, or neural functions.

So, the universal transducer can be exemplified as a biological operator describing animal or intelligent life consisting of highly organized systems functioning on natural cycles in order to increase its number and order. A biological transducer suggests the stimulus-control-response model converting stimulation energies into electrical impulses ending in mechanical movement. Sensory reception is provided by a network of receptors, including exteroceptors, photosensors, mechanosensors, thermosensors, chemosensors, and nocisensors answerable for recording external energy changes; interoceptors and propriceptors recording and reporting the internal state of the body. The brain as the mechanism of coordination of perceptions reads the meaning of the electrical signals, and provides through effectors an effective response, behavior, or action. Thus, an animal being is a kind of machine that translates changes in the environment (external or internal), such as light, temperature, sound, motion, odor, into nerve impulses, vital processes, interpreted by the central nervous system to effectively interact with the world. The nearest physical disadvantages of animal sensory reception consist in the limited range of energy perception and locomotion, while the range and power of a machine's sensors, as well as the methods of transportation, may be practically limitless.

Another example, if to specify the state-transition functions as the mental functions of seeing, memory, cognition, emotion, willing, and action, the mathematical structure of the universal operator can be properly rewritten so as to account for the AI agents interacting with the world. If one limits only by the functions of seeing: $W \rightarrow M$, the function of memory: $I \times M \rightarrow M$, the function of action of the machine on the world: $I \times M \rightarrow A$, and the effect function: $A \times W \rightarrow W$, in which, the sets of inputs I and outputs A are specialized stimuli $(I, A) \subset W$, then a special model of AI can be deduced. In such a logical model of AI, the machine state is identified with a set of sentences in the predicate calculus, and the world is accounted for in mathematical terms of objects, functions, and relations (Nilsson, 1991).

In fact, since the mind not only senses, perceives, memorizes, and wills, but also imagines images, reasons thoughts, and experiences emotions, the full set of the most vital functions of an intelligent machine should include the higher mental functions of mentation (computation or reasoning) together with perception and action. That is, the essence of an intelligent entity consists in its potency of semantic knowledge processing, its ability to adequately represent the world and power to effectively reason upon all kinds of realities, that is, the power of ontological reasoning comprising formal logical reasoning. Assigning the named attributes to the VA, in conclusion, we consider the content of its knowledge base, or what knowledge domains, regions, and realms it covers, or which subject areas, learning fields, branches of knowledge, and disciplines it will operate.

The Encyclopedic Knowledge Base of the Virtual Aristotle

As far as the EIS SUO seeks a comprehensive taxonomy and description of things together with their explanation and prediction, the Virtual Aristotle as its computing application comes out as a programmed machine capable to represent, explain, and predict all the world's phenomena. As a result of the definition, its knowledge unit must be designed as an ency clopedic knowledge base carrying an exhaustive inventory of the possible types of entities and relationships in reality. In other words, the encyclopedic knowledge base of the VA is targeted to represent the total world of knowledge as a one universe of discourse within a standard system of entity classes. For many years, computing scientists and engineers have being doing their best to create intelligent artifacts but with little success, mainly because of absence of the standard knowledge structure codifying the most essential information about the world. And the conceptual core of such standard is nothing else but the oldest science developed by the brightest minds as the science of Thing, Being, or Entity, its basic classes and relationships. If scientific knowledge domains are the sciences of special realms of the world, nature, mind, society, technology, is a universal science deals of ontology with the topmost classes of things, their categories, kinds, relations, axioms, and definitions, applicable to in any areas of knowledge and domains of interest. Whence, there comes the definition of universal formal ontology as the infinite theory of reality, total and all-embracing, containing all its kinds and levels: physical, biological, mental, cultural, or computational (or

Figure 15. The knowledge level of the Virtual Aristotle

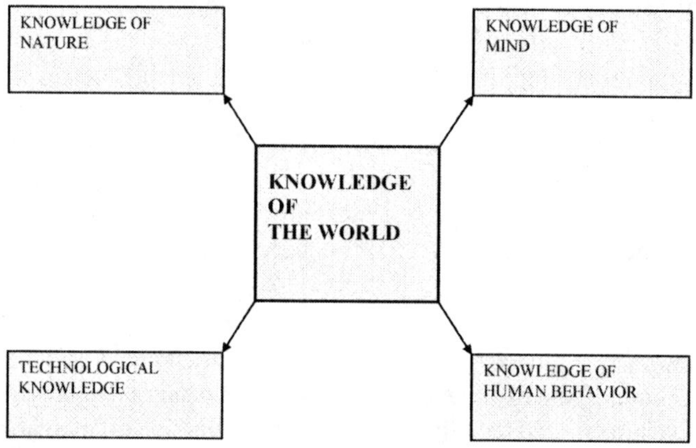

Figure 16. The relationships of knowledge domains

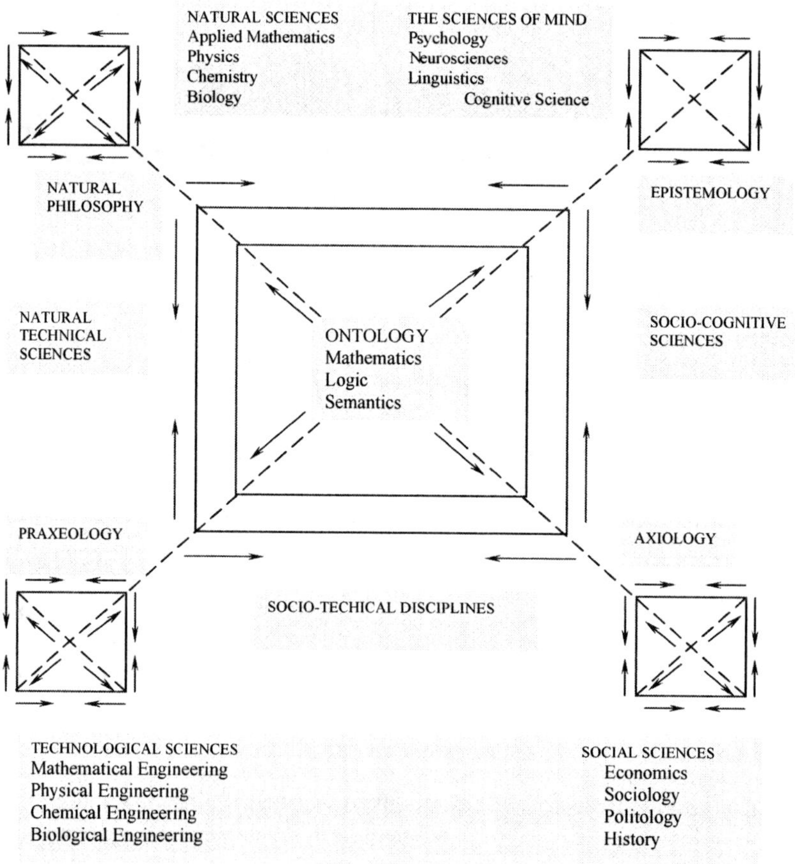

virtual) realities. So the role of the standard ontology is no more or no less than to organize the whole world of knowledge as a single universe of discourse integrating the knowledge of nature, minds, society, and technology within the general ontological learning about the world (see Figure 15).

In terms of the subject fields, the organization of world knowledge reflecting the architecture of reality comes up as the networks of sciences of natural processes, cognition, human actions and human practice, as illustrated in Figure 16. Altogether, the content of knowledge domains is distributed across interconnected disciplines as follows:

- *The knowledge of the world of nature* supplied by nature-committed sciences: applied mathematics, physics, chemistry, and biology, added by their combinations—mathematical physics, physical chemistry, mathematical biology, physical biology, chemical physics, and so forth.
- *The knowledge of the world of mind* given by psychology, neuroscience, linguistics, and cognitive science and their intersections, cognitive psychology, cognitive neuroscience, cognitive linguistics, and so forth.
- *The knowledge of world of human behavior* acquired by the network of social sciences— economics, sociology, government and politics, history, political economics, social history, and so forth.
- *The knowledge of the world of technology* including all possible types of material embodiments of scientific principles for constructing the means—machines, tools, techniques, and processes—to change or control the natural world. The art of doing technological artifacts involves both mathematical engineering, physical (nuclear, electrical, electronics, optical, mechanical) engineering, chemical engineering, biological engineering, and also all possible cross-disciplinary technological sciences, as cybernetics, control theory, computer science, mechanical bioengineering, information science, bionics, systems engineering, and so forth.

Presently, the world's collective knowledge can be found in the largest and most comprehensive learning repository, the *New Encyclopaedia Britannica*, collected according to novel principles of knowledge organization. Namely, instead of a hierarchical division, *the circle of learning* is proposed such that none of the network is an absolute beginning or ending, for "'no point is either a beginning or an end, and lines can be drawn from any branches of knowledge to any other" (Adler, 1994, Circle of Learning). The second principle is that "the whole world of knowledge is a single universe of discourse" [Adler, 1994, Knowledge…]. Or, there are no any principal differences between the humanities (philosophy, history, language, literature, and the fine arts) and the sciences regardless of their dissimilarities in methodology and criteria of validity. Altogether, the Encyclopedia covers both all of what is known about the universe ("*what we know* about the knowable universe") and the means and tools by which we get the world knowledge ("*whereby we know* which we know"). The latter comprises the branches of knowledge: logic; mathematics; the sciences (physics, biology, the social sciences with psychology, linguistics, technological sciences, Medicine with its associate disciplines); history and the humanities; philosophy; knowledge processing subjects, or information sciences. The former includes all what we know about matter and energy, Earth, life, human life, human society, art, technology, religion, and the history of mankind, which specific contents being arranged as follows (Adler, 1994, Outline of Knowledge):

- **Matter and Energy:** Atoms (atomic nuclei and elementary particles, structure and properties, nucleus and elementary particles); energy, radiation, and matter (chemical elements, compounds, and reactions; heat, thermodynamics, and the nonsolid states of matter; the solid state of matter; mechanics of particles, rigid bodies and deformable bodies—elasticity, vibrations, and flow; electricity and magnetism; waves and wave motion—water, sound, electromagnetic waves); the universe (the cosmos, galaxies, stars, and the solar system).
- **Earth:** Properties, structure, and composition (the planet Earth, physical properties, structure and composition, minerals and rocks); envelope (the atmosphere, the hydrosphere, weather and climate); surface feature (physical features, geomorphic features); history (origin and development, the geological record, the geological time).
- **Life:** Living things, nature and diversity (properties, origin and evolution, division); vital processes, molecular basis (chemicals, photosynthesis, metabolism, the nitrogen cycle, molecular mechanisms); organisms, structures and functions (cells, form and function, regulation and integration, support and covering, nutrition, exchange, transport and elimination, reproduction and sex, development and heredity); organisms, behavioral responses (nature and regularities, development and capacities); the biosphere (features, populations and communities, disease and death, ecosystems).
- **Human Life:** Stages (evolution, heredity, races); the human organism (the body, health, diseases, medicine and care); human behavior and experience (human nature and experience, attention, sensation, and perception, internal states, capacities and dispositions, learning and thinking, the self and personality).
- **Human Society:** Social groups (peoples and cultures of the World, cultural development, cultural components and institutions, language and communication); social organization and social change (structure, change, status, populations); the production, distribution, and utilization of wealth (economic concepts and systems, the consumer, the market, and pricing, organization, income, macroeconomics, growth and planning); politics and governments (theory, institutions, processes, international relations); law (philosophies, systems, and practices, public law, private law); education (aims, organization).
- **Art:** Art in general (theories and classification, experience and criticism of works of art, characteristics of the arts in particular cultures); the particular arts (literature, theatre, motion pictures, music, dance, architecture, garden and landscape design, and urban design, sculpture, drawing, painting, printmaking and photography, arts of decoration and functional design).
- **Technology:** Nature and development, elements, major fields: agriculture and food production, construction, transportation, information processing and communication systems, the major industries technology, the urban community technology, military technology, earth and space exploration technology).

- **Religion:** Religion in general (knowledge, understanding, institutions, and practices), the particular religions.
- **The History of Mankind:** Peoples and civilizations of ancient Southwest Asia, North Africa, and Europe; of medieval Europe, North Africa, and Southwest Asia; of East, Central, South, and Southeast Asia; of Sub-Saharan Africa to 1885; of Pre-Columbian America; the modern world to 1920; the world since 1920.

To represent the whole human learning with scientific knowledge as a total structure, we need to arrange this according to a unifying classification of entity kinds and relationships. For, admitting *the circle of learning* at the concrete levels of things, human knowledge as a whole, totality, or one microcosm, is rather arranged in a hierarchical order. Namely, in a descending scale from higher, fundamental ontological classes to lower, less fundamental levels of things, physical, biological, sociological, or technological worlds, with their specific domains, realms, and regions. As it is shown in Figure 16, universal ontology, enhanced with its conceptual means and tools, pure mathematics (set theory, algebra, calculus, and topology), logic of things, and general semantics, constitutes the vital part of the whole world of knowledge, thus giving it one structure, fundamental framework, and organizational scheme. As a consequence, the EIS SUO by furnishing the rudimentary system of knowledge standards as a general template and reference ontology embodies itself at the knowledge level as the heart and soul of the encyclopedic knowledge base of the Virtual Aristotle Machine (USECS, 2005; see also Supplement). In other words, having the standard universal ontology gives us the unique possibility to create a Virtual Reasoning Encyclopedia dealing with the world's collective knowledge.

Conclusion

We formulated the principles of designing the NL driven knowledge and reasoning applications, like as NL question/answering systems, domain reasoning engines, representation tools, web search mechanisms, and many other knowledge artifacts, to be empowered with human language understanding. But most important practical consequence of the standard formal ontology of major semantic classes comes to be the natural language encyclopedic intelligence, titled the Virtual Aristotle, whose progenitor had envisioned an "instrument" that "could accomplish its own work, obeying or anticipating the will of other" more than 2200 years ago. Due to the algorithmic (formal) description of natural language within the general ontological meaning framework (underlying semantic context), we could establish the principles for designing encyclopedic knowledge and reasoning machines as a new class of intelligent information (computer) systems understanding human natural language. Empowered to learn, acquire, process, accumulate, and communicate the world knowledge in NL forms, the Virtual Aristotle Machine is crucial to contribute to the advent of Global Intelligent Cyberspace, or World Wide Intelligent Web, the engine of intellectual knowledge societies.

References

Abdoullaev, A. (1997). Creating machines with universal intelligence. In *Proceedings of the IASTED International Conference on Intelligent Information Systems,* (pp. 115-122). CA, Los Alamitos: the IEEE Computer Society Press.

Adler, M. (1994). The circle of learning, outline of knowledge and guide to the Britannica, Propaedia. *The new encyclopaedia Britannica.* Chicago: Encyclopaedia Britannica, Inc.

Adler, M. (1994). Knowledge become self-conscious, outline of knowledge and guide to the Britannica, Propaedia. *The new encyclopaedia Britannica.* Chicago: Encyclopaedia Britannica, Inc.

Chen, P. (1976). The entity-relationship model – Toward a unified view of data, *ACM Transactions on Database Systems*, March *1*(1) 9-36.

Davis R., Shrobe, H., & Szolovits P. (1993). What is a knowledge representation? *AI Magazine, 14*(1), 17-33.

Friedland, N., & Allen, P. (2004). *The Halo pilot: Towards a digital Aristotle.* Retrieved April 25, 2005, from http://www.projecthalo.com

Lenat, D. (1998). *Leveraging CYC for HPKB intermediate-level knowledge and efficient reasoning.* Retrieved April 25, 2006, from http://www.cyc.com/hpkb/proposal-summary-hpkb.html

Mizoguchi, R. (1998). A step towards ontological engineering. In *Proceedings of the 12th National Conference on AI of JSAI*, pp.24-31. Retrieved April 25, 2006, from www.ei.sanken.osaka-u.ac.jp/english/step-onteng.html

Newell, A. (1982). The knowledge level. *AI Magazine, I*(1), 87-127.

Newell, A. (1990). *Unified theories of cognition.* Cambridge, MA: Harvard University Press.

Newell, A. (1993). Reflections on the knowledge level. *Artificial Intelligence* 59, 31-38.

Nilsson, N. (1991). Logic and artificial intelligence. *Artificial Intelligence*, 47, 31-56.

Ontology and taxonomies; core technologies. (2001). *Knowledge Technologies Conference* 2001, Austin USA, Retrieved April 22, 2006, from http://www.gca.org/knowledge technologies/2001/proceedings/technical.asp

Open Cyc. (2003). Retrieved April 22, 2006, from http://www.opencyc.org

Outline of knowledge and guide to the Britannica, Propaedia. (1994). *The new encyclopaedia Britannica.* Chicago: Encyclopaedia Britannica, Inc.

Reversible computing for energy and trustable computation. (2001). *The MIT AI Lab Reversible Computing home page.* Retrieved April 22, 2006, from http://www.ai.mit.edu/~cvieri/reversible.html

Roget's thesaurus. (2002). Retrieved April 22, 2006, from http://poets.notredame.ac.jp/Roget/contents.html

Simon, H. (1995) Artificial intelligence: An empirical science. *Artificial Intelligence* 77, 95-127.

Turing, A. (1950). Computing machinery and intelligence. *Mind*, Vol. LIX, No.236, 433-460.

USECS (Universal standard entity classification system). (2005). Retrieved April 22, 2006, from http://www.encyclopedic-intelligence.com

WordNet, WordNet 2.0. (2004). Retrieved April 22, 2006, from http://www.cog.sci.princeton.edu/wn/

Zadeh, L. (2001). A new direction in AI: Toward a computational theory of perceptions, *AI Magazine*, *22*(1), 73-84.

Chapter XIII

Reality Classification System: A Product Line of the EIS UFO

Introduction

The urgent necessity of a true knowledge society is mainly caused by the fact that the natural unity of world knowledge is badly destroyed by the modern system of scientific, formal research and education. For example, the harmful consequences of the greenhouse effect were predicted and explained by S. Arrenhuis. Only because the fragmented learning could not be available to the right people in the right place at the right time, the phenomenon should be reopened again after a 100 years bringing almost irreparable consequences of worldwide reach.

Division of labor and increasingly narrowing specialization are making knowledge irreparably fragmented and scattered, and thus producing a confusing multitude of disunited scientific and technological communities and associations, often duplicating the same R&D programs even within the same community, like the Semantic Web projects in the EU Framework Programs.

Due to modern information technologies and particularly the Net, there is a sea wealth of information recorded in gigantic databases. This may be mastered only by more advanced information technology, knowledge machines with a built-in common ontology framework, a single code of fundamental standards, principles, rules, and laws, suggesting a broad, integrated model of all the world things. This may alter the whole knowledge production system in an essential way, ending up in a radically new type of human society, the intellectual knowledge society, driven by the (reality)-centered semantic technology.

The major objective of standard ontology is to represent the world within a single framework of things, meeting the deep intellectual desire of the human mind for a single organic body of human knowledge and learning. Two big natural desires seem to contend for the soul of human beings: the never satiable greed for material wealth and pleasure, and the never satisfied thirst for knowledge. Which desire will dominate may determine not only human personality but the whole future of human race. The current social structure is established on the never satisfied human wishing to acquire, possess and dominate: avarice, acquisitiveness, and possessiveness. Feeling unfulfilled and dissatisfied with where humanity is heading and what it is, people increasingly recognize the need for a radically new social organization, the society of knowledge and intelligence-based economy, politics, and culture. The sophisticated social structure is intended to be the sort of advanced social system mostly aiming to provoke the alertness of mind, to stimulate an unrestricted intellectual curiosity, to enkindle the spirit of inquiry, to arouse the desire to know about everything, and thus motivating a mass participation of people in creative intellectual activities, engineering, or artistic or literary, or scientific.

By using most advanced information technologies, the knowledge society is expected to advance and boost the power of intelligent thought and perception of reality, where the economic growth and social progress and cultural dynamics are caused by the world-aware knowledge artifacts, systems, technologies, processes, services, and products, the substantive base of an intellectual civilization, or intelligent world. And encoding human knowledge in such real life intelligent applications, using unifying world models and formal ontology languages, might be a crucial contribution to creating advanced knowledge society technologies and real worldsemantic machines.

USECS®, Unified Standard Entity Classification System

To pursue the great intellectual cause of knowledge systems, we proposed a unique ontological product, the core knowledge base of any large scale effective semantic machine, the World Categorization System as the universal standard entity classification system, or USECS®, produced as a downloadable Adobe Acrobat 7.0 document. To flesh out its ontological skeleton, the rich content of the largest online lexical reference, WordNet 2.1, was readjusted according to the UFO's global schema of all things. Besides, the number of upper taxonomies and Web ontologies are included, which also can be encompassed provided that the categorical schemes used are properly sorted out.

Besides its designated trade name, the USECS® product line also might be distributed under the exchangeable labels:

- World Encyclopedia Vocabulary
- World Stock List
- World Knowledge Directory (Tree)

- Universal Scheme of Things
- Universal Directory of Meanings
- Meaning Searcher, or Sense Finder
- Human Knowledge Digest
- Global Catalog of Things
- Universal Directory of Entities
- Global Nomenclature

As a single global classification system, the USECS® is targeted to typecast and define all the key things in the world, where all the meanings of words/terms (as the real content of conceptions or the essence of things) are integrated into one semantic body. Thus this ontology ware is virtually the semantic reference system serving as an all-entity classification, the world directory of entities and relationships, or a standard code of basic meanings and definitions, supported by the UFO foundation. Such a global catalogue or directory or reference of things is targeted to contribute to the grand cause of knowledge world by expanding the comprehension of human language and verbal intelligence and so increasing the breadth, largeness, and scope of human cognition and intellection. All higher cognitive processes, such as search, knowing, thinking, decision making, problem solving, and language involve the profound understanding of verbal meanings; for to see the deep meaning of words and their associations is to have the substantial knowledge of things and their interconnections.

By its design, the whole content of the Global Catalog of Things is organized into two interrelated parts: Encyclopedic Knowledge Reference, or World Encyclopedia Digest, and the Catalogue of the World, or the Universal Directory of Entities. The Knowledge Reference is constructed as encyclopedic knowledge base dealing with the whole range of human knowledge as one totality and whole. Like general encyclopedias, it outlines various departments, fields, and domains of the universe of knowledge and learning by giving their main meanings and summary formulations open to further extension by relevant details and specific facts.

Guided by the general standard model, the World Catalogue (or Directory) splits the universe into few fundamental classes of things (substances, states, changes, and relationships) making up the skeletal frame of multilevel reality studied in diverse areas of knowledge, fields of learning, or universes of discourse. Following this guidelines, the Universal Directory of Entities comes in four complementary parts:

- The substantive catalog of objects
- The state inventory of conditions, qualities and quantities
- The full listing of changes, actions, activities, processes, and events
- The comprehensive catalogue of relationships

The USECS is tended to be designed as a lattice-ordered structure of entities limited by the universal upper bound (Entity, Thing, or Being) and the universal lower bound (Nonentity,

Nothing, or Nonbeing). All the world entities are arranged into two distinguishing levels: generic classes of things (substance, state, change, and relation taken in general, as mass nouns) and specific classes of things (objects, properties, events, and specific connections, as count nouns). The user can navigate it as a distributive lattice of things of a finite length by ascending from the bottom element of Nothing (or some specific term) to the top generalization in several steps; or vice versa, by descending from general things to special domain meanings.

The USECS also suggests the meaningful Web search by explaining the senses of search words or topics, namely, what sort of entity they represent and how the terms are featured in the common classification of semantic classes. Representing the comprehensive hierarchy of entities and relationships, it identifies the basic words in natural language with their major meanings, semantic lines of descent, and offspring. Equipped with the meaning of the search word or the structure of the search subject or topic, the user can perform intelligent online quest for data, facts, or knowledge. Unlike some comprehensive dictionaries and encyclopedias, all the meanings of terms in the USECS follow the standard formula of defining having a rigid pattern of definition structure.

As the universal schema and single hierarchy of world entities, the USECS underpins the knowledge references, classifications, and light ontologies such as Britannica's subjects, *Webster's Dictionary*, the Great Ideas, Web Directories, the UN classifications of industry and products, Wikipedia's categories as well as the International Press Telecommunications Council Subject Reference System. Thus the USECS, as the base ontology, suggests the guidelines for constructing specific ontologies, models, schemas, vocabularies, terminologies, and catalogs in any domain of knowledge or practice or business.

A Single Hierarchy of Things of Every Kind and Sort

Implicitly or explicitly, the assumption of a single hierarchy underlies all profound reference works: encyclopedias (as *Encyclopedia Britannica*, the Outline of Knowledge), thesauri (*as Roget's thesaurus*), large lexical databases (as the Princeton WordNet 2.1), and general taxonomies (from Yahoo hierarchy to upper ontologies and Web ontology languages).

For example, the largest human knowledge reference, *Encyclopedia Britannica*, tends to organize the world's collective knowledge as a single universe of discourse following the natural order of particular things: **substance** (matter and energy, from atoms to the universe), **earth** (globe), **life**, **human life**, **human society**, and **culture** (art, technology, religion, the history of mankind, and the branches of knowledge).

The WWW classification schemas involve the subject categories such as Arts & Humanities & Literature, Business & Economy, Computers & Internet, Games, Health, Home, Kids & Teens, News & Media, Recreation & Sports, Reference, Regional, Science, Shopping, Society, and World.

Roget's thesaurus splits up all English lexicon into several classes of words as expressing the hierarchy of basic categories: abstract relations (existence or being or entity, relation, quantity, order, number, time, change, and causation); space; matter; the intellectual faculties; the voluntary powers; the sentiment and moral powers.

Discarding the principle of multiple hierarchies, the online lexical database WordNet 2.1 also puts entity first in the hierarchy of words (nouns) having as its heterogeneous subordinates:

- The realms of physical entity (thing, object, causal agent, substance or matter, and physical process)
- The realms of abstract entity or abstraction (psychological feature, attribute, group, relation, communication, quantity, otherworld, and set)

Still the above classifications reflect a habitual prejudice to divide everything into concrete entities (physical objects, agents, or processes, completed sometimes with groups, collections of concrete things) and abstractions (classes, states, qualities, quantities, and relations) or simply organizing information by pragmatic topics. Such widely practiced supposals have been the principal obstacles in describing the universe of learning as a whole totality.

The Rationale for the All-Thing Catalog

In the great human enterprise of knowing and learning, one of the hardest tasks has been to determine the body of words in terms of which the language can effectively represent reality, express the mental experiences, and communicate information. Having such an all-embracing system of words to which one can refer for standard definitions of significant terms has always been considered as the most sought-after resource. But the main barrier to achieve the encyclopedic inventory of senses consisted in the wanting of the assumption to view the entire macrocosm of knowledge as a single universe of discourse. That is, the whole undertaking demanded the universal schema capable to represent everything in a single hierarchical order of things, from Everything to fundamental, domain-independent classes to domain-specific, particular individuals, to Nothing, as in:

Thing, Entity, or Being (Everything, the World, or the Universe) > Anything >

> Something > Nothing (Nonentity, Nonexistence, Nonbeing)

So to benefit the uniform catalog of things is possible only if our accounting of reality affords an exhaustive categorization of things applicable in every area of the universe, in all spheres of existence, at every grade and level of being. As a great reward, within the unifying scheme of world entities, the major knowledge and language resources (encyclopedias, lexical taxonomies, semantic thesauri, Web directories, search engine subject indices, upper ontologies, domain terminologies, languages, and nomenclatures) will become integrated in a single consistent system of terms. In order to approach the target, the encyclopedic source of meanings should be constructed as addressing all significant information about natural, mental, social, cultural, and virtual domains. To meet this tall task, the universal directory of all the basic names of world things is created in the form of the world knowledge wordbook. In its substance, such a general nomenclature, terminology, or language of things is nothing but the digest (epitome, compendium, outline, abstract, or synopsis) of human learning covering all key domains of reality, subject areas, and fields of study.

Giving the systematized description of general entities, with kinds and instances, by defining their language, terms, names, and expressions, the USECS appears as an essential tool in transforming the World Wide Web into the Meaningful Web of words and constructions, a digital embodiment of human knowledge. Besides, providing the common language of things, it also can be viewed as a complement to the ISO international standards supplying the common reference frame and technological language for specific things: materials, products, technologies, processes, and services (http://www.iso.org/iso/).

Being drawn on the unified model of reality and aimed toward unity of standards, the entire content is divided into two parts:

1. The world catalogue of general classes of things such as **substance** (substances and objects), **state** (states, conditions, qualities, quantities), **change** (changes, events, and processes), and **relationship** (relationships, connections, associations) with the standard formulas of defining their child classes, kinds, varieties, and instances.
2. The encyclopedic knowledge digest designed as a source of essential knowledge covering the main points of any significant subject area, domain of interest, or field of study.

USECS and the World Wide Intelligent Web

There is now an urgent necessity to rationalize both the web keyword search and the Web directories browsing. The World Wide Web or the Web is an information space of software agents (servers, search engines, spiders, browsers, proxies, multimedia players) and resources (documents, graphics, sounds or audios, and videos or animation) identified by URI's (Uniform Resource Identifiers) and interconnected by hypertext links via the Internet protocols (as the HTTP - hypertext transfer protocol). There are two ways to navigate the information space of Internet sites providing textual documents, images, animation and sound: either by browsing the Web Directories—organizing the Web content as a tree-like hierarchy of subjects or topics and subtopics, or by means of the Search Engines (software programs)—entering keywords or key phrases or a piece of text. As a result, the keyword searches, basic or advanced, retrieve all the occurrences of the keyword in the Web pages that are listed in the search engine index.

Despite such sophisticated tricks of trade as relevancy rankings, using statistics and heuristics, refined search, Boolean operators, and indexing metatags, both the subject-based queries and the keyword search engine enquiries are often ended up with a multitude of irrelevant information, promptly bringing a jumble of useless URLs.

Considering this, the World Directory of Things as the Meaning Finder suggests the meaningful Web search by explaining the senses of search words or topics; namely, what sort of entity they represent and how the terms are featured in the common classification of semantic classes. Representing the comprehensive hierarchy of entities and relationships, the Universal Directory of Meanings identifies all the basic words in natural language with their major meanings, semantic lines of descent, and offspring. Equipped with the meaning of the search word or the structure of the search subject or topic, the user will be capable to perform more relevant and coherent online quest for data, facts, or knowledge. Unlike most basic dictionaries and encyclopedias, using different ways and techniques of definitions for various sorts of words, the Universal Directory follows a single standard uniform formula for defining key words in a language, so that a primary meaning is enlarged and extended by such a variety of special meanings where the basic sense stays as the determining content and underlying element.

What the User Can Get from USECS

Nowadays there are many special services and applications offered at one's fingertips, from electronic banking services to funny AI agents. But someone hardly ever has been proposed to have the solid and integral content that affords:

- What classes of things can be in the world and what kinds of relationships the things may have to each other
- All-embracing hierarchical framework of terms and meanings
- Discovering what is behind and beyond any significant words or notions
- Deepening the consciousness (awareness, knowingness, cognizance, understanding) of realities with their connections
- The outlines of human knowledge about the world, covering virtually all subject areas and domains
- One of the richest vocabularies and lexicons of real world things
- The sense finder and meaning searcher complement for meaningful Web search

How to Use the World Entities Reference

To browse the World Directory content of entities or relationships covering the references and meanings of key words in a language and navigate through the search results, the user can apply either the Find Dialog Box or the Search PDF Window. Pressing Ctrl + F (Windows) or Command + F (Mac OS) opens the Find toolbar. Pressing Shift + Ctrl + F or Shift + Command + F (Mac OS) opens the Search PDF Window (for more details, see Adobe Essentials, search for words in a PDF document). The whole procedure asks for the following simple steps. Click the button Search; set a search preference, basic search options or advanced search options while looking for central senses and primary meanings employ a case-sensitive and whole words only options. CAPITALIZE your search, write in capital letters, in case you are looking for general things or classes such as ENTITY, PROCESS, SUBSTANCE, OBJECT, CHANGE, ACTION, RELATION, MATTER, MATERIAL, CAUSE, BEING, EVENT, LIFE, ANIMAL, HUMAN, PERSON; or knowledge domains and subject areas, such as Science, Mathematics, Language, and so forth.

While searching on a topic or subject, the user gets as the results the outlines or summary of the subject: its central meaning, basic concepts and principles, classification, content, and relationships with other subjects.

When looking for key words, the user will get their basic senses, semantic relatives, and family lines traced as deep as from the universal term of entity, as in:

DATABASE < INFORMATION < MESSAGE < COMMUNICATION < SOCIAL RELATION < RELATION < ENTITY

or

SEARCH ENGINE < COMPUTER PROGRAMM < SOFTWARE PROGRAM < COMPUTER CODE < CODING SYSTEM < WRITING < WRITTEN COMMUNICATION < COMMUNICATION < RELATION < ENTITY.

Under the headword DATABASE, its offspring or special types are enumerated in boldface or italic (list, listing; electronic, online, computer database; subdata base), all with their issues as well. Having constructed the most general meanings and senses, the user can proceed with searching the Internet putting into service the search engines such as Ask Jeeves, Yahoo, and Google, or the web directories as Open Directory Project, Yahoo Directory, or Google Directory, which are inbuilt features of the World Directory. Additionally, for *Encyclopedia Britannica*'s online subscribers, there is the opportunity to browse the rich electronic content by Britannica's categories and subjects.

A close utility of having the USECS is that the user can create a domain classification in any field of knowledge, kind of activity, or sort of practice. To show how to create and develop domain ontologies, there are use cases in the fields of:

- Medicine
- Business (Commerce)
- The World Countries

Audience

To meet the human thirst for encyclopedic knowledge and intelligence, the content is particularly committed to digesting, systematizing, and encapsulating the human knowledge into a naturally ordered scheme retaining the most essential things. As the world knowledge digest and all-meaning language resource, the USECS content is addressing all people desiring to know: students and scientists, programmers and engineers, intellectuals and laymen, politicians and artists, presidents and housewives, managers and workers, technologists and theoreticians. Its audience includes all who seek to understand the world and its worlds of words from a single outlook of reality, in one standard comprehensive way, within the unifying schema of things. While having at his fingertips the universal directory of meanings and definitions, the user can perform a meaningful searching for specific information on requested words, phrases, subjects or topics, occurring in the wider context of Web pages. The existing possibility of enlarging the basic content and meaning of entity terms with specific facts and information makes the USECS a valuable contribution in building the Intelligent World Wide Web, the information foundation of Global Knowledge Society.

Contents

Open Directory Project http://dmoz.org/

Yahoo! Directory http://dir.yahoo.com/

Google Directory http://directory.google.com/

Britannica Categories and Subjects http://www.britannica.com/eb/subject

International Standard Industrial Classification of All Economic Activities (ISIC) http://unstats.un.org/unsd/cr/family2.asp?Cl=17

Central Product Classification (CPC) http://unstats.un.org/unsd/cr/family2.asp?Cl=16

The United Nations Standard Products and Services Code® (UNSPSC®) http://www.unspsc.org/

UK Standard Industrial Classification of Economic Activities (UK SIC) http://www.statistics.gov.uk/methods_quality/sic/default.asp

Wikipedia http://wikipedia.org/

International Press Telecommunication Council Subject Reference System http://www.iptc.org/NewsCodes/0.0/ documentation/SRS-doc-Guidelines_3.pdf

UFO and Upper Ontologies

In accordance with the EIS UFO, the fundamental kinds of reality compose the principal ontological categories in the universe of discourse, which are substances and objects (spatial and non-spatial or abstract), states (attribute, property, quality, or quantity), changes (actions and processes), and relationships. While all sorts of specific objects, qualities, events, and associations make up their instances, the UFO's classes provide the knowledge standards for all kinds of knowledge domains and subject fields. Below it is recapitulated how the relevant classes of extant lexical taxonomies and upper level schemas can be unified and merged within the EIS SUO.

Entity, Thing, or Being: The Universal Class of all Kinds of Things

- Being, Existence, Entity, Thing (the universal class) and Nonbeing, Nonexistence, Nonentity, Nothing (the null class)
- The Basic Kinds of Entity: Substance or Object; State or Property and Quality and Quantity; Change, Action or Process; Relation, Relationship, Association, and Connection

Substance

- Substrate, stuff: The class of substances (objects, material or nonmaterial, spatial or non-spatial, physical or mental, concrete or abstract)
- **WordNet 2.0:** Entity {thing; causal agent or cause; substance or matter; object or physical object (whole or unit, natural object, living or animate thing (plant, flora; animal, fauna; person, human being)); artifact or artifact)}, group or grouping, psychological feature {cognition, knowledge (mind, head, brain, psyche or nous, content or mental object, lexis, vocabulary, information, place, and public knowledge, history)}, abstraction (set)
- **WordNet 2.1:** Entity {physical entity (thing; object, physical object; substance, matter; causal agent), abstract entity (group, grouping; set)}
- **Roget's Thesaurus:** Matter
- **EuroWordNet:** 1st order entity (origin, form, composition, function)
- **EDR:** Subject or human (person, animal, body or group, supernatural being), matter (thing or concrete object, objective subject)
- **SUMO:** Physical object (self-connected object, collection, agent, region), abstract object (set or class, graph)
- **CYC:** Individual (spatial thing), mathematical or computational thing (set or collection), groups, parts of objects, composition of substances, agents, organizations, actors, materials, devices, construction, food, clothing)

- **DOLCE:** Aggregate (amount of matter, arbitrary collection), object (physical object, mental object, social object), feature (relevant part, place)}
- **Sowa:** Objects
- **Knowledge Machine Component Library:** Entities (things, or objects)

WebKB-2: Entities (spatial objects and non-spatial objects)

OWL: Classes and individuals.

State

- The class of properties, qualities, and quantities
- **WordNet 2.0:** State, phenomenon (state of matter), abstraction (attribute (property, quality, shape or form), quantity and amount or measure or quantum), possession, psychological feature (cognition, knowledge (ability, inability, attitude)
- **WordNet 2.1:** Abstract entity {attribute (state; quality, property; shape, form; human nature, character, personality); measure, quantity, amount}
- **Roget's Thesaurus:** Quantity, number
- **EuroWordNet:** 2nd order entity (static situation and situation component, condition, existence, quantity)
- **EDR:** Event/occurrence (condition/state)
- **SUMO:** Abstract object (quantity, attribute, internal)
- **CYC:** Intangible individual (static situation, attribute value)
- **DOLCE:** Quality (color, size, shape, smell), quality region (qualia), occurrence (state)
- **KM Component Library:** Properties, states
- **OWL:** Properties (monadic)

Change

- The class of actions, activities, and processes
- **WordNet:** Event (happening, occurrence or natural event), phenomenon (natural phenomenon, consequence or effect, process), act, human action or human activity, abstraction (relation, change), psychological feature (cognition, knowledge, process; feeling; motive, motivation or need)
- **WordNet 2.1:** Entity {physical entity (cause; process, physical process), abstract entity (psychological feature (cognition, knowledge; motivation, motive, need; event)}
- **Roget's Thesaurus:** Change
- **EuroWordNet:** 2nd order entity (situation type, dynamic, situation component, cause, experience)

- **EDR:** Event or occurrence (phenomenon, action or deed, movement, change in state or change in condition)
- **Sowa:** Processes
- **SUMO:** Physical process (internal change, intentional change, dual object change, shape change, motion)
- **CYC:** Partially intangible (event), doing, transformations, changes of state, transfer of possession, movement, emotion, perception, waves
- **DOLCE:** Occurrence (dynamic events, processes and accomplishments)
- **KM Component Library:** Events (actions and processes)

Relation

- The class of relations
- **WordNet 2.0:** Abstraction (relation, time, space), state (relationship, relationship)
- **WordNet 2.1:** Entity {abstract entity (relation; communication; attribute (time, space, or infinite, temporal property)}
- **Roget's Thesaurus:** Relation, causation, order, time, space
- **EuroWordNet:** 2nd order entity (situation type, static, relation, situation component, communication)
- **EDR:** Location, locale, place, time
- **SUMO:** Abstract object (relation, proposition, attribute, relational)
- **CYC:** Mathematical object (relation, formal), temporal relations (times and dates), spatial relations
- **DOLCE:** Quality (position and place, time interval), quality region (space, time)
- **KM Component Library:** Entity (object)-to-entity (object) relations, event-to-event relations, event-to-object relations
- **WebKB-2:** Entity (object)-to-entity (object) relations
- **OWL:** Set-theoretical properties (relations)

The Encyclopedic Knowledge Base of Virtual Aristotle

Accounting
Agriculture
Algae
Amphibians
Angiosperms
Annelids
Arachnids
Arithmetic
Arthropods
Aschelminths
Atmosphere
Atoms
Automata theory
Bacteria
Beauty
Behavior
Biblical Literature
Biology
Birds
Blood
Broadcasting
Buddhism
Business Law
Capitalism
Cat
Censorship
Childhood
Christianity
City
Civilizations
Cnidarians
Combinatorics
Communication, a practice
Computer Science

Aesthetics
Alcohol
Algebra
Analysis (mathematical)
Animal
Anthropology
Architecture
Artefact, artifact
Art
Astronomy
Atomic Nuclei
Attention
Automation
Banks
Behavior
Beverage
Biochemistry
Biosphere
Birth
Brain
Bryophytes
Building Construction
Business organization
Cancer
Cell
Chemistry
Chordates
Circulation and Circulatory Systems
Citizen
Climate
Color
Combinatorial Geometry
Communism
Computers

Constitution

Continental Landforms

Courage

Cryptology

Custom and Convention

Dance

Decorative Arts and Furnishings

Desire

Disease

Document

Drawing

Earth

Earth Sciences

Education

Electromagnetic Radiation

Emotion

Energy Converters

Epistemology

Evolution

Exercise and Physical Conditioning

Family and Kinship

Farming

Fishes

Food

Freedom

Galaxy

Genetics and Heredity

Geomorphic Processes

Government budget

Growth and Development, Biological

Habit

Health

History, human

Hypothesis

Human Body

Human Society

Ice

Immunity

Continent

Cosmos

Crime and Punishment

Culture

Cyberspace, the Internet

Death

Democracy

Digestion and Digestive Systems

Doctrines and Dogmas

Drugs

Duty

Earthquakes

Economics

Electricity

Electronics

Endocrine Systems

Engineering

Ethics

Excretion

Exploration

Family Law

Ferns and other lower vascular plants

Flatworms; Fylum Platyhelminthes

Forestry

Fungi

Geochronology

Geography

Good

Gravitation

Gymnosperms

Happiness

Hinduism, religious pluralism

Honor

Human Being

Human Life

Hydrosphere

Ideology

Induction

Industrial Engineering and Production Management

Industries

Infectious Diseases

Information Processing and Information Systems

Inheritance and Succession

Insurance

Intergumentary Systems

Intelligence, natural

Intelligence, artificial

Intelligence, agency

International Trade

International Law, Relations, Trade

Islam

Judaism

Judgment

Judicial and Arbitrational System

Justice

Labor

Language

Law, natural

Law, social

Learning, animal

Learning and Cognition, human

Legal Systems

Libraries

Life

Light

Linguistics

Literature

Logic

Love

Machine

Magnetism

Mammals

Man

Mapping

Marketing and Merchandising

Market

Materials Science

Mathematics

Matter

Measurement System

Mechanics

Medicine

Memory

Mental Disorder

Metabolism

Metaphysics

Mind

Minerals

Modernization and Industrialization

Monerans

Money

Monarchy

Mollusks

Moss Animals

Muscles and Muscle Systems

Music

Myth and Mythology

Names

Nature

Navigation

Nebula

Nervous Systems

Number

Numerical analysis

Nutrition

Ocean

Occultism

Oligarchy

Opinion

Optics

Organism

Pain

Painting

Peoples and Cultures

Perception

Philosophy

Physical Sciences

Plants

Pleasure

Poisons and Poisoning

Population

Printing/Publishing

Probability

Progress

Propaganda

Protists

Prudence

Public administration

Public works

Radar

Reaction

Relativity

Reproduction and Reproductive System

Respiration and Respiratory System

Revolution

Rites and Ceremonies

Rocks

Science

Semantics

Set Theory

Sign and Symbol

Slavery

Social Sciences

Soils

Soul

Space

Sponges

Star

State

Steppe

Supportive and Connective Tissues

Taxation

Personality

Photosynthesis

Physics

Plate Tectonics

Poetry

Political Systems

Police

Principle

Procedural Law

Prophesy

Property Law

Protozoa

Psychology

Public opinion

Punishment

Radiation

Reasoning

Religion

Reptiles

Resources

Rhetoric

Rivers

Scholarship

Sculpture

Sensory Reception

Sex and Sexuality

Sin

Sleep and Dreams

Society

The Solar

Sound

Speech

Sport

Star Clusters

Statistics

Subatomic or Elementary Particles

Surveying

Teaching

Technology
Telecommunication Systems
Telescopes
Temperance
Theatre
Theology
Thermodynamics
Thought
Time
Tools
Topology
Tort
Truth
Tyranny and Despotism
Tissue and Organ Transplants
Transportation
Tree
Trigonometry
The Universe
Virtue and Vice
Virus
War
Wealth
Will
Wisdom
Work and Employment
Writing

About the Author

From 1975 to 1981, **Azamat Abdoullaev** is a postgraduate and research associate at the USSR Academy of Sciences, the Institute of Physics (Moscow, Russia). In 1988, he received a scientific degree in physics and mathematics (the PhD equivalent) conferred by the USSR Academy of Sciences, the Lebedev's Institute of Physics, Moscow. In 1989, a year latter, he submitted a doctorate dissertation in the philosophical sciences, which was published as a book, 'Introduction into Information World', by the USSR Academy of Sciences. The work pioneered the role of global ontology for building (encyclopedic) intelligent systems.
From 1983 to 1991, the author was a research scientist at the Institute of Scientific and Technical Information at the USSR Academy of Sciences and the Government Committee in Scientific and Technical Information. From 1993 up to date, Director and Chief Research Scientist of a Russian company, the first business corporation entirely engaged in research and development of Computer Intelligence Framework enabling a new class of knowledge society artifacts: *large scale intelligent systems with encyclopedic knowledge, general reasoning and natural language power driven by standard global ontology and semantics technology.*
Dr. Abdoullaev established EIS Encyclopedic Intelligent Systems Limited Company in Cyprus to futher the emerging knowledge market of ontological technologies and participate in the European Union programs for building knowledge-based economies. EIS Ltd. has been incorporated with the major aim to develop and manufacture advanced knowledge applications and universal intellectual technologies:

- Semantic web global ontology
- Encyclopedic knowledge products

- General-purpose intelligent software programs, systems, and packages
- Internet-based general reasoning applications
- Global web trading reasoning systems and intelligent business semantic applications
- Knowledge intensive embedded systems

The projected knowledge product lines are to be designed as the realization of the author's pioneering research work on the fundamental principles of knowledge processing machines and general intelligent software systems, which have been published in several books and technical articles and promoted via the Internet sites:

http://www.eis.com.cy ('ONTOLOGY, SEMANTIC TECHNOLOGY, and KNOWLEDGE SOCIETY')

In the field of Encyclopedic Intelligence Systems, for many years, the research web site dedicated to the standard global ontology and encyclopedic intelligence has been occupying the top slot in major web directories and search engines such as Yahoo, Google, and Ask Jeeves.

Dr. Abdoullaev is the creator of USECS, Universal Standard Entity Classification System, known as ''the World Directory of Things,'' the core knowledge base for any large scale real world knowledge system or semantic machine.

Index